QUANTUM OPTICS

XIII SUMMER SCHOOL ON

QUANTUM OPTICS

2-8 September 1985, Frombork, Poland

editors

J FIUTAK
J MIZERSKI

World Scientific

Published by

World Scientific Publishing Co Pte Ltd.
P. O. Box 128, Farrer Road, Singapore 9128
242, Cherry Street, Philadelphia PA 19106-1906, USA

Library of Congress Cataloging-in-Publication Data

Quantum optics.

Papers presented at a meeting held on the 2-8 September 1985 at Gdańsk, Poland.
Edited by J. Mizerski and J. Fiutak.
1. Quantum optics — Congresses. I. Mizerski, J. II. Fiutak, J.
QC446.15.Q83 1986 535 86-1673
ISBN 9971-50-097-3
ISBN 9971-50-098-1 (pbk)

Printed in Singapore by Kim Hup Lee Printing Co. Pte. Ltd.

PREFACE

The 13th Summer School on Quantum Optics was organized by the University of Gdańsk in the historical city of Frombork, in the first week of September 1985. This School was visited by 17 lecturers from Europe and Canada and about 50 participants from Poland and Eastern Europe. All participants had an opportunity to listen to very interesting lectures devoted to experiments, techniques, theoretical ideas and computations concerning various aspects of the light interaction with matter as well as laser physics. A significant part of the lectures is devoted to the investigations of the atomic and molecular collisions by means of laser spectroscopy methods (Campos, Andersen, Burnett, Le Gouët and Ferrante). The theoretical studies of the atomic and molecular structures and atom-atom interactions are represented by the lectures of Baylis, Pascale and Rudzikas. The new applications of lasers to high resolution spectroscopy of atoms and molecules are demonstrated in lectures given by Steudel, Rostas and Niemax. The recently discovered effect of diffusion of laser-excited atoms in gases is described theoretically and illustrated with experimental results in the lectures by Nienhuis and Gel'mukhanov. The state-of-art and recent developments for white-light emission of gaseous lasers are presented by Telle. The lecture by Wódkiewicz raises an interesting question of optical testing of quantum mechanics. Finally, the possibility of the production of squeezed-states in resonance fluorescence is considered by Tanaś.

The interesting contents of all these lectures have stimulated close and friendly contacts among all physicists attending the School. We think this fact to be also of a great importance.

J. Fiutak

J. Mizerski

CONTENTS

QUANTUM OPTICS

CORRELATION AND RELATIVISTIC EFFECTS IN ATOMS AND SMALL MOLECULES

W.E. Baylis *
Service de Physique des Atomes et des Surfaces, Centre d'Etudes Nucléaires de Saclay, 91191 Gif-sur-Yvette Cedex, France.

1. Introduction

1.1 - Scope and purpose

Total energies calculated in *ab initio* treatments of moderately heavy atoms are typically many thousands of hartrees **, but the energies probed by collisions or spectroscopy are rarely more than extremely small differences, amounting usually to no more than a fraction of a hartree. Often, in order to make useful computations of optical-collisional phenomena such as the far-wing pressure broadening of spectral lines, one needs accuracies to a few wavenumbers (cm^{-1}), i.e. to 10^{-5} hartrees or so.

Ab initio calculations of systems with heavy atoms practically never attain the required accuracies of roughly a part in 10^9, but instead, one attempts to compute small differences more directly by freezing closed-shell cores or by replacing the cores by pseudo-or model-potentials (Pascale, 1985). Obviously at the level of precision required, relatively small and subtle effects can be important. Two of the most important such effects are correlation and relativity ; they form the subjects of the present chapter.

* On leave until the end of 1985 from the Department of Physics, University of Windsor, Windsor, Ontario, Canada N9B 3P4.

** 1 hartree is the atomic unit (a.u.) of energy : 1 hartree = 27.2116 eV = 219 474.6 cm^{-1}.

The emphasis will be on correlation and on the use of a core-polarization model to represent it. First we will try to develop an understanding and "feeling" for what is meant by interelectronic correlation ; we will investigate a simple oscillator model for important contributions to the correlation in both free and interacting atoms. Next we will show how such a model can be derived formally from first principles as a correction to the frozen-core approximation, and we note that past work along these lines has neglected important exchange and interelectronic repulsion terms.

Calculations on atoms are used to illustrate the importance of correlation and relativistic effects and the success which can be achieved with a polarization model in combination with a multiconfiguration treatment for atoms with more than one electron in the valence shell. Finally, we will extend the simple oscillator model to interacting atoms and thereby find that there is an important core-polarization correction which reduces the molecular dispersion interaction.

1.2 - Isotropic oscillator in an electric field

We begin by looking at a harmonically bound charge in the presence of a uniform electric field. This is a problem often encountered in elementary classes on quantum mechanics. It is an attractive problem because analytical results are obtained quite easily. We will find it useful for illustrating correlation effects and for providing a model for the polarization of atomic cores by valence electrons and external charges.

Let the charge q be isotropically and harmonically bound to the origin with a force constant $m\omega^2$, where m is the mass of the charge. In the presence of a uniform electric field $\vec{F}$ the system has a hamiltonian

$$H = \frac{p^2}{2m} + \frac{1}{2} m\omega^2 r^2 - q \vec{r}.\vec{F} , \qquad (1)$$

where $\vec{p} = - i\hbar \vec{\nabla}$ is the momentum operator. If one displaces the origin to $\vec{r_0}$

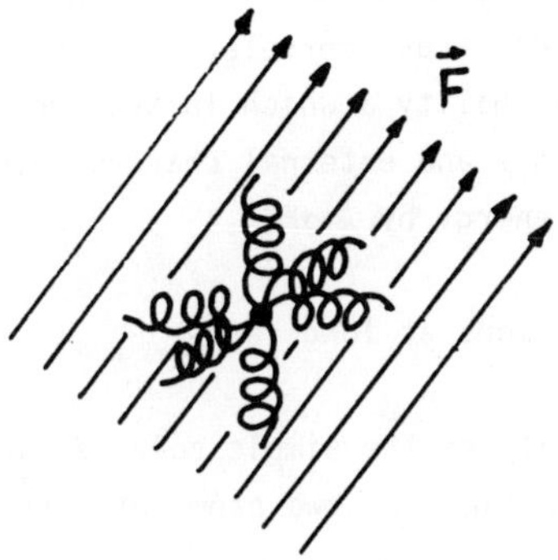

Fig. 1 - Harmonically bound charge q in a uniform electric field $\vec{F}$.

$$\vec{r} = \vec{r}\,' + \vec{r}_o \tag{2}$$

where

$$\vec{r}_o = q\,\vec{F}/(m\omega^2) \quad , \tag{3}$$

the terms linear in $\vec{r}\,'$ cancel and the hamiltonian (1) takes the form of that for an isotropic oscillator with no external field, plus a constant :

$$H = \frac{p^2}{2m} + \frac{1}{2} m\omega^2 r'^2 - \frac{1}{2} \alpha F^2 \tag{4}$$

where

$$\alpha = q^2/(m\omega^2) \tag{5}$$

The average dipole induced by the field is

$$q\langle\vec{r}\rangle = q\,\langle\vec{r}\,'\rangle + q\,\vec{r}_o = q\,\vec{r}_o = \alpha\,\vec{F} \tag{6}$$

so that the term α is readily identified as being the dipole polarizability of the harmonically bound charge.

Consequently, the effect of the external field $\vec{F}$ on the oscillator is (i) to induce a dipole moment $\alpha\vec{F}$ and (ii) to shift all energy levels by the same amount, $-\frac{1}{2}\alpha F^2$. If we think of the oscillator as representing an atomic core and the field $\vec{F}$ as representing the instantaneous field arising from valence electrons, the results illustrate the most important effects of the correlation between the core and valence electrons. They suggest that one might model these correlation effects by endowing the core with a dipole polarizability α which in the presence of a field $\vec{F}$ due to the valence electrons and external charges develops a dipole moment $\alpha\vec{F}$ and lowers the total energy by $\frac{1}{2}\alpha F^2$.

1.3 - Interatomic interactions at long range

As an example of the utility of the simple polarization model suggested above, consider the interactions of two atoms at long range. The total hamiltonian

$$H = H_A + H_B + V_{AB} \tag{7}$$

is a sum of the hamiltonians H_A and H_B of the isolated atoms and the interaction V_{AB}. The problem is to find a practical expression for V_{AB}. A useful approach - one which forms the basis of pseudopotential calculations - is to consider the separate contributions of the valence electrons and the closed-shell cores of atoms A and B. One expands V_{AB} as follows :

$$V_{AB} = V(e_A\text{-}e_B) + V(\text{core}_A\text{-}\text{core}_B) + V(e_A\text{-}\text{core}_B) + V(e_B\text{-}\text{core}_A) \tag{8}$$

where $V(e_A\text{-}e_B)$ is the interaction between the valence electrons of atom A and those of atom B, $V(e_A\text{-}\text{core}_B)$ is the interaction of the valence electrons of A with the core electrons of B, etc. To minimize subtraction error, the monopole part of the direct Coulomb interaction is omitted from each term in (8) ; the sum of such monopole contributions vanishes, of course, if one or both atoms is neutral.

The last three terms in (8) represent interactions with core electrons. They can be most simply calculated if perturbations of the cores are ignored : only overlap effects with the frozen cores would be included. However, such an approximation is severe at long range, because it eliminates a substantial fraction of the Van der Waals interaction. Indeed if one of the atoms has no valence electrons outside its closed-shell core, it would miss the entire interaction occurring beyond the region of overlap. The polarization model discussed in the previous section suggests that corrections can be added which account for the most important core perturbations. These corrections take the form

$$-\frac{1}{2}\alpha_A\,(F_{AB}^2 + 2\,\vec{F}_{AB}\cdot\vec{F}_{Ae_A}) - \frac{1}{2}\,\alpha_B\,(F_{BA}^2 + 2\,\vec{F}_{BA}\cdot\vec{F}_{Be_B}) \qquad (9)$$

where α_A is the dipole polarizability of the closed-shell core A, $\vec{F}_{AB}$ is the electric field appropriately averaged over core A which arises from atom B (core plus valence electrons), $\vec{F}_{Ae_A}$ is the corresponding field due to the valence electrons of A, and α_B, $\vec{F}_{BA}$, $\vec{F}_{Be_B}$ are the analogous expressions for core B. Note that terms like $-\frac{1}{2}\alpha_A\,F^2_{Ae_A}$ and $-\frac{1}{2}\alpha_B\,F^2_{Be_B}$ are already effectively included in H_A and H_B, respectively.

Calculations based on (8) with the correction (9) can predict long-range Van der Waals coefficients much more easily and usually with higher accuracy than *ab initio* calculations (Baylis 1969, 1978).

2. Derivation of V_{pol}

2.1 - Second-order corrections to frozen-core approximation

As attractive and intuitive as the polarization model is, a derivation of the model from first principles is desirable so that its limitations and possible extensions can be seen. Although the basic derivation has been given before (Bottcher and Dalgarno 1974 ; Peach 1983), previous work has ignored exchange effects, which are important in the region of overlap with the core.

To make our derivation more concrete, we consider the particular case of an atom with a spherically symmetric closed-shell core c and valence electrons v, but most of the derivation is easily applied to more complex systems. The total hamiltonian $\mathcal{H}$ is written as the sum

$$\mathcal{H} = H_c + H_v + \Delta V \tag{10}$$

where H_c is the hamiltonian of the ion core in the absence of any valence electrons, H_v is the hamiltonian for the valence electrons and comprises both the kinetic-energy operator T_v and an effective potential V_{eff} which we want to determine, and

$$\Delta V = U - V_{eff} \tag{11}$$

is the difference between the true potential U and V_{eff}.

In the special case of a nonrelativistic treatment of an atom, we would have

$$\mathcal{H} = \sum_i (T_i - \frac{Z}{r_i}) + \sum_{i<j} r_{ij}^{-1} \tag{12}$$

where the sums are over all electrons, i.e. with i referring either to a valence electron ($e \in v$) or to a core electron ($e \in c$), $T_i = -\frac{1}{2}\nabla_i^2$ is the kinetic-energy operator of the ith electron in atomic units (a.u.), r_{ij} is the length of $\vec{r}_{ij} = \vec{r}_i - \vec{r}_j$, and Z is the charge on the nucleus (also in a.u.). The separation indicated in (10) is achieved with

$$H_c = \sum_{i \in c} (T_i - \frac{Z}{r_i}) + \sum_{i<j \in c} r_{ij}^{-1} \tag{13}$$

and

$$H_v + \Delta V = T_v + U \tag{14}$$

where

$$T_v = \sum_{i \in v} T_i \tag{15}$$

$$U = - \sum_{i \in v} \frac{z}{r_i} + \sum_{i<j \in v} r_{ij}^{-1} + U_{pol} \tag{16}$$

$$U_{pol} = \sum_{i \in v} (\sum_{j \in c} r_{ij}^{-1} - \frac{Z - z}{r_i}) \tag{17}$$

and z is the residual charge on the core (Z-z = no. of core electrons).

Formally, the equations we need to solve are just those of steady-state perturbation theory. We write eigenvalue equations for both the total system with hamiltonian $\mathcal{H}$ and the "unperturbed" system with hamiltonian $H_c + H_v$:

$$(\mathcal{H} - \varepsilon_n)\,\psi_n = 0$$
$$(H_c + H_v - \varepsilon_n^{(o)})\,\psi_n^{(o)} = 0 \tag{18}$$

and take the "unperturbed" eigenfunctions to be given by simple products of eigenfunctions representing the ground state ϕ_o of the core and state ψ_n of the valence shell :

$$\Psi_n^{(o)} = \phi_o\,\psi_n \quad ; \quad \varepsilon_n^{(o)} = \epsilon_o + E_n$$
$$H_c\,\phi_o = \epsilon_c\,\phi_o \quad ; \quad H_v\,\psi_n = E_n\,\psi_n \tag{19}$$

As usual in perturbation theory, we choose normalizations

$$\langle\psi_n^{(o)} \mid \Psi_n^{(o)}\rangle = \langle\Psi_n^{(o)} \mid \Psi_n\rangle = 1 \tag{20}$$

and develop expansions for the energy shift $\Delta\varepsilon_n \equiv \varepsilon_n - \varepsilon_n^{(o)}$ in powers of ΔV by interating the relation

$$\langle\Psi_m^{(o)}|\,(\mathcal{H} - H_c - H_v)\,|\Psi_n\rangle = (\varepsilon_n - \varepsilon_m^{(o)})\,\langle\Psi_m^{(o)} \mid \Psi_n\rangle = \langle\Psi_m^{(o)}|\,\Delta V\,|\Psi_n\rangle \tag{21}$$

with both $n = m$ and $n \neq m$. However, whereas in perturbation theory one knows ΔV and seeks $\Delta\varepsilon_n$, we now look for the ΔV which makes the energy differences $\Delta\varepsilon_n$ vanish. Knowing ΔV we easily find V_{eff}, our model potential, in terms of U.

From the first-order expression for $\Delta\varepsilon_n$,

$$\Delta\varepsilon_n^{(1)} = \langle\,\Psi_n^{(o)}|\,\Delta V\,|\Psi_n^{(o)}\rangle \tag{22}$$

the condition $\Delta\varepsilon_n^{(1)} = 0$ gives

$$\langle\phi_o \psi_n| V_{eff}^{(1)} |\phi_o \psi_n\rangle = \langle\phi_o \psi_n| \mathcal{U} |\phi_o \psi_n\rangle .$$

Thus, a simple choice for $V_{eff}^{(1)}$ is the operator on the valence-electron space given by*

$$V_{eff}^{(1)} = \langle\phi_o | \mathcal{U} | \phi_o\rangle \qquad (23)$$

This is of course just the frozen-core approximation, and contributions to $\varepsilon_n^{(1)}$ can only arise from higher-order corrections $\Delta V_{eff} \equiv V_{eff} - V_{eff}^{(1)}$.

From the second-order expression

$$\Delta\varepsilon_n^{(2)} = \sum_{m \neq n} \frac{\langle\psi_n^{(o)}| \Delta V |\psi_m^{(o)}\rangle \langle\psi_m^{(o)}| \Delta V |\psi_n^{(o)}\rangle}{\varepsilon_n - \varepsilon_m} \qquad (24)$$

and the condition $\Delta\varepsilon_n^{(1)} + \Delta\varepsilon_n^{(2)} = 0$, we similarly obtain the polarization correction to (23) :

$$\Delta V_{eff}^{(2)} \equiv V_{pol} = \langle\phi_o| \mathcal{U}_{pol}\, G\, \mathcal{U}_{pol} |\phi_o\rangle \qquad (25)$$

where the Green's operator is

$$G = - \sum_{c \neq o} |\phi_c\rangle (\epsilon_c - \epsilon_o + \mathcal{L}_v)^{-1} \langle\phi_c| \qquad (26)$$

with the sum extending over all excited states c of the core.

* A more general solution to (22) is $V_{eff}^{(1)} = \langle \phi_o| \mathcal{U} |\phi_o\rangle + i\mathcal{L}_v A$ where $\mathcal{L}_v$ is valence-shell liouvillean (see below) and A is an arbitrary hermitean operator.

Here, $\mathcal{L}_v$ is the liouvillean for the valence electrons, defined by $\mathcal{L}_v\, U = -\, U\mathcal{L}_v^{+} \equiv [H_v, U]$. In deriving (25), one notes that $\langle\phi_o|\ (U - V_{eff}^{(1)})\ |\phi_c\rangle$ vanishes if c = o and is identical to $\langle\phi_o|\ U_{pol}\ |\phi_c\rangle$ otherwise.

The division of electrons into core and valence classifications is useful when the core excitation energies $\epsilon_c - \epsilon_o$ are large (and core distortions consequently small) compared to those of the valence shell. Under these circumstances, G (26) can be expanded in a power series of $\mathcal{L}_v/(\epsilon_c - \epsilon_o)$:

$$G = \sum_{k=o}^{\infty} \sum_{c \neq o} |\phi_c\rangle \frac{\mathcal{L}_v^{\,k}}{(\epsilon_o - \epsilon_c)^k} \langle\phi_c| \tag{27}$$

The lowest-order (k = o) term in the expansion (27) is independent of the energy of the valence electrons and gives the so-called "static" contribution. Higher-order (k > o) terms give the "dynamic" or "nonadiabatic" corrections (Dalgarno, Drake and Victor, 1968).

Often the wave functions of the core states ϕ_o and ϕ_c will be represented by single Slater determinants of one-electron orbitals, and the various ϕ_c will correspond to one, two, or more virtual excitations from the closed-shell ground state ϕ_o. Then U_{pol} will have nonvanishing matrix elements only between states which differ at most in a single orbital. In particular, the second-order polarization correction V_{pol} (25) will have contributions only from those ϕ_c representing a single virtual excitation from ϕ_o.

2.2 - Exchange contributions

There is a problem with the derivation above of the frozen-core potential $V_{eff}^{(1)}$ (23) and its polarization correction V_{pol} (25) : as in the works of Bottcher and Dalgarno (1974) and Peach (1983), the valence electrons are treated as being distinguishable from the core electrons. Consequently, all core-valence exchange effects are missing. To correct the formulation, we need to antisymmetrize the wave functions $|\phi\psi\rangle$ for exchanges between electrons in the core and those in the valence shell : we replace $|\phi\psi\rangle$ by $|A_{cv}\phi\psi\rangle$ where A_{cv} is the core-valence antisymmetrizer

$$A_{cv} = \binom{N_c+N_v}{N_c}^{-1/2} \sum_{p\epsilon p(c,v)} (-)^p \, p \qquad (28)$$

and where the sum extends over all permutations p which interchange a number of valence electrons with the same number of core electrons, without regard to ordering among either the valence electrons selected or among the core electrons with which they are interchanged ; N_c and N_v are the total numbers of core and valence electrons, respectively ; $\binom{n}{m}$ is the binary coefficient $n!\,[m!(n-m)!]^{-1}$; and $(-)^p$ represents the parity of the exchange : $(-)^p = +1\ (-1)$ for interchanges of an even (odd) number of electrons. Note that there is one such permutation (namely the identity) which interchanges zero electrons, there are $N_c\, N_v$ which interchange one, $[N_c(N_c-1)/2][N_v(N_v-1)/2] \equiv \binom{N_c}{2}\binom{N_v}{2}$ which interchange two, etc., so that the total number of perturbations on the R.H.S. of (28) is

$$\sum_k \binom{N_c}{k}\binom{N_v}{k} = \binom{N_c+N_v}{N_c} \quad . \qquad (29)$$

Note also that the total antisymmetry operator A for all $N_c + N_v$ electrons can be written as the product

$$A = A_{cv}\, A_c\, A_v \qquad (30)$$

where A_c (A_v) is the antisymmetrizer for the core (valence) electrons by themselves.

The replacement of $|\phi\psi\rangle$, in which ϕ and ψ are individually antisymmetric, by the fully antisymmetric $|A_{cv}\,\phi\psi\rangle$ is not a serious complication. Assuming that the core and valence wave functions are orthogonal, we find that the only matrix elements affected are those of the two-electron terms r_{ij}^{-1} in U_{pol} which couple a core electron, $i \epsilon c$, with a valence one, $j \epsilon v$, and the effect on these terms is simply to add an exchange factor :

$$\sum_{i\epsilon c, j\epsilon v} \langle A_{cv}\, \phi\psi | r_{ij}^{-1}\; A_{cv}\, \phi'\psi' \rangle = \sum_{i\epsilon c, j\epsilon v} \langle \phi\, \psi | r_{ij}^{-1} | \sum_{p\epsilon p(c,v)} (-)^p\, p\, \phi\psi \rangle$$

$$= \sum_{i\epsilon c, j\epsilon v} \langle \phi\psi | r_{ij}^{-1}\, (1 - \mathbb{P}_{ij})\; | \phi'\psi' \rangle \tag{31}$$

where $\mathbb{P}_{ij}$ interchanges electrons i and j. Note furthermore that ϕ and ϕ' can differ by at most a single virtual excitation and that only those terms in the Slater determinants ϕ and ϕ' contribute which have the ith electron in the excited orbital.

The result of correcting the expressions of section 2.1 for the indistinguishability of core and valence electrons is thus to add core-valence exchange to the frozen-core interaction $V_{eff}^{(1)}$ and to correct the polarization term V_{pol} for such exchange. Of course this exchange is only important where the overlap of the valence electrons with the core is significant.

2.3 - Multipole expansion and limiting forms

Calculations of matrix elements of U_{pol} (17) is facilitated by use of the familiar multipole expansion

$$r_{ij}^{-1} = \sum_{l=0}^{\infty} \frac{r_<^l}{r_>^{l+1}}\, P_l\, (\hat{r}_i . \hat{r}_j) \tag{32}$$

where $r_<$ ($r_>$) is the smaller (larger) of r_i and r_j. Only the l = 0 part contributes to the direct frozen-core interaction when the core is spherically symmetric :

$$\langle \phi_0 | \sum_{i\epsilon c} r_{ij}^{-1} - \frac{Z - z}{r_j} | \phi_c \rangle = \int_0^\infty dr\, D_0(r)\, (\frac{1}{r_<} - \frac{1}{r_j}) = - r_j^{-1}\, (Z-z)\, \chi_0\, (r_j) \tag{33}$$

where $D_0(r)$ is the radial number density of the N_c = Z-z electrons in the core state ϕ_0 and $\chi_0(r_j)$ is the core screening function

$$(Z - z)\, \chi_o (r_j) = \int_{r_j}^{\infty} dr\, D_o(r)\, (1 - \frac{r_j}{r}) \quad ; \quad \chi_o(o) = 1. \tag{34}$$

Results of (34), (31), (23) and (16) are easily combined to give the frozen-core potential

$$V_{eff}^{(1)} = \sum_{i<j \in v} r_{ij}^{-1} - \sum_{j \in v} r_j^{-1} \left[z + (Z-z)\, \chi_o(r_j) \right] + V_{ex} \tag{35}$$

where V_{ex} is the usual nonlocal core-valence exchange potential when the core is in its ground state (see section 2.2 and eq. (31)) :

$$V_{ex} = - \sum_{i \in c, j \in v} \langle \phi_o |\, r_{ij}^{-1}\, \mathbb{P}_{ij}\, | \phi_o \rangle \, . \tag{36}$$

The polarization correction V_{pol} (25) to the frozen-core potential also contains both a local direct contribution and a nonlocal exchange part. Substituting (32) into $\mathcal{U}_{pol}$ (17) and the result into (25), we find for the direct part

$$V_{pol.dir} = - \sum_{c \neq o} \sum_{\substack{l_{ij} \\ l'_{i'j'}}} \langle \phi_o | \frac{r_<^l}{r_>^{l+1}} P_l\, (\hat{r}_i . \hat{r}_j)\, | \phi_c \rangle\, (\varepsilon_c - \varepsilon_o + \mathcal{L}_v)^{-1}$$

$$\times \langle \phi_c | \frac{r'^{\,l'}_<}{r'^{\,l'+1}_>} P_{l'}\, (\hat{r}_{i'} . \hat{r}_{j'})\, | \phi_o \rangle \tag{37}$$

Let J_c, M_c be the eigenvalues of the total angular momentum and its projection on the quantization axis of the core state ϕ_c. The corresponding eigenvalues for the core ground state ϕ_o are 0,0. If we expand the Legendre polynomials P_l in spherical harmonics C_{lm} (Brink and Satchler, 1968), we can apply the Wigner-Eckart theorem

$$\langle \phi_o | \frac{r_<^l}{r_>^{l+1}} C_{lm}\, (\hat{r}_i)\, | \phi_c \rangle = \begin{pmatrix} 0 & l & J_c \\ 0 & m & M_c \end{pmatrix} \langle \phi_o || \frac{r_<^l}{r_>^{l+1}} C_l\, (\hat{r}_i) || \phi_c \rangle \tag{38}$$

where the reduced matrix element $\langle\phi_o|| \frac{r_<^l}{r_>^{l+1}} C_l ||\phi_c\rangle$ is written with the Edmonds' normalization. We obtain, after applying an orthogonality relation for 3-j symbols,

$$V_{pol.dir} = - \sum_{\substack{c\neq o\\ l,m}} (2l+1)^{-1} \sum_{ij\ i'j'} \langle\phi_o|| \frac{r_<^l}{r_>^{l+1}} C_l(\hat{r}_i)|| \phi_c\rangle$$

$$\times C^*_{lm}(\hat{r}_j)\ (\epsilon_c-\epsilon_o\ +\mathcal{L}_v)^{-1}\ C_{lm}(\hat{r}'_j)\quad \langle\phi_o|| \frac{r'^l_<}{r'^{l+1}_>} C_l(\hat{r}_{i'})||\phi_c\rangle^* \delta_{J_c,l}\delta_{M_c,o} \tag{39}$$

The "static" or k = o part of (39) (see (27)) is

$$V^{(o)}_{pol.dir} = -\frac{1}{2} \sum_{jj'} \sum_{l} a_l\ (r_j,r_{j'})\ P_l\ (\hat{r}_j\cdot\hat{r}_{j'}) \tag{40}$$

where the function a_l is

$$a_l(r_j,r_{j'}) = 2 \sum_{\substack{c\neq o\\ i,i'}} \frac{\langle\phi_o|| \frac{r_<^l}{r_>^{l+1}} C_l(\hat{r}_i)|| \phi_c\rangle \langle\phi_o|| \frac{r'^l_<}{r'^{l+1}_>} C_l(\hat{r}_{i'})|| \phi_c\rangle^*}{(2l+1)(\epsilon_c - \epsilon_o)} \times \delta_{J_c,l}\,\delta_{M_c,o} \tag{41}$$

The multipole expansion terms a_l simplify in the limiting cases of large and small r_j and $r_{j'}$: let r_o be a characteristic core radius ; for $r_j \gg r_o$ and $r_{j'} \gg r_o$ a_l has the asymptotic form

$$a_l(r_j,r_{j'}) \sim (r_j\ r_{j'})^{-(l+1)}\ \alpha_l \tag{42}$$

where α_l is the static 2^l-pole polarizability of the core defined by

$$\alpha_l = 2 \sum_{c\neq o} \frac{|\langle\phi_o|| \sum_i r_i^{\,l}\, C_l(\hat{r}_i)||\, \phi_c\rangle|^2}{(2l+1)\,(\epsilon_c-\epsilon_o)}\, \delta_{J_c,l}\, \delta_{M_c,o} \tag{43}$$

with the index i summed over all core electrons and c over all excited core states. On the other hand, in the limit of small r_j and $r_{j'}$,

$$a_l(r_j,r_{j'}) \xrightarrow{r_j,r_{j'}\to o} 2(r_j\, r_{j'})^l \sum_{c\neq o} \frac{|\langle\phi_o||\sum_i r_i^{-(l+1)}\, C_l(\hat{r}_i)||\, \phi_c\rangle|^2}{(2l+1)\,(\epsilon_c-\epsilon_o)}\, \delta_{J_c,l}\delta_{M_c,o} \tag{44}$$

The monopole term clearly vanishes at long range : $\alpha_o = 0$; but at short range it is not generally zero. Its size in the limit $r_j \to 0$ and $r_{j'} \to 0$ can be found by comparing the total binding energy of the ion core of nuclear charge Z with that of the isoelectronic ion core with nuclear charge Z-1. The effect of exchange is to reduce the interaction in the region $r_j \lesssim r_o$ or $r_{j'} \lesssim 0$. It will have no effect at long range (see (42) and (43)).

2.4 - Dynamic corrections

Contributions from terms in the expansion (27) with $k > 0$ give dynamic corrections to the static ($k = 0$) terms considered above. The kth correction is

$$V^{(k)}_{pol.dir} = \sum_{c\neq o} \langle\phi_o|\, \mathcal{U}_{pol}\, |\phi_c\rangle \frac{\mathcal{L}_v^k}{(\epsilon_o-\epsilon_c)^{k+1}} \langle\phi_c|\, \mathcal{U}_{pol}|\, \phi_o\rangle \tag{45}$$

and asymptotically (r_j and $r_{j'} \gg r_o$), where the exchange becomes negligible, one has simply

$$V_{pol}^{(k)} \sim (-)^{k+1} \sum_{l} \beta_l^{(k)} \sum_{m} \left[\sum_{j} \frac{C_{lm}^{*}(\hat{r}_j)}{r_j^{l+1}} \right] \mathcal{L}_v^{k} \left[\sum_{j'} \frac{C_{lm}(\hat{r}_{j'})}{r_{j'}^{l+1}} \right] \tag{46}$$

where

$$\beta_l^{(k)} = \sum_{c \neq o} \frac{|\langle\phi_o|| \sum_i r_i^l \, C_l \, (\hat{r}_i)|| \, \phi_c\rangle|^2}{(2l+1) \, (\epsilon_c - \epsilon_o)^{k+1}} \, \delta_{J_c,1} \, \delta_{M_c,o} \tag{47}$$

For the lowest-order (k=1) dynamic correction, the fact that $V_{pol.dir}^{(k)}$ and $\langle\phi_o| \, U_{pol} \, |\phi_c\rangle$ can be taken to be real functions obeying $\nabla_j^2 \langle\phi_o| \, U_{pol} | \, \phi_c\rangle = 0$, $r_j \neq 0$, allows us, after integration by parts of $\langle\psi_n|V_{pol.dir}^{(1)}| \, \psi_n\rangle$, to write simply

$$V_{pol.dir}^{(1)} = \frac{1}{4} \sum_{c \neq o} (\epsilon_c - \epsilon_o)^{-2} \sum_{j \in v} \nabla_j^2 \, (|\langle\phi_o| \, U_{pol}| \, \phi_c\rangle|^2) \tag{48}$$

which reduces asymptotically to

$$V_{pol}^{(1)} \sim \frac{1}{2} \sum_{l} (l+1) \, (2l+1) \, \beta_l^{(1)} \sum_{j} r_j^{-(2l+4)} \tag{49}$$

We note, in agreement with Peach (1983), the lack of any k = 1 correction, asymptotically, to the static cross terms ($j \neq j'$ in eq. (40)).

The next order (k = 2) does contain a correction to the cross terms, however. To calculate it, we note

$$\mathcal{L}_v^2 \, f = [H_v, \, [H_v, f] \,] \tag{50}$$

where in our case $H_v = -\frac{1}{2} \sum_{j \in v} \nabla_j^2 + V_{eff}$ and asymptotically

$f = \sum_{j \in v} r_j^{-l-1} \, C_{lm}(\hat{r}_j)$. Consequently $\vec{\nabla}_j \vec{\nabla}_{j'} \, f = \delta_{jj'} \, \vec{\nabla}_j \vec{\nabla}_j \, f$ and $\nabla_j^2 \, f = 0$ so that

$$\mathcal{L}_v^2 f = \sum_j (\vec{\nabla}_j f).(\vec{\nabla}_j V_{eff}) + \sum_j (\vec{\nabla}_j \vec{\nabla}_j f) : \vec{\nabla}_j \vec{\nabla}_j \tag{51}$$

and the operator for the second-order dynamic correction takes the asymptotic form

$$V_{pol}^{(2)} \sim \sum_{1jj'} \beta_1^{(2)} \Big[\vec{Q}_1^{(1)} (\vec{r}_j,\vec{r}_{j'}).\vec{\nabla}_j V_{eff} + \overset{\leftrightarrow}{Q}{}_1^{(2)} (\vec{r},\vec{r}_{j'}) : \vec{\nabla}_j \vec{\nabla}_j \Big] \tag{52}$$

where the vector and symmetric dyad functions are defined by

$$\vec{Q}_1^{(1)} (\vec{r},\vec{r}') = \vec{\nabla} \frac{P_1(\xi)}{(rr')^{1+1}} = \frac{1}{(rr')^{1+1}} \cdot \frac{1}{r} \Big[(1+1)\, \hat{r}\, P_1(\xi) + (\hat{r}\, \xi - \hat{r}')\, P_1'(\xi) \Big] \tag{53}$$

$$\overset{\leftrightarrow}{Q}{}_1^{(2)} (\vec{r},\vec{r}') = \vec{\nabla}\vec{\nabla} \frac{P_1(\xi)}{(rr')^{1+1}} = \frac{1}{(rr')^{1+1}} \cdot \frac{1}{r^2} \Big[(1+1)(1+2) P_{1+1}(\xi)(\hat{r}'\hat{r}' - \hat{\varepsilon}_1\hat{\varepsilon}_1) -$$

$$P_{1+1}'(\xi)(\hat{\varepsilon}_2\hat{\varepsilon}_2 - \hat{\varepsilon}_1\hat{\varepsilon}_1) + (1+1)\, |\hat{r}\times\hat{r}'|\, P_{1+2}'(\xi)(\hat{r}'\, \hat{\varepsilon}_1 + \hat{\varepsilon}_1\, \hat{r}')\Big] . \tag{54}$$

In the definitions (53) and (54), $\xi \equiv \hat{r}.\hat{r}'$, differentiation $\vec{\nabla}$ is only with respect to $\vec{r}$, a prime on a Legendre polynomial indicates a derivative with respect to its argument, and the orthogonal unit vectors $\hat{\varepsilon}_1$ and $\hat{\varepsilon}_2$ are given by

$$\hat{\varepsilon}_2 = \frac{\hat{r}' \times \hat{r}}{|\hat{r}' \times \hat{r}|} \quad , \qquad \hat{\varepsilon}_1 = \hat{\varepsilon}_2 \times \hat{r}' \quad . \tag{55}$$

One easily verifies

$$\overset{\leftrightarrow}{Q}{}_1^{(2)} (\vec{r},\vec{r}') : \vec{1} = \nabla^2 \frac{P_1(\xi)}{(rr')^1} = 0 \tag{56}$$

where $\overset{\leftrightarrow}{1} \equiv \hat{x}\,\hat{x} + \hat{y}\,\hat{y} + \hat{z}\,\hat{z}$ is the unit dyad. Also, using the values $P_1(1)=1$ and $P_1'(1) = \frac{1}{2}\, 1\, (1+1)$, one finds the rather simple diagonal ($\vec{r} = \vec{r}'$) expressions

$$\vec{Q}_l^{(1)}(\vec{r},\vec{r}) = -(l+1)\,\hat{r}\,r^{-2l-3}$$

$$\overleftrightarrow{Q}_l^{(2)}(\vec{r},\vec{r}) = (l+1)(l+2)\,.\,\frac{1}{2}(3\,\hat{r}\,\hat{r} - \vec{1})\,r^{-2l-4} \tag{57}$$

Since V_{eff} has the asymptotic form (see (35))

$$V_{eff} \sim -z\sum_j r_j^{-1} + \frac{1}{2}\sum_{\substack{j,k\\ j\neq k}} r_{jk}^{-1} , \tag{58}$$

the vector contribution to $V_{pol}^{(2)}$, which involves

$$\vec{\nabla}_j V_{eff} \sim z\,\vec{r}_j\,r_j^{-3} - \frac{1}{2}\sum_{\substack{k\\ k\neq j}} \frac{(\vec{r}_j-\vec{r}_k)}{r_{jk}^{3}} , \tag{59}$$

contains 2^l-pole corrections varying at large distances as $z\,r_{j'}^{-l-1}\,r_j^{-l-4}$ and as $r_{j'}^{-l-1}\,r_j^{-l-1}\,r_{jk}^{-3}$. Although both terms are of the same order, the latter (but not the former) have been ignored by Peach (1983).

The leading contribution to $V_{pol}^{(2)}$ at long range nevertheless arises, due to the dominance of radial derivatives of the valence wave function ψ at long range, from the radial part of the dyad term. By using $(1-\xi^2)\,P'_{l+1}(\xi) = (l+1)\left[P_l(\xi) - \xi\,P_{l-1}(\xi)\right]$, we find from (54) that

$$\overleftrightarrow{Q}_l^{(2)}(\vec{r},\vec{r}') : \hat{r}\,\hat{r} = \frac{(l+1)\,(l+2)}{r'^{\,l+1}\,r^{l+3}}\,P_l(\hat{r}.\hat{r}') \tag{60}$$

which asymptotically and on the average gives as the leading term in $V_{pol}^{(2)}$ the effective potential

$$V_{pol}^{(2)} \sim \sum_{ljj'} \beta_l^{(2)}\,d_2\,\frac{(l+1)(l+2)}{r_{j'}^{l+1}\,r_j^{l+3}}\,P_l\,(\hat{r}_j.\hat{r}_{j'}) \tag{61}$$

where d_2 is a constant roughly equal to $2\,|E_n|/N_v$ where E_n is the eigenenergy of the valence shell (see (19)) and N_v is the number of valence electrons.

Collecting results for the asymptotic form of the polarization potential, we can write

$$V_{pol} \sim -\frac{1}{2} \sum_{jj'l} (r_j r_{j'})^{-l-1} P_l(\hat{r}_j \cdot \hat{r}_{j'}) (\alpha_l - r_j^{-2} \delta_l) \qquad (62)$$

where the correction δ_l is

$$\delta_l = (l+1) \left[(2l+1) \beta_l^{(1)} \delta_{jj'} + 4 (l+2) \beta_l^{(2)} d_2\right] \quad . \qquad (63)$$

Note that the notation $\gamma_l \equiv \frac{1}{2} \beta_l^{(2)}$ is common in the literature (see for example Peach 1983) and that the approximation $d_2 \simeq 2 |E_n|/N_v$ is exact for $N_v = 1$. Indeed for $N_v = 1$, the simple results (57) can be used to give V_{pol} correct through terms of order r^{-2l-5} : one finds $d_2 = 2 |E_n| - (2l+5) (l+2)^{-1} z/r$, in agreement with Peach (1983) *.

2.5 - A suggested form of the polarization potential

The perturbation expansion (27) of the resolvant operator G (26) is useful for obtaining limiting forms and constraints on the polarization potential V_{pol}, but its poor convergence properties, particularly at small r, and the difficulty of calculating higher-order terms make it unsuitable for most numerical calculations. Indeed as seen in the previous section, uncertainties in the second-order corrections to the dipole polarizability are large enough to make it pointless to include either higher-order dipole corrections or higher-order multipole polarizabilities beyond the static quadrupole contribution.

* Unfortunately, for $N_v > 1$, the r^{-2l-5} terms given by Peach (1983) are missing contributions both from the interelectronic repulsion (see above) and from non-radial derivatives of ψ (see (52) and (54)).

Let us refer back to the general expression for V_{pol}, which itself arises from a second-order perturbation expression for V_{eff}, in order to seek an appropriate form for a polarization potential. From (25) and (26) we have

$$V_{pol} = \sum_{\substack{c \neq o \\ j,j' \in v}} v_{oc}^{(j)} (\epsilon_o - \epsilon_c - \mathcal{L}_v)^{-1} v_{co}(j') \tag{64}$$

where the core matrix elements, including exchange effects, can be written (see (17) and (31))

$$v_{oc}^{(j)} = \langle \phi_o | \sum_{i \in c} r_{ij}^{-1} (1 - \mathbb{P}_{ij}) | \phi_c \rangle \quad . \tag{65}$$

A multipole expansion as in section 2.3 allows us to write (64) as a sum of multipole contributions for each pair (jj') of valence electrons :

$$V_{pol} = \sum_{l} \sum_{jj' \in v} V_{pol}^{(l)} (jj') \tag{66}$$

where

$$V_{pol}^{(l)} (jj') = \sum_{m,c} v_{oc}^{(l,m)} (j') (\epsilon_o - \epsilon_c - \mathcal{L}_v)^{-1} v_{oc}^{(l,m)} (j)^* \tag{67}$$

with

$$v_{oc}^{(l,m)} (j) = (2l+1)^{-1/2} \sum_{i \in c} \langle \phi_o || \frac{r_<^l}{r_>^{l+1}} C_l(\hat{r}_i) (1 - \bar{\mathbb{P}}_{ij}) || \phi_c \rangle \times C_{lm}^*(\hat{r}_j) \delta_{J_c,l} \delta_{M_c,o} \tag{68}$$

in which the interchange operator $\bar{\mathbb{P}}_{ij}$ differs from $\mathbb{P}_{ij}$ in that it operates only on radial and spin coordinates, not on the angular ones.

The form of (67) suggests that rather than write $V_{pol}^{(l)}$ (jj') as a sum over k of terms

$$V_{pol}^{(l)(k)}(jj') = \sum_{m,c} v_{oc}^{(l,m)} (j') \frac{\mathcal{L}_v^k}{(\epsilon_o - \epsilon_c)^{k+1}} v_{oc}^{(l,m)} (j)^* \tag{69}$$

a better adapted approximation without the convergence problems of (69) might be obtained by simply replacing the liouvillean $\mathcal{L}_v$ in (67) by an effective potential $\mathcal{L}_{eff}$ to give

$$V_{pol}^{(1)}(jj') \simeq \sum_{m,c} V_{oc}^{(1,m)}(j')\,(\epsilon_o - \epsilon_c - \mathcal{L}_{eff})^{-1}\, V_{oc}^{(1,m)}(j)^*$$

$$= V_{pol}^{(1)(o)}(jj')\left[1 + \mathcal{L}_{eff} / (\epsilon_c - \epsilon_o)\right]^{-1} \quad . \tag{70}$$

The effective liouvillean $\mathcal{L}_{eff}$, which in general may be a nonlocal potential, should obviously be chosen so as to reproduce the low-order dynamic correction $V_{pol}^{(1)(k)}(jj')$, k = 1, 2, ..., to the static $V_{pol}^{(1)(o)}(jj')$ polarization terms. In the asymptotic limit, which is the only case in which the dynamic corrections are known to have a simple analytic form, one finds from (62) and (70)

$$\frac{\mathcal{L}_{eff}}{\epsilon_c - \epsilon_o} \sim \sum_{\substack{s \le 2 \\ t \ge 0}} b_{s,t}\, r_j^{-s}\, r_{j'}^{-t} \tag{71}$$

where the coefficient of the leading asymptotic term is

$$b_{2,0} = \delta_1 / \alpha_1 \tag{72}$$

However, whereas the asymptotic expansions of $V_{pol}^{(1)(k)}(jj')$, $k > 0$, diverge, that of $\mathcal{L}_{eff}$ (71) gives well-behaved results when substituted into (70), even for small r_j and $r_{j'}$. Indeed, the asymptotic form of $V_{pol}^{(o)}$, eqs. (40) and (43), together with the expansion (71) might be used at all r if the expansion is truncated after the s = t = 2l+1 term to ensure the correct short-range limiting form (44).

3. Atomic calculations

3.1 - Relativistic and correlation effects in the valence shell

Direct effects of relativity are most important for those electrons with a significant probability of moving with velocities approaching the speed of light. These, of course, are the electrons of tightly bound core orbitals, and one might well imagine relativistic effects are negligible, even in heavy atoms, for the weakly bound valence electrons. It may in fact be true that the direct relativistic effects are small for valence electrons, but the indirect effects can be very important.

The indirect effects arise from direct effects on the low-lying core states together with the self-consistency of all electrons on the atom. Thus the 1s electrons become more tightly bound and spatially "contracted" due to direct relativistic effects. All the other s orbitals, which are constrained to be orthogonal to the 1s orbital, also contract and become more tightly bound, but mainly through the effects of self-consistency with the 1s orbital and with each other, rather than as a result of direct relativistic effects. The less penetrating d and f electrons, on the other hand, experience a more effectively screened Coulomb attraction of the nucleus, and consequently the d and f orbitals tend to dilate and become less tightly bound.

To see how large relativistic effects can be on valence electrons, compare the nonrelativistic Hartree-Fock (HF) results with the relativistic ones (RHF) in tables 1 and 2. Relativistic effects are seen to increase the ionization energy of neutral mercury by more than 20 % (table 1) and to decrease the oscillator strength (a dimensionless measure of the strength of the radiative transition) of the resonance line from 3 to 2. Of course mercury, with its nuclear charge of Z = 80, has relatively large relativistic effects because its innermost electrons move at characteristic speeds $\alpha Zc \sim (Z/137)c$ which approach the speed of light c. The point here is that in such atoms, all electrons are affected, not just those of

Table 1 - Ionization energy $E\,[Hg^+(6\ ^2S_{1/2})] - E\,[Hg\ (6\ ^1S_0)]$ of mercury.

Method	Energy (a.u.)	Reference
Experiment	0.3836	Moore (1971)
HF	0.2510	Baylis (1977)
RHF	0.3152	Migdalek and Baylis (1985a)
MCRHF	0.3399	"
MCRHF-CP	0.3674	
(MCRHF-ACP	Exact)	"

Table 2 - Oscillator strengths of the resonance $(6\ ^1S_0 - 6\ ^1P_1)$ and intercombination $(6\ ^1S_0 - 6\ ^3P_1)$ lines of Hg.

Method	$f\ (^1S_0 - {}^1P_1)$	$f\ (^1S_0 - {}^3P_1)$	Reference
Experiment	1.18 ± 0.08	0.0247 ± 0.0004	Lurio (1965) Deech and Baylis(1971)
HF	3.03	0	Desclaux (1972)
RHF-LS	1.99	~ 0	Desclaux and Kim(1975)
MCRHF	1.53	~ 0.012	Desclaux and Kim(1975)
MCRHF-CP	1.247	0.0200	Migdalek and Baylis (1985a)
MCRHF-ACP	1.174	0.0254	"

the most tightly bound orbitals. Indeed, the fractional change in ionization energy or average orbital radius is often larger for the valence shell than for the $1s^2$ shell (Baylis, 1977).

A comparisons of the RHF results in tables 1 and 2 with experimental values reveals that relativistic effects account for only a fraction of the error in the HF values. The remaining part is due almost entirely to electron correlation *, that is, to the response of electrons to deviations from the average charge distribution. An important example of correlation is the polarization of the remaining electron distribution when one electron is at $\vec{r}$ (see fig. 2a), as discussed in part 1 of this chapter.

In single-configuration HF or RHF calculations, each electron moves in a potential arising from the average distribution of all other electrons. Consequently such calculations include none of the correlation effects.

(a)

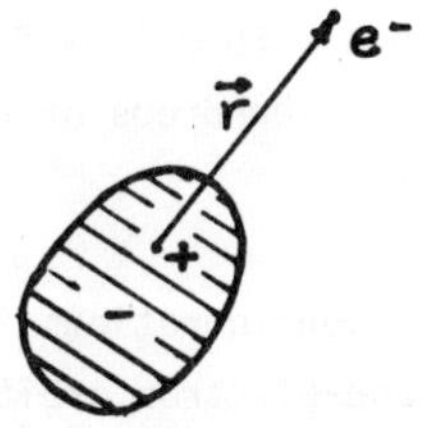

(b)

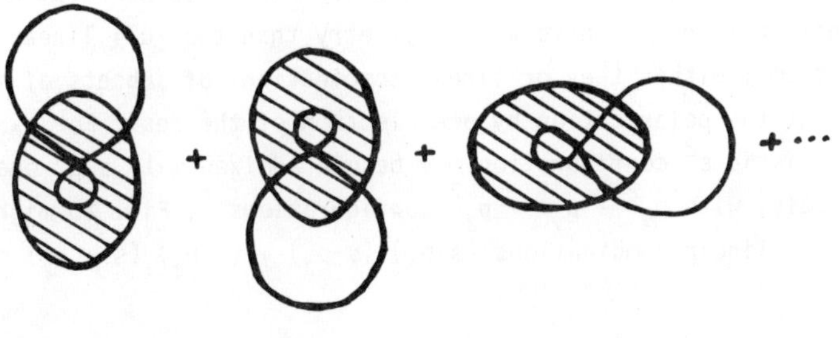

Fig. 2 - Example of correlation. (a) Polarization of remaining electron distribution when one electron is at $\vec{r}$. (b) The representation of such correlations in a multiconfiguration or configuration interaction treatment.

* Some authors define correlation to be the difference between the single-configuration HF values and experiment, but I prefer a more physical definition which avoids questions such as what if any relativistic effects are included in the HF calculation.

They would, for example, predict none of the Van der Waals interaction between neutral atoms. *Ab initio* calculations of correlation effects must allow for configuration mixing : the total N-electron wave function is written as a linear combination of Slater determinants, each of which represents a single configuration. In a configuration-interaction (CI) calculation, the one-electron orbitals in each determinant are fixed, perhaps by a previous computation, and only the coefficients of the linear combination are varied. On the other hand, in multi-configuration (MC) calculations there is one set of one-electron orbitals out of which all configurations are constructed, and these orbitals are self-consistently optimized as the best mixture of configurations is found. As consequence of the orbital optimization, MC calculations with only a few configurations can often produce results which require hundreds or even thousands of fixed configurations in a CI calculation.

Fig. 2b shows in principle now the various configurations in an MC or a CI calculation can represent correlation. A one-electron orbital is shown together with the electron distribution of the remaining electrons in three possible configurations which would contribute to correlation. Individual configurations may have lower symmetry than the full linear combination of them ; either they or linear combinations of subsets of them may represent the polarization by one electron of the rest. For example, a nonrelativistic s^2 configuration may become admixed with a p^2 one, or more precisely, with $p_x^2 + p_y^2 + p_z^2$ configurations *. Fig. 2b might represent the linear combinations $(s+p_z)\ (s-p_z) + (s-p_z)\ (s + p_z) + (s-p_x)\ (s+p_x) + \ldots$

* In atomic calculations, one usually integrates over angular variables and sums over azimuthal quantum numbers analytically so that the existence of lower symmetry components of "a" p^2 configuration tend to be hidden. Their presence is none-the-less important for understanding the physics in configuration mixing.

3.2 - RHF and MCRHF calculations

RHF (Dirac-Fock) calculations of atoms have been available now for more than a decade (Desclaux 1975), even though the very existence of stable Dirac-Fock solutions for systems with more than one electron in a central field continues to be debated (see, for example Sucher 1985 and references therein). Only recently has it become clear that the boundary conditions imposed on the numerical one-electron orbitals in atomic RHF treatments act effectively as projection operators on positive-energy space and prevent mixing with the negative-energy continuum of the eigenvalue spectrum of the Dirac hamiltonian. RHF calculations of molecules still pose major problems, but the spherical symmetry of atoms permit straightforward applications including all the dominant relativistic effects.

The major remaining problem in accurate atomic calculations is how to determine reliable estimates of the correlation effects. Multiconfiguration RHF (MCRHF) calculations can be made, but these tend to be considerably more complex than their nonrelativistic counterparts. In part, the added complexity is due to the multiplicity of relativistic (nlj) orbitals compared to the nonrelativistic (nl) ones. There are also problems with how rapidly MC expansions converge. These will be addressed in the following section.

3.3 - Problems with the MC approach

It is convenient to split the total correlation into three parts : intravalence, valence-core, and intra-core. The additional configurations added in MC or CI approaches can be derived from the initi al configuration by virtual excitations. Those configurations which can be derived solely by virtual excitations from the valence shell will be considered to contribute to intravalence correlation, whereas those others which can be derived from one or more virtual excitations from the valence shell plus one or more from the core will be taken as part of valence-core correlation, and the rest will contribute to intracore correlation.

The correlated wave function ψ_m will be a linear combination of Slater determinants Φ_n, each representing a configuration :

$$\psi_m = C_o \Phi_m + \sum_{n \neq m} C_n \Phi_n \tag{73}$$

where Φ_m is the principal configuration and corresponds to the uncorrelated wave function. The size of the coefficients C_n can be estimated by first-order perturbation theory :

$$C_o \simeq 1 \quad , \quad C_n \simeq \frac{\langle \Phi_n \mid \sum_{i<j} r_{ij}^{-1} \mid \Phi_m \rangle}{E_m - E_n} \tag{74}$$

If only excitations from the valence shell are involved, a few low-lying configurations Φ_n with good overlap with Φ_m will usually dominate, and coefficients C_n for other configurations will be small in comparison. When core excitations are involved, however, there will usually be many configurations, all with similar energy differences E_m-E_n and similar overlaps with Φ_m. The convergence of (73) is then usually poor.

The valence-core correlation differs from the intracore correlation in two important respects : it is much larger, and it depends strongly on the state of the atom. It is the part of the correlation which usually causes the greatest error in *ab initio* calculations.

An approach which has proved its value in a large numbers of applications to both one-electron spectra (Migdalek and Baylis, 1978, 1979a-e, 1981, 1982, 1984a) and two-electron spectra (Migdalek and Baylis, 1984b, 1985a, b and c) is to use an MCRHF calculation in which only configurations contributing to intervalence correlation are admixed and to approximate the rest of the correlation, mainly valence-core contributions, by a core-polarization (CP) model.

3.4 - Core-polarization model

In part 2 of this chapter, we showed that the main corrections to a frozen-core calculation are valence-core correlation terms which represent the polarization of the core by the electric fields of the valence electrons. Here we consider a simple model which gives a dipole polarization potential of the form suggested in section 2.5. Higher multipole polarizabilities have not been considered.

To allow for both dynamic effects and the finite size of the core, the effective electric field at the core due to the jth valence electron at $\vec{r}_j$ is taken to be

$$\vec{F}_j = \frac{\vec{r}_j}{(r_j^2 + r_o^2)^{3/2}} \tag{75}$$

where r_o is a parameter corresponding to the "size" of the core. The total polarization potential is then

$$\begin{aligned} V_{CP} &= -\frac{1}{2}\alpha\,(\sum_j \vec{F}_j)^2 \\ &= -\frac{1}{2}\alpha \sum_j F_j^2 - \alpha \sum_{i<j} \vec{F}_i.\vec{F}_j \end{aligned} \tag{76}$$

where α is the static dipole polarizability of the core. In a HF-CP calculation, the terms $-\frac{1}{2}\alpha\, F_i^2$ can be added directly to the one-electron potentials. They always act to lower the eigenenergies. The cross or dielectric terms $-\alpha\, \vec{F}_i.\vec{F}_j$ can be easily calculated with the interelectronic repulsion r_{ij}^{-1}. Their effect can either raise or lower energies, depending on whether the angular average $\langle \hat{r}_i.\hat{r}_j \rangle$ is negative or positive (see section 3.7).

3.5 - The core-polarization correction to the dipole-moment operator

The interaction of an atom with a uniform external field $\vec{F}_{ext}$ is given by $-\vec{d}.\vec{F}_{ext}$ where $\vec{d}$ is the dipole-moment operator

$$\vec{d} = -\sum_i \vec{r}_i \,, \tag{77}$$

and where the sum extends over all electrons. If the core of the atom is frozen, the sum can be restricted to valence electrons, but then a core-polarization correction should be added. The correction arises from a modification to V_{CP} (76) which must be made in the presence of the external field :

$$V_{CP} = -\frac{1}{2}\alpha(\vec{F}_{ext} + \sum_j \vec{F}_j)^2$$

$$= -\frac{1}{2}\alpha F^2_{ext} - \frac{1}{2}\alpha(\sum_j \vec{F}_j)^2 - \vec{d}_c.\vec{F}_{ext} \tag{78}$$

where

$$\vec{d}_c = \alpha \sum_j \vec{F}_j \tag{79}$$

is the dipole moment induced in the core by the valence electrons. The first term $-\frac{1}{2}\alpha F^2_{ext}$ shifts all levels equally and is of little importance here. The second term in (78) is the usual core-polarization potential (76). The last term must be added to the interaction of the valence dipole with the external field. Consequently the total dipole operator for the frozen-core atom is

$$\vec{d} = -\sum_{j \in v} \vec{r}_j + \vec{d}_c$$

$$= -\sum_{j \in v} \vec{r}_j \left[1 - \frac{\alpha}{(r_j^2 + r_o^2)^{3/2}}\right] \tag{80}$$

Since the dipole moment induced in the core by the valence electrons is in the opposite direction to the dipole moment of the valence electrons themselves, the effect of core polarization (and hence of valence-core correlation) is to reduce the atomic dipole moment, and hence to reduce the interaction with the external field. Of course the same interaction, $-\vec{d}.\vec{F}_{ext}$, is also responsible for electric-dipole transitions, and consequently the strengths of such transitions tend to be decreased by core polarization.

3.6 - Modified oscillator-strength sum rules

Analytic expressions have long been known (Vinti 1932) for a number of the sums

$$S_k = \sum_{a \neq o} (E_a - E_o)^k f_{ao} \tag{81}$$

where

$$f_{ao} = \frac{2}{3} (E_a - E_o) \, |\langle a| \vec{d} \, |o\rangle|^2 \tag{82}$$

is the absorption oscillator strength for the transition a ← o. These "sum rules" are useful in calculating and checking a number of results. Of course they become modified in the presence of core polarization. Thus if we put

$$\lambda_j = \alpha \, (r_j^2 + r_o^2)^{-3/2} \tag{83}$$

so that the dipole-moment operator (80) is :

$$\vec{d} = - \sum_{j \in v} \vec{r}_j \, (1 - \lambda_j) \, , \tag{84}$$

we find

$$S_0 = \frac{1}{3} \langle \, [\vec{d}_j \; [H,\vec{d}] \;] \, \rangle_o$$

$$= N_v \langle (1 - \lambda_1)^2 - \frac{2}{3} r_1 \, (1 - \lambda_1) \, \lambda_1' + \frac{1}{3} (r_1 \, \lambda_1')^2 \rangle_o \tag{85}$$

$$S_1 = - \frac{2}{3} \langle \, [H,\vec{d}\,]^2 \, \rangle_o \tag{86}$$

$$S_{-1} = \frac{2}{3} \langle d^2 \rangle_o$$

$$= \frac{2}{3} N_v \langle \, r_1^2 (1-\lambda_1)^2 \rangle_o + \frac{2}{3} N_v \, (N_v - 1) \, \langle \hat{r}_1 \cdot \hat{r}_2 \, (1-\lambda_1) \, (1-\lambda_2) \rangle \tag{87}$$

where $\langle \; \rangle_o$ indicates the expectation value in the ground state 0 and

$\lambda_j' = d\lambda_j/dr_j$. It is easily seen that if core polarization is neglected ($\lambda_j = 0$), the usual results $S_0 = N_v$, $S_1 = + \frac{2}{3} \langle(\sum_j p_j)^2\rangle$, and $S_{-1} = \frac{2}{3} \langle(\sum_j r_j)^2\rangle$ are obtained. The presence of core polarization tends to increase the effective number of electron S_0 and to decrease S_{-1}.

3.7 - MCRHF-CP calculations

Energies and oscillator strengths have been calculated for the mercury and cadmium isoelectronic sequences as well as for ytterbium. In each case, the closed-shell ion core was frozen, a limited MC approach was used to determine the intravalence correlation, and valence-core correlation was approximated by the polarization potential V_{CP} (76) and the correction (80) to the dipole-moment operator. Values of the core polarizability were taken from tabulated values (Fraga et al. 1976) and the core size parameter, r_0, was set equal to the average radius of the outermost (nlj) core orbital in some calculations (MCRHF-CP), and in others adjusted to reproduce energies relative to those of the closed-shell core (MCRHF-ACP). For the more highly charged members of the Hg and Cd isoelectronic sequence, the ionization energies required have not been measured, and consequently, only the version with fixed r_0 (MCRHF-CP) could be run.

The number of configurations included in the MC treatment of the valence shells was minimal : for the Cd and Hg isoelectronic sequences, the ground $^1S_0$ state mixed the lowest available s^2, $p^2_{1/2}$, and $p^2_{3/2}$ configurations whereas the $^1P_1$ and $^3P_1$ states combined only the configurations $s_{1/2}\ p_{1/2,3/2}$ necessary for intermediate coupling. In the case of Yb, which has a $4f^{14}$ core instead of the $4f^{14}\ 5d^{10}$ core of Hg, 6p5d configurations were added to the P states. Results confirm the ability of the core-polarization model to represent most of the valence-core correlation. When r_0 could be adjusted to match experimental ionization energies, calculated oscillator strengths were generally within or quite close to given experimental error limits. In a few exceptional cases for multiply charged ions where agreement with experiment was less satisfactory, it may be argued that the beam-foil measurements concerned failed to account adequately for cascading. The reader is referred to the original articles (Migdalek and Baylis, 1984b, 1985 a-c) for further details.

Note that our polarization model is relatively simple : there is a single size parameter r_0 for both direct terms $-\frac{1}{2}\alpha F_j^2$ and dielectric terms $-\alpha \vec{F}_i.\vec{F}_j$. Although the form of the potentials is that suggested in section 2.5, the asymptotic results (62) and (63) indicate that the dielectric terms should have a smaller parameter r_0 than the direct terms. Results, especially for the P states of the Cd sequence, indicate that the added flexibility of a separate size parameter for the dielectric term might permit a useful extension and refinement of our calculations.

The importance of the dielectric term is revealed by comparing the energy shifts caused by core polarization in the $^1P_1$ states with those in the $^3P_1$ states : the $^1P_1$ states often drop in energy more than the ground $^1S_0$ states whereas the $^3P_1$ states are lowered by relatively small amounts. The difference is due to the dielectric term. A simple calculation (Migdalek and Baylis, 1985c) shows that the average value of the cosine of the angle between the two valence electrons, $\langle\hat{r}_1.\hat{r}_2\rangle$, is positive for the $^1P_1$ state and negative for the $^3P_1$ state. In other words, the valence electrons are more often on the "same side" of the atom in the $^1P_1$ state and on "opposite sides" in the $^3P_1$ state. As a result, the dielectric term reinforces the effects of the direct polarization term in the $^1P_1$ state but largely cancels them in the $^3P_1$ state. Thus the dielectric term decreases the $^1P_1$ - $^3P_1$ fine-structure splitting and thereby increases the intermediate coupling and hence the $^1S_0$-$^3P_1$ oscillator strength (Migdalek and Baylis, 1985a).

4. Atom-atom interactions

4.1 - Oscillator model revisited

In the first part of this chapter, an oscillator model was used to suggest a polarization model for effects of valence-core correlation. The model was developed formally and more completely in part 2 as perturbation-theory corrections to the frozen-core approximation. In section 3, the utility of the model was demonstrated in applications to the two-electron spectra of some heavy atoms and ions. Now we return to the oscillator model to see what might be learned about atom-atom interactions.

Consider two isotropic harmonic oscillators with mass, charge, and natural angular-frequency given by (m_i, q_i, ω_i), i = 1, 2, in the presence of uniform electric fields F_i, and coupled to each other by the dipole-dipole interaction. The hamiltonian is

$$H = \sum_{i=1,2} \left(\frac{P_i^2}{2m_i} + \frac{1}{2} m_i \omega_i^2 r_i^2 - q_i \vec{r}_i . \vec{F}_i\right) + q_1 q_2 \vec{r}_1 . (1-3\hat{R}\hat{R}) . \vec{r}_2 \, R^{-3} \tag{88}$$

where $\vec{R}$ is the center of oscillator 2 with respect to oscillator 1. If we take $R = R\hat{z}$, the Schrödinger equation separates trivially into x, y, and z parts. The final separation into normal modes is effected by orthogonal transformations of the form

$$\begin{pmatrix} \sqrt{m_1}\, x_1 \\ \sqrt{m_2}\, x_2 \end{pmatrix} = \begin{pmatrix} \cos\theta & -\sin\theta \\ \sin\theta & \cos\theta \end{pmatrix} \begin{pmatrix} \xi_+ \\ \xi_- \end{pmatrix} \tag{89}$$

Details are left to the reader. We quote here only the shift ΔE in energy of the system ground state, through terms varying as R^{-6} ($h = e = 1$) :

$$\Delta E = -\frac{3}{2} \frac{\omega_1\omega_2}{\omega_1+\omega_2} \frac{\alpha_1\alpha_2}{R^6} + \vec{F}_1 . \overleftrightarrow{\Phi} . \vec{F}_2 \frac{\alpha_1\alpha_2}{R^3} - \frac{1}{2}\alpha_1 F_1^2 - \frac{1}{2}\alpha_2 F_3^2$$

$$- \frac{1}{2}\left(\alpha_1 \vec{F}_1 . \vec{\Phi}^2 . \vec{F}_1 + \alpha_2 \vec{F}_2 . \overleftrightarrow{\Phi}^2 . \vec{F}_2\right) \frac{\alpha_1\alpha_2}{R^6} \tag{90}$$

where $\alpha_i = q_i^2/(m_i\omega_i^2)$ is the polarizability of oscillator i, and the dyads $\overleftrightarrow{\Phi}$ and $\overleftrightarrow{\Phi}^2$ are given by

$$\overleftrightarrow{\Phi} = \overleftrightarrow{1} - 3\hat{R}\hat{R} \quad , \quad \overleftrightarrow{\Phi}^2 = \overleftrightarrow{\Phi}.\overleftrightarrow{\Phi} = \overleftrightarrow{1} + 3\hat{R}\hat{R} \tag{91}$$

The first term on the RHS of (90) is the long-range Van der Waals interaction between the oscillators. The significance of the other terms is clarified if we specify the fields $\vec{F}_1$ and $\vec{F}_2$ for a simple model such as the alkali (1) - noble gas (2) interaction. The oscillators then represent the ion core of the alkali and the closed-shell noble-gas atom, and the fields $\vec{F}_1$ and $\vec{F}_2$ represent the effective average field at the alkali core due to the alkali valence electron at r and the field at the noble-gas atom arising from both the alkali valence electron and the positively charged alkali core, respectively. For F_2, we consider only the leading term in the expansion in inverse powers of R :

$$\vec{F}_1 = \vec{r}\,(r^2 + r_o^2)^{-3/2} \quad , \quad \vec{F}_2 = \vec{r}.\overleftrightarrow{\Phi}\, R^{-3} \tag{92}$$

$$\Delta E = -\frac{3}{2}\,\frac{\omega_1\omega_2}{\omega_1+\omega_2}\,\frac{\alpha_1\alpha_2}{R^6} - \frac{1}{2}\,\alpha_1\,F_1^{\,2} - \frac{1}{2}\,\frac{\alpha_2}{R^6}\,\vec{r}\,.\,\overleftrightarrow{\Phi}^2\,.\,\vec{r}\,(1-\lambda)^2 \tag{93}$$

where $\lambda = \alpha\,(r^2 + r_o^2)^{-3/2}$ is the same expression as in section 3.6 ; as there, it enters to reduce the dipole moment $-\vec{r}$ of the valence electron. Here it reduces the main Van der Waals interaction between the alkali valence shell and the noble-gas atom. Its presence does not appear to have been recognized previously in atom-atom interactions.

4.2 - The Van der Waals coefficient

In addition to the core polarization term $-\frac{1}{2}\,\alpha_1\,F_1^{\,2}$, equation (93) contains a long-range interaction with an expectation value $-\,C_6\,R^{-6}$ where the Van der Waals coefficient C_6 is :

$$C_6 = \frac{3}{2}\,\frac{\omega_1\omega_2}{\omega_1+\omega_2}\,\alpha_1\alpha_2 + \alpha_2 < [r(1-\lambda)]^2 > . \tag{94}$$

Of course (94) contains no dynamics effects because in its derivation the position $\vec{r}$ of the valence electron was taken as a constant. One can check the presence of the factor $(1-\lambda)$ and calculate dynamic corrections by a direct calculation of C_6 from second-order perturbation theory :

$$C_6 = \sum_{a,b} \frac{|\langle a|\ \vec{d}_1\ |o\rangle \ . \ \overleftrightarrow{\Phi} \ . \ \langle b|\ \vec{d}_2\ |o\rangle^{\ 2}}{E_a(1) - E_o(1) + E_6(2) - E_o(2)} \tag{95}$$

where a and b are summed over excited states of atoms 1 and 2, respectively. Because the unperturbed hamiltonians of the noninteracting atoms are spherically symmetric and invariant under inversion, (95) can be reduced to

$$C_6 = \frac{2}{3} \sum_{a,b} \frac{|\langle a|\vec{d}_1|o\rangle|^2 |\langle b|\vec{d}_2|o\rangle|^2}{E_a(1)-E_o(1)+E_b(2)-E_o(2)}$$

$$= \frac{3}{2} \sum_{a,b} \frac{f_{ao}(1)\ f_{bo}(2)}{[E_a(1)-\ E_a(1)][E_b(2)-E_o(2)][E_a(1)+E_o(1)+E_b(2)-E_o(2)]} \tag{96}$$

where $f_{ao}(1)$ and $f_{bo}(2)$ are absorption oscillator strengths (see (82)). If the excitation energies of atom 2 are generally large compared to those of the important transitions in atom 1, one can expand the factor

$$[E_a(1)-E_o(1) + E_b(2)-E_o(2)]^{-1} \simeq [E_b(2)-E_o(2)]^{-1}\ (1 - X + x^2 - \ldots) \tag{97}$$

where $X = [E_a(1)-E_o(1)]/[E_b(2)-E_o(2)]$ in an expansion analogous to that for the resolvant operator (27). The result can be expressed directly in terms of the sums S_k (81) :

$$C_6 \simeq \frac{3}{2}\ [S_{-1}(1)\ S_{-2}(2) - S_0(1)\ S_{-3}(2) + S_1(1)\ S_{-4}(2) - \ldots\ ]$$

$$\simeq \frac{3}{2}\ S_{-1}(1)\ \alpha_2 - 3\ S_0(1)\ \beta_2^{(1)} + 6\ S_1(1)\ \gamma_2 - \ldots \tag{98}$$

where we have identified the static dipole polarizability and its first- and second-order dynamic corrections as

$$\alpha_2 = S_{-2}(2) \quad , \quad \beta_2^{(1)} = \frac{1}{2} S_{-3}(2) \ , \text{ and } \gamma_2 \equiv \frac{1}{2} \beta_2^{(2)} = \frac{1}{4} S_{-4}(2) \qquad (99)$$

By using the oscillator-strength sum rules corrected for core polarization (section 3.6), one confirms the presence of the reduction factor $(1-\lambda)$ in (94), but in addition one generalizes the result to include dynamic corrections for atom 2 and to allow for several valence electrons on atom 1. The core-core term does not appear in (98) ; it would arise in first-order perturbation theory as a contribution to the polarization potential $-\frac{1}{2}\alpha_1 F_1^{\,2}$ from that part of F_1 due to the instantaneous dipole $\vec{d}_2$ in atom 2.

4.3 - Future work

The basic idea of a core-polarization model is not new. Polarization potentials were used many years ago to improve hand-cranked Hartree computations (Biermann, 1943). The emphasis in this chapter has been to improve the theoretical foundation of the model, to demonstrate its use both in MCRHF calculations of atoms and in pseudopotential and frozen-core calculations of small molecules, and to investigate possible refinements and extensions in its applications.

More work needs to be done. In particular, the short-range part of V_{pol} should be determined. How important is the exchange (section 2.2) ? the monopole contribution (section 2.3) ? Answers to such questions may be needed before core-polarization methods can make useful calculations requiring accurate valence wave functions at the nucleus, as for example, in determinations of electro-weak interactions (Martenson-Pendrill 1985 ; Dzuba et al. 1985 and references therein).

The static contribution to V_{pol} can be calculated in a relativistic version of coupled-HF theory (Johnson, Kolb and Huang 1983 and references therein). In addition molecular calculations should be made which include the reduction in valence multipole moments by core polarization (sections 3.5, 4.1 and 4.2). Such work will provide useful improvements to a model whose continued use in accurate atomic and molecular calculations seems assured, both with pseudopotential approaches and with *ab initio* methods like MCRHF in which the correlation contribution is restricted to the intravalence part where MC treatments excel.

References

. Baylis W.E. : J. Chem. Phys. 51, 2665 (1969).

. Baylis W.E. : J. Phys. B 10, L583 (1977).

. Baylis W.E. : in Progress in Atomic Spectroscopy, part A, ed. by W. Hanle and H. Kleinpoppen (London : Plenum) (1978).

. Biermann L. : Z. Astrophys. 22, 157 (1943).

. Bottcher C. and Dalgarno A. : Proc. R. Soc. A 340, 187 (1974).

. Dalgarno A., Drake G.W.F. and Victor G.A. : Phys. Rev. 176, 194 (1968).

. Deech J.S. and Baylis W.E. : Can J. Phys. 49, 90 (1971).

. Desclaux J.P. : Int. J. Quantum Chem. 6, 25 (1972).

. Desclaux J.P. : Comput. Phys. Commun. 9, 31 (1975).

. Desclaux J.P. and Kim Y.K. : J. Phys. B 8, 1177 (1975).

. Dzuba V.A., Flambaum V.V., Silvestrov P.G. and Shushkov O.P. : J. Phys. B 18, 597 (1985).

. Fraga S., Karwowski J. and Saxena K.M.S. : Handbook of Atomic Data (Amsterdam : Elsevier) (1976).

. Johnson W.R., Kolb D. and Huang K-N : Atom. Data Nucl. Data Tab. 28, 333 (1983).

. Lurio A. : Phys. Rev. 140, 1505 (1965).

. Martenson-Pendrill A.M. : Phys. Rev. Lett. 54, 1153 (1985).

. Migdalek J. and Baylis W.E. : J. Phys. B 11, L497 (1978).

. Migdalek J. and Baylis W.E. : J. Phys. B 12, 113 (1979a).

. Midgalek J. and Baylis W.E. : J. Quant. Spectrosc. Radiat. Transfer 22, 113 (1979b).

. Migdalek J. and Baylis W.E. : J. Quant. Spectrosc. Radiat. Transfer 22, 127 (1979c).

. Migdalek J. and Baylis W.E. : Can. J. Phys. 59, 769 (1981).
. Migdalek J. and Baylis W.E. : Can. J. Phys. 60, 1317 (1982).
. Migdalek J. and Baylis W.E. : Phys. Rev. A 30, 1603 (1984a).
. Migdalek J. and Baylis W.E. : J. Phys. B 17, L459 (1984b).
. Midgalek J. and Baylis W.E. : J. Phys. B 18, 1533 (1985a).
. Migdalek J. and Baylis W.E. : J. Phys. B 18, in press (1985b).
. Migdalek J. and Baylis W.E. : Phys. Rev. A 32, in press (1985c).
. Moore C.E. : Atomic Energy Levels NSRDS-NBS N° 35 (Washington, DC : US Govt Printing Office) (1971).
. Pascale J. : (Chapter in this volume) (1986).
. Peach G. : in Atoms in Astrophysics, ed. by P.G. Burke, W.B. Eissner, D.G. Hummer, and I.C. Percival (London : Plenum) (1983).
. Sucher J. : Phys. Rev. Lett. 55, 1033 (1985).
. Vinti J.P. : Phys. Rev. 41, 432 (1932).

USE OF PSEUDOPOTENTIALS IN ATOM-ATOM (OR MOLECULE) COLLISIONS

Jean Pascale

Service de Physique des Atomes et des Surfaces, Centre d'Etudes Nucléaires de Saclay, 91191 Gif-sur-Yvette Cedex, France

1. Introduction

Knowledge of interactions betweens ions, atoms or molecules is fundamental for interpretating or predicting collisional processes which may occur under various circumstances. The aim of this paper is to demonstrate the usefulness of using semiempirical effective interactions (more particularly, emphasis will be put on the pseudopotential approach) in the study of atom-atom (or molecule) collisions. Therefore, we will not review in this paper the numerous theoretical works devoted to various atom-atom (ion or molecule) systems and using a semiempirical effective interaction approach. Instead, we would like to show that if the semiempirical effective interactions are carefully defined, their use in molecular-structure calculations and in collision problems can give quite accurate results. These can be of comparable or even higher quality than those obtained from more sophisticated approaches. We will limit our examples to one-electron systems, while extension to more than one valence electron does not present any particular difficulty in principle. The alkali-metal (M) -atom - rare gas systems, which may be reduced to interactions between a one valence electron only and two closed-shell cores are certainly the most appropriate systems to be studied using an effective interaction approach. Moreover, these systems are easy to handle in laboratories, and so a large amount of experimental data has been produced up to now, leading to numerous confrontations between experiment and theory. In this paper, we will consider the M-atom-He systems as a first example. For these systems, recent molecular-structure calculations have been carried out using an l-dependent semiempirical pseudopotential approach (Pascale, 1983a) and they have been tested against numerous experimental data in extensive calculations of cross sections for intra- and -inter-doublet transitions in the M-atom in collisions with He (Pascale, 1983 b, c, and to be published ; Kimura and Pascale, 1985). Our second example will concern the M-H_2 systems, for which semiempirical pseudopotential molecular-structure calculations have been performed very recently (Rossi and Pascale, 1985) using a one-electron two-center model (in which the molecular nature of H_2 is accounted for). The results of these calculations are quite encouraging and we foresee the use of the pseudopotential approach in future studies of some reactive scattering processes.

2. Pseudopotential molecular-structure calculations

A - Generalities

Let us consider the interaction between two atoms or ions. Generally, one has to solve the Schrödinger equation describing the whole system within the Born-Oppenheimer (BO) approximation, in which the nuclear and electronic wave functions are separable. Therefore, one needs to solve the electronic Schrödinger equation,

$$H_{el}\ \psi_{el} = E_{el}(R)\ \psi_{el}\ , \tag{1}$$

for any fixed distance R between the two nuclei, in order to obtain the electronic energy $E_{el}(R)$. In Eq. (1), H_{el} is the full electronic Hamiltonian which contains all the one-electron kinetic-energy operators and all the Coulombic interactions involving the electrons. Then, the adiabatic potential energy E(R) of the system for a given state is defined as the sum of $E_{el}(R)$ and of the Coulombic repulsion between the nuclei.

A.1 - Full electron *ab initio* methods

For solving Eq. (1) by *ab initio* methods one makes no approximation, at any stage of the calculations, which would require either the use of some experimental data or other input not determined from fundamental principles. Generally, in order to solve Eq. (1), a variational procedure is used. It consists of defining an approximate electronic wave function ψ_a depending on several parameters, and in varying these parameters to minimize the expectation value

$$\langle\psi_a|H_{el}|\psi_a\rangle = E_a , \qquad (2)$$

where E_a is the approximate electronic energy. Obviously, the accuracy of such calculations will depend upon the flexibility of ψ_a. Usually, ψ_a may be obtained within the Hartree-Fock self-consistent-field (HF-SCF) approximation, in which each electron moves in the average potential due to the other electrons of the system. Then correlations between the electrons may be accounted for by configuration interaction (CI) calculations (some aspects of these *ab initio* methods are discussed by W.E. Baylis elsewhere in this volume). To illustrate the complexity of such calculations, when a high accuracy of the results is expected, let us consider the all-electron CI calculations of Saxon et al. (1977) for the NaAr system. In these calculations, where the eleven electrons of Na and the eighten electrons of Ar are considered, ψ_a was expanded in terms of orthonormal configuration state functions (CSF). Each CSF was written as a linear combination of slater determinants (SD). And each SD was built from an orthonormal basis set of one-electron wave functions expanded in terms of Slater-type orbitals (STO), including STO's with large values of the orbital quantum number l to account correctly for dispersion forces. To obtain the adiabatic potential energies of the $X^2\Sigma^+$, $B^2\Sigma^+$ and $A^2\Pi$ states, 2568 (for the $^2\Sigma^+$ states) or 4235 (for the $^2\Pi$ state) CSF's were included in the calculations. However, in order to reduce the computation time, only interatomic valence correlation energies were considered in these calculations. The results showed large deviations of the CI calculations from the SCF calculations and improved the agreement between experiment (York et al., 1975) and previous semiempirical pseudopotential calculations (Baylis, 1969 ; Pascale and Vandeplanque, 1974). However, the results showed also some deficiencies of the CI calculations, due mainly to the neglect of intra-atomic valence correlation energies.

Because the computer time and computer storage increase rapidly as the number of basis functions increases, the need of introducing some simplifications in the treatment of large systems becomes manifest. Simplifications in the calculations, without loss of accuracy, is the purpose of using effective interactions.

A.2 - *Ab initio* effective potential methods

The basic idea of these methods is that only the outermost valence electrons of the atoms (or ions) constituting the molecular system are important in determining the chemical bonding. Thus, the full-electron calculations are simplified by separating the electrons of the system into valence and core electrons. An effective potential is defined to describe the interaction between a valence electron and the core electrons. Then, in CI calculations the core electrons are "frozen" and only the correlations between the valence electrons are considered. There has been a quite large number of reviews on *ab initio* effective interaction methods (see for example, Weeks et al., 1969 ; Bardsley, 1974 ; Kahn et al., 1976 ; Barthelat et al., 1977 ; Dixon and Robertson, 1978) to which we refer the reader for details. Here, we just survey the effective interaction approach by considering again the example of the NaAr system studied by Laskowsky et al. (1981). In this approach, Na is considered as a core A and one valence electron, and Ar as a core B and eight valence electrons. The electronic Hamiltonian of the system is written as,

$$H_{el} = \sum_{i=1}^{N} \left[-\frac{1}{2} \nabla_i^2 + \sum_x \left(-\frac{Z_x}{r_{ix}} + v_{ix} \right) + \sum_{j>i} \frac{1}{r_{ij}} \right] \qquad (3)$$

where $N = 9$ and Z_x is the nuclear charge of the core x (A or B). The problem is then to determine an effective potential v_{ix} which describes the interaction between a valence electron e_i^- and the core x. Defining the one-electron Hamiltonian as :

$$h_{ix} = -\frac{1}{2}\nabla_i^2 - \frac{Z_x}{r_{ix}} + v_{ix} , \qquad (4)$$

the effective potential v_{ix} is such that the valence electron wave function φ_{ix} satifies the Schrödinger equation :

$$h_{ix}\varphi_{ix} = \varepsilon_{ix}\varphi_{ix} \qquad (5)$$

where ε_{ix} is the valence electron energy. Provided that φ_{ix} is orthogonal to all the core orbitals, in order to satisfy the Pauli principle, v_{ix} can be defined as a non-local operator in terms of both the valence and core orbitals (note that the orthogonality of φ_{ix} to all the core orbitals implies that the radial part of φ_{ix} oscillates in the core region). However, such a definition of v_{ix} is not very useful for molecular-structure calculations because all the core orbitals have then to be considered in the basis function set. In the pseudopotential approach, first one restricts the non-locality of v_{ix} to the angular momentum dependence, defining

$$v_{ix} = \sum_l v_{ix}^{(l)}(r_{ix})\mathcal{P}_l^x , \qquad (6)$$

where $\mathcal{P}_l^x$ is an l-projector on the center x. Second, for each l-symmetry, the local potential $v_{ix}^{(l)}(r_{ix})$ is defined such that the radial pseudo-wave function of the valence electron in its lowest state has no node, but is identical to the true valence wave function at large distance r_{ix}. With such a definition v_{ix}, the orthogonality constraints of φ_{ix} to all the core orbitals are implicitely satisfied, and the core orbitals can be safely ignored in the molecular-structure calculations. In the *ab initio* pseudopotential-CI calculations of Lakowsky et al. (1981) for NaAr, all the correlations among the nine valence electrons were considered. However, because their basis set of Gaussian functions did not include large values of l, their calculations failed to predict correctly the dispersion energies in spite of the large computational effort which was required (several thousands of CSF's were used). These calculations, as well as those of Saxon et al. (1977), illustrate clearly the considerable amount of work needed in CI-*ab initio* methods to predict correctly the dispersion energies. To overcome these difficulties, an *ab initio* method has been proposed recently by M. Hliwa et al. (1985) for the calculation of dispersion energies. This method, prompted by the perturbative approach of Bottcher and Dalgarno (1974), does not require the definition of a cut-off radius as in the semiempirical model potential or pseudopotential approaches described below ; but it is limited to dipolar terms. While the method seems promising, its application to NaAr in particular is not fully convincing (due to an insufficient knowledge of the core-core interaction, a scaling of the theoretical results to an experimental ground-state potential curve was necessary for improving the agreement between experiment and theory).

A.3 - Semiempirical effective potential methods (model potential (MP) and pseudopotential (PP) approaches)

In these methods the effective potentials discussed above are determined by modeling fully or partly the interactions in order to reproduce accurately some experimental data. In that sense these methods are semiempirical.

In order to illustrate these approaches, let us consider again the example of an alkali-metal (M)-atom-rare gas system. Because both the M-ion and the rare gas have closed-shell structures, the problem of determining the interaction between the M-atom and the rare gas is reduced to a three-body problem (namely a core A representing the M-ion, the valence electron e^-, and a core B representing the rare gas, as in Fig. 1 , in which the two-body interactions are described by effective potentials. An effective Hamiltonian describing the three-body system may be written as

$$H_{eff} = H_{el} + V_{AB}(R) \quad , \tag{7}$$

with

$$H_{el} = -\frac{1}{2}\nabla^2_{\vec{r}_x} + V_A + V_B + V_{CT}(\vec{r}_x, \vec{R}) \ . \tag{8}$$

V_x, $x \equiv A$ or B and V_{AB}. are effective potentials describing, respectively

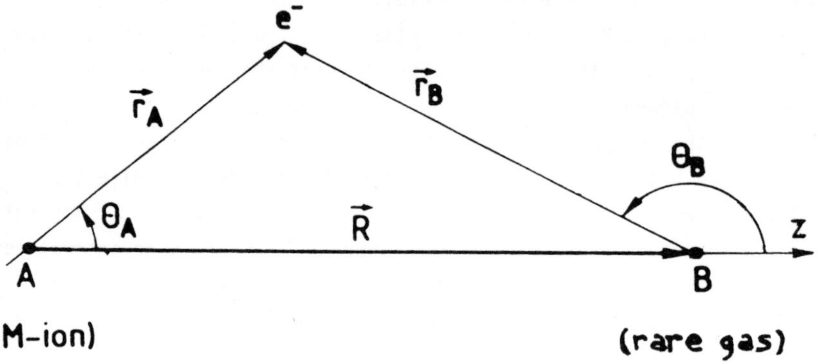

Fig. 1 - The three-body model describing the alkali-metal-atom-rare gas interaction.

the e^--x and core-core interactions. V_{CT} $(\vec{r}_x, \vec{R})$ is a term (the so-called cross term defined later on) which must be included in the calculations in order to have the correct behavior of the adiabatic potential energies at large internuclear distances R. This term depends both of the position vector $\vec{r}_x$ of e^- with respect to the core x, and of the position vector $\vec{R}$ of B with respect to A (see Fig. 1). As before, the electronic energies $E_{el}(R)$ are obtained by solving the one-electron Schrödinger equation (1) corresponding to the effective electronic Hamiltonian (Eq. (8)) and the adiabatic potential energies are then :

$$E(R) = E_{el}(R) + V_{AB}(R) \ . \tag{9}$$

The determination of V_{AB} may cause a real problem in some cases. Usually, V_{AB} may be modeled to fit some reliable experimental scattering data, when available.

Let us discuss here the effective potential V_x describing the e^--x interaction. Its determination has become rather standard and depends only on the approach chosen : model potential or pseudopotential. In general, V_x is separated into a short-range part and a long-range part :

$$V_x = V_x^{SR} + V_x^{LR}(r_x) \tag{10}$$

The long-range part may be expressed as a local potential $V_x^{LR}(r_x)$, a sum of polarization terms and of the Coulombic interaction between e^- ans the net charge of the core x. The polarization terms are relatively well-known at large values of r_x (see, for example, Buckingham, 1967 ; Peach, 1983), but their extension to small values of r_x require to the introduction of analytic cut-off functions which are rather arbitrarily defined. The short-range part, V_x^{SR}, is now defined such that the valence electron wave function satisfies the Schrödinger equation,

$$\left[-\frac{1}{2}\nabla^2_{\vec{r}_x} + V_x^{SR} + V_x^{LR}(r_x) \right] \varphi(\vec{r}_x) = \varepsilon\, \varphi(\vec{r}_x) \tag{11}$$

ε is the valence electron energy which can be negative (bound state) or positive (continuum state). When x is an M-ion, V_x^{SR} is usually modeled to reproduce correctly the nl-energy levels of the M-atom ; when x is a rare gas, V_x is modeled to reproduce low-energy e^--x elastic scattering experimental data (usually, the lowest l-wave phase shifts and the scattering length). Now, the distinction between pseudopotential (PP) and model potential (MP) approaches comes essentially from the definition of V_x^{SR}, or more exactly from the conditions imposed on $\varphi(\vec{r}_x)$. To be more specific let us consider first the case where x is an M-ion ($x \equiv A$).

If one puts $V_A^{SR} = 0$ and $\varepsilon = \varepsilon_{nl}$, the experimental energy of a given nl level of the M-atom, and integrates the radial Schrödinger equation corresponding to Eq. (11) inward from infinity :

$$\left[-\frac{1}{2}\frac{d^2}{dr_A^2} - \frac{l\,(l+1)}{2\,r_A^2} + V_A^{LR}(r_A) \right] R_{nl}(r_A) = \varepsilon_{nl}\, R_{nl}(r_A)\ , \tag{12}$$

one finds the following results : $R_{nl}(r_A)$ behaves correctly at large r_A-values, but may oscillate for $r_A \to 0$ because $V_A^{LR}(r_A)$ is attractive and the effective potential $\frac{l(l+1)}{2r_A^2} + V_A^{LR}(r_A)$ may support several bound states for the lowest values of l. In the PP approach, one defines V_A^{SR} in terms of projectors onto the angular momentum l as discussed in Sec A.1,

$$V_{Al}^{SR} = \sum_{l=0}^{\infty} V_{Al}^{SR}(r_A)\mathcal{P}_l^A \qquad (13)$$

Then, for each l-symmetry, the local potential $V_{Al}^{SR}(r_A)$ is introduced in Eq. (12) and modeled so that the pseudo-radial wave function $R_{nl}^{PP}(r_A)$ has no node for the lowest nl level. Let $l_{max}-1$ be the largest value of l corresponding to a real bound state of the core. Then, one finds that for $l < l_{max}$ $V_{Al}^{SR}(r_A)$ has to be repulsive to satisfy the Pauli principle. For $l \geq l_{max}$, $V_{Al}^{SR}(r_A)$ has to be attractive to take into account the incomplete screening of the nucleus by the core electrons for small r_x-values. Usually it is enough to take $V_{Al}^{SR}(r_A) = V_{Al_{max}}^{SR}(r_A)$ for $l > l_{max}$, so that the summation in Eq. (13) is limited only to a few terms. In the MP approach, $V_{Al}^{SR}(r_A)$ is introduced in Eq. (12) and modeled so that the model potential-radial wave function $R_{nl}^{MP}(r_A)$ has the correct number of nodes (n-l-1 nodes). Then virtual bound states are predicted, which may correspond closely to the real bound states of the core. Usually, a local potential (in the sense that it does not depend on l) can be found, which allows one to fit closely the experimental energy levels ε_{nl} of the M-atom. The non-locality of the e^--A interaction, as discussed in Sec. A.2, is then reflected by the conditions that the valence electron wave function $\varphi^{MP}(\vec{r}_A)$ must be orthogonal to the wave functions of the virtual bound states.

In the case where x is a rare gas atom (X ≡ B) the definition of V_B^{SR}, in the PP or MP approach, is equivalent to above by considering the Levinson's theorem (see, for example, de Alfaro and Regge, 1965) for the e^--x elastic scattering : the number n_l of real bound states of the core x corresponds to an l-wave phase shift $\eta_l(E)$ such that

$$\eta_l\ (E = 0) = n_l\ \Pi\ , \qquad (14)$$

where E is the incident electron energy. Therefore, when solving the radial Schrödinger equation similar to Eq. (12) but with ε_{nl} replaced by E, in order to determine $\eta_l(E)$, a local potential is introduced in a manner equivalent to above. In the PP approach an l-dependent short-range potential $V_{Bl}^{SR}(r_B)$ is modeled so that $\eta_l^{PP}(E)$ fits the experimental l-wave phase shift η_l^{exp} (E) (modulo Π) at low energies, and satisfies

$$\eta_l^{PP}\ (E = 0) = 0 \tag{15}$$

For $l < l_{max}$ (where l_{max} is defined above), $V_{Bl}^{SR}(r_B)$ is repulsive to simulate the Pauli principle, and for $l \geq l_{max}$ it is attractive and, $V_{Bl}^{SR}(r_B) = V_{Bl_{max}}^{SR}(r_B)$. In the MP approach, a local short range potential is modeled such that $\eta_l^{MP}(E)$ fits $\eta_l^{exp}(E)$ at low energies, and satisfies Eq. (14) ; moreover, $\varphi^{MP}(\vec{r}_B)$ must satisfy the requirement that it is orthogonal to the wave functions of the virtual bound states of the same symmetry predicted by the effective potential V_B. An attractive potential $V_B^{SR}(r_B)$, independent of l, is generally determined, which should predict virtual bound states as close as possible of the real ones. However, this condition is difficult to realize since V_B does not contain any Coulombic term, and therefore makes the position of the virtual bound states very sensitive to $V_B^{SR}(r_B)$. Two procedures have been proposed : $V_B^{SR}(r_B)$ may be modeled by fitting $-\eta_l^{exp}(E)$ at low energies, with the constraint that $\varphi^{MP}(\vec{r}_B)$ is orthogonal to HF core orbitals of the same symmetry when solving the radial Schrödinger equation (see, for example, Philippe et al., 1979) ; or $V_B^{SR}(r_B)$ may be modeled by fitting $\eta_l^{exp}(E)$ over a larger domain of energies than above (G. Peach, 1982).

Given the interactions V_X and the cross term $V_{CT}(\vec{r}_X, \vec{R})$ defining the electronic Hamiltonian (Eq. 8), the electronic energies $E_{el}(R)$ and the electronic wave functions $\psi_{el}(\vec{r},\vec{R})$ are usually obtained by solving the one-electron Schrödinger equation (Eq. 1) with standard variational procedures. But, whereas in the PP approach $\psi_{el}(\vec{r},\vec{R})$ has only to tbe expanded over a basis set of valence orbitals centered on the M-atom the MP approach requires that the basis set includes also orbitals describing the virtual bound states of the M-ion and of the rare gas atom. Therefore, the PP approach appears to be more efficient for molecular-structure calculations over the MP approach when the number of core electrons increases. In particular, for two-valence electrons systems the MP approach may give rise to instabilities of $E_{el}(R)$ at small R (see, Mo et al., 1985) due to the presence of virtual bound states, while the PP approach avoids these problems in principle.

In the following sections we present recent applications of the PP approach to molecular-structure calculations of M-atom-He (Pascale, 1983a) and M-atom-H_2 systems (Rossi and Pascale, 1985).

B - Recent applications of the PP approach

B.1 - M-atom-He systems. The interaction between the M-atom and He is described by the above three-body model (see Fig. 1).

B.1.a - The interactions. The effective interaction V_x between e^- and the core x is separated as in Eq. (10) into a short-range part V_x^{SR} and a long-range par V_x^{LR}. For V_x^{SR} we use an l-dependent pseudopotential (see Eq. 13), where the local potential $V_{xl}(r_x)$ is of Gaussian-type :

$$V_{xl}(r_x) = C_{xl} \exp(- D_{xl} r_x^2) \tag{16}$$

and C_{xl} and D_{xl} are parameters. For V_x^{LR}, we use a local potential,

$$V_x^{LR}(r_x) = - \frac{Z_x}{r_x} - \frac{1}{2} \frac{\alpha_{d_x}}{(r_x^2 + d_x^2)^2} - \frac{1}{2} \frac{\alpha'_{q_x}}{(r_x^2 + d_x^2)^3} \tag{17}$$

Z_x is the net charge of the core ; α_{d_x} is the experimental static dipole polarizability, and

$$\alpha'_{q_x} = \alpha_{q_x} - 6 \beta_x + 2 \alpha_{d_x} d_x^2 , \tag{18}$$

where α_{q_x} is the experimental static quadrupole polarizability ; β_x is a dynamical correction to the static dipole polarizability and d_x is a cut-off radius to be adjusted. In the case of the M-ion, β_A was considered as a parameter. Then the parameters of the effective interactions were determined by fitting some experimental data. In the case of the e^--M-ion interaction, the parameters were obtained by reproducing the experimental energies levels of the M-atom (we have used the values determined by Bardsley, 1974). For the e^--He interaction, the parameters were obtained by fitting experimental l-wave phase shifts (s, p, and d waves) of Williams (1979), as well as the value of the scattering length, namely 1.18 au (atomic units) determined by O'Malley et al. (1979).

The core-core interaction $V_{AB}(R)$ was taken as

$$V_{AB}(R) = V_{AB}^{SR}(R) - \frac{1}{2}\,\frac{\alpha_{d_B}}{(R^2 + d_B^2)^2} - \frac{1}{2}\,\frac{\alpha''_{q_B}}{(R^2 + d_B^2)^3} \tag{19}$$

where $\alpha''_{q_B} = \alpha_{q_B} + 2\,\alpha_{d_B}\,d_B^2$, and the short-range part of the potential was modeled as

$$V_{AB}^{SR}(R) = a\,\exp\,(-\,b\,R) \tag{20}$$

Then the parameters a and b were adjusted by fitting the experimental $X^1\Sigma^+$ potential curves of the M-ion-He systems (Inouye et al., 1979) obtained from elastic scattering experiment in the 0.5-4 keV energy range. We refer to Pascale (1983a) for a tabulation of the various parameters used in the calculations.

Finally, in order to define fully the one-electron Hamiltonian, we need to determine the cross-term $V_{CT}(\vec{r}_x,\vec{R})$ which results from the polarization of He by both e^- and the M-ion. This term is important to obtain the correct behavior of the adiabatic potentials at large internuclear distances :

$$E_i(R) - E_i(\infty) = -\,\frac{\alpha_{d_B}}{R^6} < r_A{}^2\,(1 + P_2\,(\hat{r}_A)>_i + \frac{3\beta_B}{R^6} \tag{21}$$

where $<\,>_i$ denotes the average value with respect to the wave function of the M-atom in the state i. The expression of $V_{CT}(\vec{r}_x,\vec{R})$ is well-known asymptotically (see, for example, Peach, 1983). For any position $(\vec{r}_x,\vec{R})$ one may use :

$$V'_{CT}(\vec{r}_B,\vec{R}) = \frac{\alpha_{d_B}\,\xi_B}{(R^2 + d_B^2)\,(r_B^2 + d_B^2)} + \frac{1}{2}\,\frac{\alpha''_{q_B}\,(3\,\xi_B^2 - 1)}{(R^2 + d_B^2)^{3/2}\,(r_B^2 + d_B^2)^{3/2}} \qquad (22)$$

whith is consistent with the polarization terms used in Eqs. (17) and (19), where $\xi_B = \hat{r}_B.\hat{R}$ ($\hat{r}_B = \vec{r}_B/r_B$). But since this expression is only valid when the electronic charge densities of the two atoms do not overlap, it was proposed (Pascale, 1983a) to use

$$V_{CT}(\vec{r}_B,\vec{R}) = V'_{CT}(\vec{r}_B,\vec{R})\; f_c\left(\frac{R}{r_A}\right) \qquad (23)$$

where the additional cut-off function in R/r_A forces V_{CT} smoothly to zero for $r_A \geq R$.

B.1.b - Molecular-structure calculations. The electronic wave function $\psi_e(\vec{r}_A,R)$ was expanded over a large basis set of STO for the M-atom, where $\vec{R}$ is chosen as the quantization axis. The non-linear parameters of the STO's were optimized in order to reproduce accurately the ionization energies of the excited levels up to the first nG level (with an accuracy generally better than 2.5 10^{-4} au) ; they were kept constant for all R-values. The electronic energies (up to the first nG level of the M atom) were obtained from the diagonalization of the electronic hamiltonian (Eq. (8)) for each value of the projection M_L of the total orbital momentum $\vec{L}$ (equal to that of the valence electron, in the present case). The calculations were performed for all the M-atom-He system in the range R = 2 - 50 a.u.

B.1.c - Results and discussion. To illustrate the accuracy of the PP calculations (J. Pascale, 1983a), let us consider the case of the NaHe and CsHe systems. Fig. 2 shows the comparisons of the PP calculations with the $X^2\Sigma$ and $A^2\Pi$ potential curves obtained from far-wing intensity measurements of the resonance line of Na broadened by He (Havey et al., 1982) and using the quasi-static approximation to analyze the data. The PP results are in excellent agreement with experiment, while the well depths predicted by the SCF-HF calculations of Krauss et al. (1971) and the MP

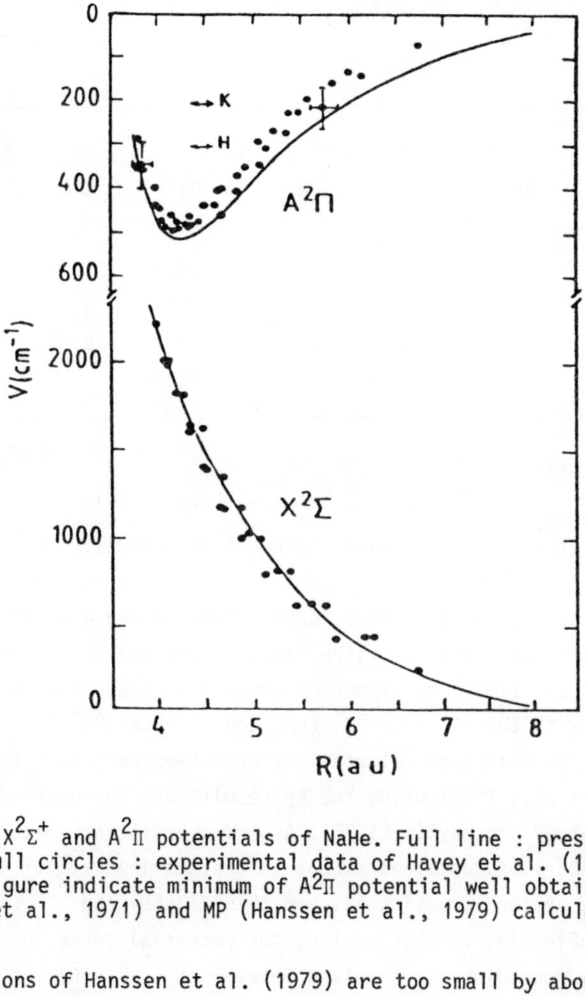

Fig. 2 - $X^2\Sigma^+$ and $A^2\Pi$ potentials of NaHe. Full line : present PP calculations. Full circles : experimental data of Havey et al. (1982). The arrows in the figure indicate minimum of $A^2\Pi$ potential well obtained by HF-SCF (Krauss et al., 1971) and MP (Hanssen et al., 1979) calculations.

calculations of Hanssen et al. (1979) are too small by about a factor of two. The PP results confirmed also the far-red wing intensity measurements of York et al. (1975) which were questioned by the MP calculations. Results for LiHe and KHe indicate the same tendency for the SCF-HF calculations or the MP calculations (using the same e-He model potential as that for NaHe) to underestimate the $A^2\Pi$ well depth (see Table I). In the case of CsHe, the PP calculations improve considerably previous l-independent pseudo-potential (liPP) calculations for both the adiabatic potential

Table I - Position Re(a.u.) and depth De (cm^{-1}) of the $A^2\Pi$ potential wells of LiHe, NaHe and KHe.

Alkali atom	Li		Na		K	
	Re	De	Re	De	Re	De
Method						
SCF-HF[a]	3.5	500	4.53	210		
MP[b]			4.58	299	5.30	190
PP[c]	3.44	1025	4.35	511	5.30	245
Experiment[d]	3.45(.08)	8.50(100)	4.4(.2)	480(50)		

a Krauss et al. (1971).
b NaHe, Hanssen et al. (1979) ; KHe, Masnou-Seeuws (1982).
c J. Pascale (1983).
d LiHe, Balling et al. (1982) ; NaHe, Havey et al. (1982).

energies and the dipole moments. This is seen Figs. 3 and 4 where the PP results are compared with previous liPP calculations and the experimental data of Ferray et al. (1980) derived from absorption measurements in the bands associated with the $^2\Sigma^+$ (6S)-$^2\Sigma^+$ (5D) and $^2\Sigma^+$ (6S)-$^2\Sigma^+$ (5D) transitions. Recently, semiclassical calculations have been performed for these bands (Visticot et al., 1985) using the PP results and the unified Franck-Condon model of Szudy and Baylis (1975). A good overall agreement between experiment and theory was obtained except in the spectral region corresponding to the avoided crossing between the $^2\Sigma^+$ (5D) and $^2\Sigma^+$ (7S) potential curves (see Fig. 3). In this region, the potential curves have to be improved, and the non-adiabatic coupling between the $^2\Sigma^+$ (5D) and $^2\Sigma^+$ (7S) molecular states has to be considered to bring experiment and theory into closer agreement. This has been shown recently by using a Landau-Zener approximation (O'Callaghan et al., 1985).

Recent PM calculations by Mason and Peach (1985) for LiHe and NaHe confirm the good agreement between experimental data and the PP calculations ;

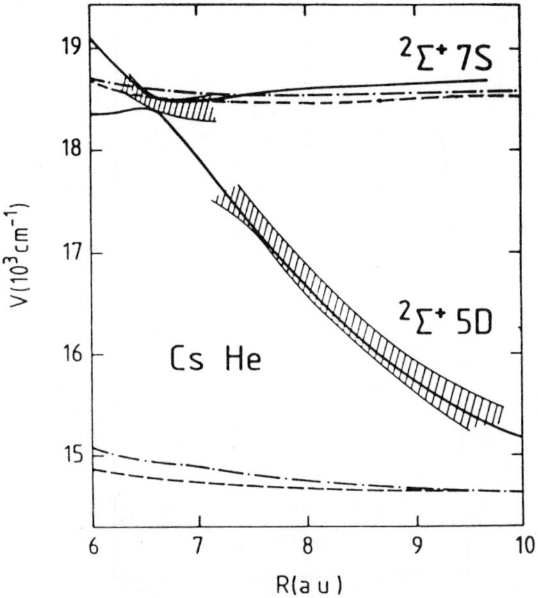

Fig. 3 - Adiabatic potential curves for the $^2\Sigma^+$ (5D) and $^2\Sigma^+$ (7S) states of CsHe. Full line : present PP calculations, dash-dotted line : liPP calculations (Pascale and Vandeplanque, 1974) ; dashed line : liPP calculations (Czuchaj and Sienkiewicz, 1979). Hatching : experimental data (Ferray et al., 1980).

their results for NaNe confirm also the good agreement of the MP calculations of Masnou-Seeuws et al. (1978) with the experimental data of Ahmad-Bitar et al. (1977). This illustrates the difficulty, in the treatment of M-atom-rare gas systems by both the PP and MP methods, of defining correctly the e^--rare gas effective interaction. Additional difficulties arise for the M-atom-heavier rare gas systems, due mainly to an inadequate knowledge of the core core interaction (for a discussion of this problem, see Pascale, 1985) and also because the calculation are more sensitive to the cut-off functions introduced in the expression of the cross-term. Therefore, more work is needed for these systems (in particular, it should be interesting also to evaluate the importance of using a PP approach rather than an liPP approach for the heavier rare gases).

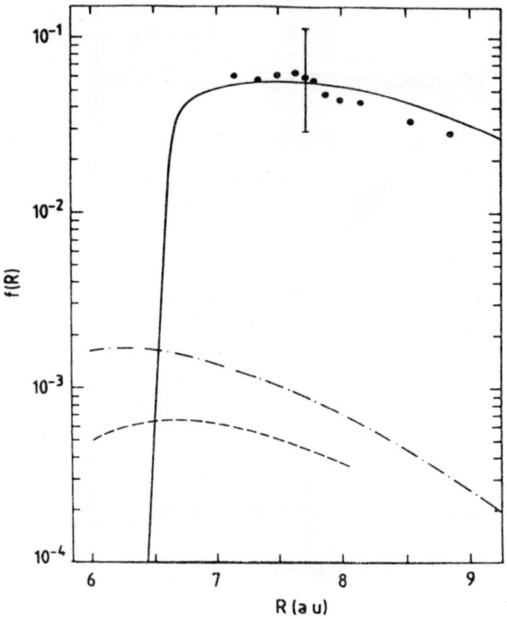

Fig. 4 - Dipole-induced oscillator strength versus R for the $^2\Sigma^+$ (6S)-$^2\Sigma^+$ (5D) transition in CsHe. Full curve : PP calculations (Pascale, 1983) ; dashed-dotted line : liPP calculations (Pascale, 1977) : dashed line : liPP calculations (Czuchaj, 1979). The dots are the experimental data (Ferray et al., 1980).

Treatment of one-electron systems by PP or MP approaches are tending to become standard techniques, in particular for systems such as M-ion-M-atom, alkali-earth ion-rare gas systems, and quite accurate adiabatic energies can be obtained. Treatment of systems as Cd-rare gases or Tl-rare gases is more difficult, but has been considered recently using a liPP approach (Czuchaj and Sienkiewicz, 1984 and 1985). Extensions of the PP and MP methods to systems with two or more valence electrons is relatively straightforward, but the treatment of the valence-electron correlations requires additional effort. Moreover, for systems with more than one electron, the PP approach should be preferred to the MP approach, as discussed above. Studies on M-atom-M-atom and alkali earth-rare gas systems are presently in progress in several laboratories. Finally, treatment of polyatomic systems by PP or MP approaches is certainly possible, but calculations are more difficult than for diatomic systems because of the many-center integrals to evaluate. In the following section, we present the first detailed calculations on M-atom-H_2 systems using a PP approach, in which H_2 has been treated as a one-center system and anisotropic terms have been defined to account for the molecular structure of H_2 (Rossi and Pascale, 1985).

B.2 - M-atom-H_2 systems. In the PP approach of these systems, a one-electron two-center model has been considered (see Fig. 5) in which the valence

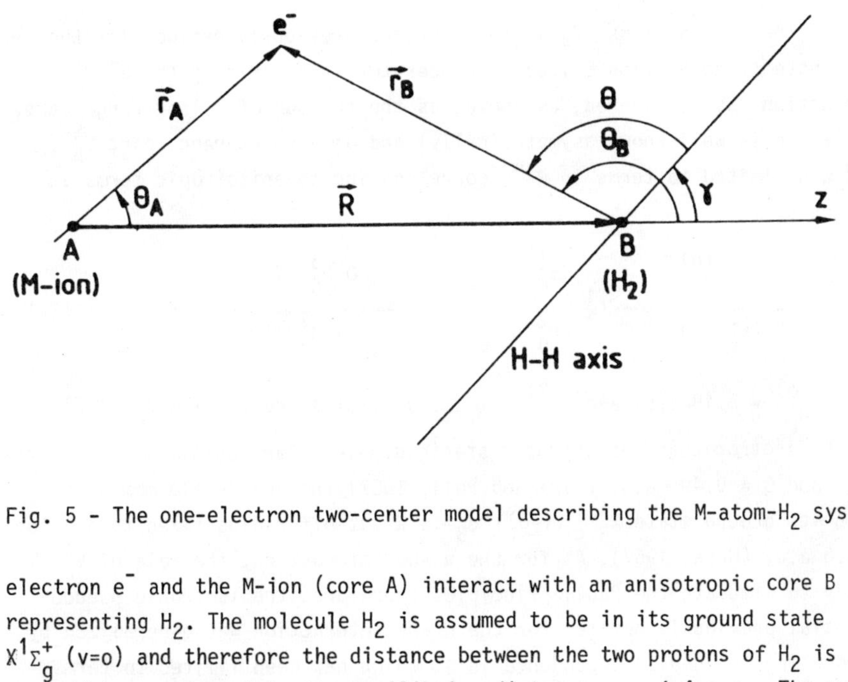

Fig. 5 - The one-electron two-center model describing the M-atom-H_2 system.

electron e^- and the M-ion (core A) interact with an anisotropic core B representing H_2. The molecule H_2 is assumed to be in its ground state $X^1\Sigma_g^+$ (v=o) and therefore the distance between the two protons of H_2 is assumed to correspond to the equilibrium distance r_e = 1.4 a.u.. The angle γ specifies the orientation of the H-H axis with respect to the core-core AB axis taken as quantization axis. The problem is now to determine the electronic energies E_{el} (R,γ) by solving the one-electron Schrödinger equation,

$$H_{el}\,\psi_{el} = E_{el}(R,\gamma)\,\psi_{el}\,, \tag{24}$$

for any given distance R and angle γ. Then, the adiabatic potential energies $E(R,\gamma)$ are obtained by adding to $E_{el}(R,\gamma)$ the core-core interaction $V_{AB}(R,\gamma)$. As for the M-He system, the one-electron Hamiltonian is defined as

$$H_{el} = -\frac{1}{2}\nabla^2_{\vec{r}_X} + V_A + V_B + V_{CT}\,(\vec{r}_X,\vec{R},\gamma) \tag{25}$$

where again the cross term V_{CT} has to be included to have the correct behavior of $E(R,\gamma)$ at large R-values.

B.2.a - The interactions. V_A is the operator previously defined for the M-He system (see Section B.1.a). The operator V_B describing the e^--H_2 interaction was considered, as above, as the the sum of a long-range part V_B^{LR} (which is well-known asymptotically) and of a short-range part V_B^{SR}. $V_B{}^{LR}$ was limited to terms in R^{-4}, corresponding to anisotropic terms in P_2 $(\cos\theta)$:

$$V_B^{LR} = -\frac{1}{2}\frac{\alpha_{d_B}^{(o)}}{(r_B^2 + d_B^2)^2} - \left(\frac{1}{2}\frac{\alpha_{d_B}^{(2)}\; r_B^2}{(r_B^2 + d_B^2)^2} + \frac{Q\; r_B^3}{(r_B^2 + d_B^2)^3}\right) P_2\,(\cos\theta) \tag{26}$$

where $\alpha_{d_B}^{(o)} = 5.18$ a.u. and $\alpha_{d_B}^{(2)} = 1.2$ a.u. (Kolos and Wolniewicz, 1967) are the isotropic and anisotropic static dipole polarizabilities, respectively, and $Q = 0.49$ a.u. (Karl and Poll, 1967) the quadrupole moment of H_2 in its ground state $X^1\Sigma_g^+$ (v=o). d_B is a cut-off radius fixed to a value of 1.6 a.u. (Hara, 1967). As for the e^--He interaction, the role of V_B^{SR} is mainly to simulate the Pauli principle. Therefore, the isotropic pseudo-potential previously defined for the e^--He interaction was generalized by introducing an angular dependence in θ, which has been limited to terms in P_2 $(\cos\theta)$ to be consistent with V_B^{LR}, and

$$V_B^{SR} = \sum_{l=o}^{\infty} V_{Bl}^{(o)} (r_B) \mathcal{P}_l^B + \sum_{l=o}^{\infty} V_{Bl}^{(2)} (r_B) \frac{1}{2} \{ P_2 \cos \theta), \mathcal{P}_l^B \} . \qquad (27)$$

The anticommutator { } ensures the Hermiticity of H_{el}, and $V_{Bl}^{(o),(2)}(r_B)$ are Gaussian-type radial operators (see, Eq. (16)). The coefficients determining V_B^{SR} were adjusted to reproduce the experimental data of Linder and Schmidt (1971) concerning the differential elastic scattering of e^- with H_2 in its ground state $X^1\Sigma_g^+$ (v=o), as well as the theoretical value 1.27 a.u. determined by Chang (1981) for the scattering length. As seen in Fig. 6, the limitation of the l-dependence of the pseudopotential to l = 0,1, as for the e^--He interaction, was sufficient to obtain good agreement with scattering experimental data.

The cross-term V_{CT} which results from the polarization of H_2 by both the point charges e^- and the M-ion is well-known asymptotically (see, for example, Buckingham, 1967), and it is taken as :

$$V_{CT}(\vec{r}_A,\vec{R},\gamma) = - \Big[\frac{\alpha_{d_B}^{(o)} \cos \theta_B}{(R^2 + d_B^2)(r_B^2 + d_B^2)} + \frac{\alpha_{d_B}^{(2)} \, r_B \, R \, (3 \cos \theta \cos \gamma - \cos \theta_B)}{2\,(R^2 + d_B^2)^{3/2} \, (r_B^2 + d_B^2)^{3/2}} \Big] f_c \left(\frac{R}{r_A}\right) \qquad (28)$$

where cut-off functions have been introduced consistently with the definition of V_B^{LR} above and that of $V_{AB}(R,\gamma)$ below ; the same additional cut-off function $f(\frac{R}{r_A})$ previously defined for the M-He systems was also used.

Finally, in order to calculate the adiabatic potential energies $E(R,\gamma)$, the core-core interaction $V_{AB}(R,\gamma)$ has to be determined. An approach similar to that used for calculating the M-ion-He interaction (based on the knowledge of the interaction asymptotically and its determination at short R-values from high-energy elastic scattering experiments) is not possible for the M-ion-H_2 systems since the scattering data are averaged over the angle γ. In order to estimate $V_{AB}(R,\gamma)$, a method has been proposed (Rossi

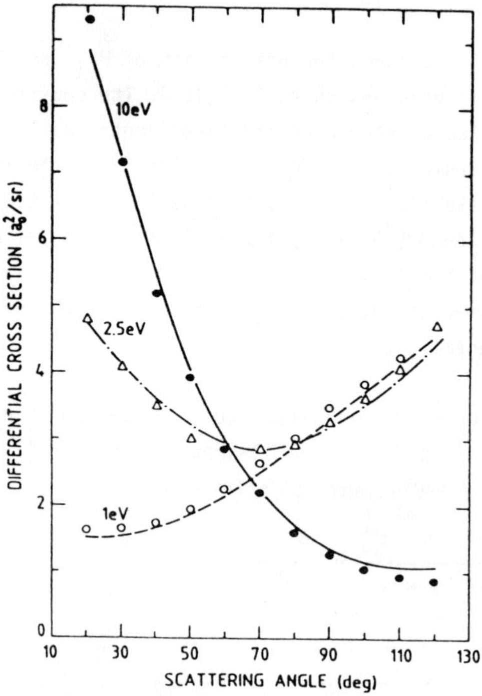

Fig. 6 - Calculated differential cross sections versus scattering angle for the elastic scattering of e^- by H_2 in its ground state $X^1\Sigma_g^+$ (v=o), for three energies as indicated in the figure. The symbols are the experimental points of Linder and Schmidt (1971).

and Pascale, 1985). It consists of writting $V_{AB}(R,\gamma)$ as,

$$V_{AB}(R,\gamma) = V_{static}(R,\gamma) + V_{induction}(R,\gamma) + V_{dispersion}(R,\gamma) \ . \qquad (29)$$

The last two terms of expression (29) can be easily obtained (see, for example, Buckingham, 1967) :

$$V_{induction}(R,\gamma) = - \frac{\alpha_{d_B}^{(o)}}{2\,(R^2 + d_B^2)^2} - \frac{\alpha_{d_B}^{(2)}\ R^2}{2\,(R^2 + d_B^2)^3}\ P_2\ (\cos\gamma) \quad , \qquad (30)$$

and

$$V_{dispersion}(R,\gamma) = -\frac{3}{2} F \left[\frac{\alpha_{d_A} \alpha_{d_B}^{(o)}}{(R^2 + d_A^2)^{3/2} (R^2 + d_B^2)^{3/2}} + \frac{1}{2} \frac{\alpha_{d_A} \alpha_{d_B}^{(2)} R^2 P_2 (\cos \gamma)}{(R^2 + d_A^2)^2 (R^2 + d_B^2)^2} \right], \tag{31}$$

where the factor F in Eq. (31) is determined by the Slater-Kirkwood formula (1931). An estimation of $V_{static}(R,\gamma)$ can be obtained by considering the first-order term of a stationary perturbative method. It consists of taking the average value, with respect to an approximate wave function describing H_2 in its ground-state $X^1\Sigma_g^+$ (v=o) (Hara, 1967), of the sum of the interactions between the M-ion and each of the particles constituting H_2. The interaction between an electron and the M-ion is taken to be the sum of V_A^{SR} previously defined and of the Coulombic interaction ; the interaction between a proton and the M-ion is taken as the sum of the Coulombic interaction and of minus $V_{Al_{max}}^{SR}$, where l_{max} has been defined above (see Chapter 2.A.3). It was found that this approach leads to an underestimation of the quadrupole moment for H_2. Then, to remedy this problem, $V_{static}(R,\gamma)$ was taken as

$$V_{static}(R,\gamma) = V'_{static}(R,\gamma) + \frac{Q R^3}{(R^2 + d_B^2)^3} P_2 (\cos \gamma) \tag{32}$$

where V'_{static} is calculated as above, but where the Coulombic part of the interaction is replaced by the first order term of its expansion in terms of Legendre polynomials P_1 (cos γ). A good overall agreement with previous *ab initio* calculations was then found for Li^+H_2 and Na^+H_2 for the two symmetries $C_{\infty v}$ and C_{2v} which were considered (Rossi and Pascale, 1985).

B.2.b - Molecular-structure calculations. The large basis set of STO's centered on the M-ion previously defined for the treatment of the M-atom-He systems was also used for the M-atom-H_2 systems. The calculations were performed for the $C_{\infty v}$ and C_{2v} symmetries. In the $C_{\infty v}$ symmetry (γ=0), the classification of the molecular states is identical to that for the M-atom-He systems, and the $^2\Sigma$, $^2\Pi$, $^2\Delta$, etc..., result from diagonalization of the electronic Hamiltonian H_{el} for each class of states. In the C_{2v} symmetry ($\gamma = \frac{\Pi}{2}$), the molecular states of the M-atom-H_2 system separate in classes $^2A_1$, $^2B_1$, $^2B_2$ and $^2A_2$, and their adiabatic energies are similarly obtained from diagonalization of H_{el} for each class of states.

B.2.c - Results and discussion. No experimental data are presently available for the M-H_2 systems for testing the validity of the PP calculations. Therefore, comparisons with accurate *ab initio* results, when available, are quite valuable. The *ab initio* calculations of Botschwina et al. (1981) for the first lowest states of NaH_2, for R_{H-H} = 1.4 a.u. corresponding to the condition of the PP calculations, are very instructive. These authors have used the restricted-Hartree-Fock (RHF)-SCF method as a first step. Then, calculations were repeated for the X^2A_1 and A^2B_2 potential curves using the paired-natural-orbitals-coupled-electron-pair-approximation (PNO-CEPA) method, which takes into account most of the electronic correlations. The *ab initio* results are shown in Figs. 7 and 8 for the C_{2v} and $C_{\infty v}$ symmetries, respectively, along with the PP results. While the X^2A_1 potential curve is little changed from the RHF-SCF results, the A^2B_2 potential curve is found more attractive when calculated using the PNO-CEPA method. Therefore, the good agreement which is observed between the PNO-CEPA and PP results (see Fig. 7), in particular for the A^2B_2 potential curve, is quite significative since the PP approach takes implicitely into account correlation effects through the use of experimental core polarizabilities. As expected, the PP adiabatic potential energies are generally lower than the energies obtained by the RHF-SCF method. For the B^2A_1 and $B^2\Sigma^+$ potential curves, however, where repulsive effects are predominant and the correlation effects less important, a good agreement is observed between PP and RHF-SCF results. This tendency of the PP approach to find (in agreement with more sophisticated calculations) adiabatic energies which are generally lower than those obtained from RHF-SCF calculations was already observed for the M-atom-He systems (see, Sec. B.1.c). The comparisons of the PP results with the HF-SCF calculations of Krauss (1968) for the LiH_2

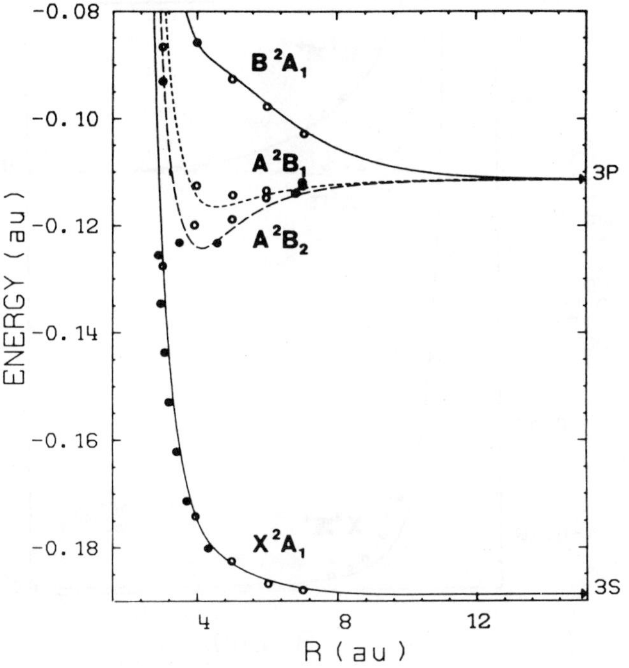

Fig. 7 - Adiabatic potential energies for the lowest states of NaH_2 in C_{2v} symmetry. PP results are compared with RHF-SCF (open circles) and PNO-CEPA (full circles) calculations of Botschwina et al. (1981).

system are consistent with these observations (see, Figs. 9 and 10), considering the *ab initio* calculations were performed for a distance R_{H-H} = 1.5 a.u. corresponding to an absolute minimum in the A^2B_2 potential curve (similar calculations for R_{H-H} = 1.4 a.u. should increase the energy values). The results of Wagner et al. (1978) for the X^2A_1 and $X^2\Sigma^+$ potential curves, obtained from multiconfiguration-self-consistent-field-optimized-valence-configurations (MCSCF-OVC) calculations, are also shown in Figs. 9 and 10. These calculations illustrate again the difficulty for *ab initio* calculations to take correlation effects correctly into account. Comparisons with the PP results show a good agreement for the X^2A_1 potential curve, while the PP $X^2\Sigma^+$ potential curve is found less repulsive.

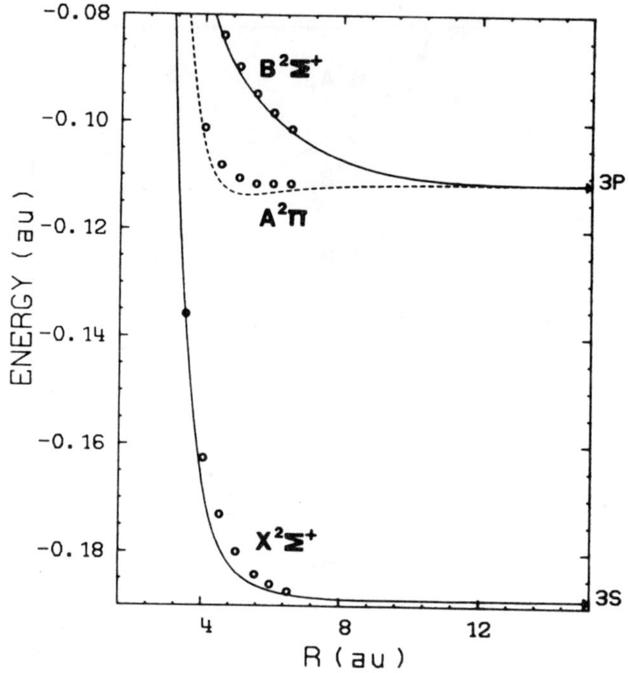

Fig. 8 - Adiabatic potential energies for the lowest states of NaH_2 in $C_{\infty v}$ symmetry. As in Fig. 7.

This good overall agreement between the PP results and available *ab initio* results (in general, considering the most sophisticated of them) allows us to have some confidence in the validity of the PP predictions for the M-atom-H_2 systems. Note that for CsH_2, preliminary calculations of Gadea et al. (1983) using an *ab initio* pseudopotential method are only in qualitative agreement with present PP calculations, the adiabatic energies obtained by this method being much lower than those obtained by the PP approach.

In view of the rather encouraging results obtained for the M-atom-H_2 systems by using the PP two-center approach, treatment of other systems (M-atom-N_2 systems, for example) by the same approach could be envisaged. However, in order to study reactive collisional problems, the present

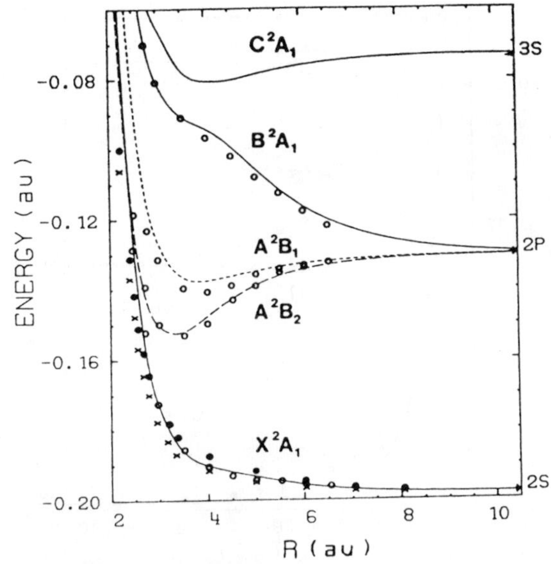

Fig. 9 - Adiabatic potential energies of the lowest states of LiH_2 in C_{2v} symmetry. The PP results are compared with HF-SCF calculations of Krauss (1968), open circles ; and with MCSCF-OVC calculations of Wagner et al. (1978) for two levels of approximation (15 OVC, crosses ; 28 OVC, full circles).

approach should be extended to a three-center model in which the interaction between a valence electron and each center is described by an l-dependent pseudopotential.

3. Applications of the PP approach to the study of atom-atom collisions

In the following, we show that the use of the PP approach in molecular-structure calculations of the M-atom-He systems allows us a systematic study (from Li to Cs) of some collisional processes involving the M-atom. First, the PP molecular-structure calculations have been applied to extensive cross-section calculations concerning intra-n^2P transitions induced in thermal or suprathermal collisions of the M-atoms with He in its ground-

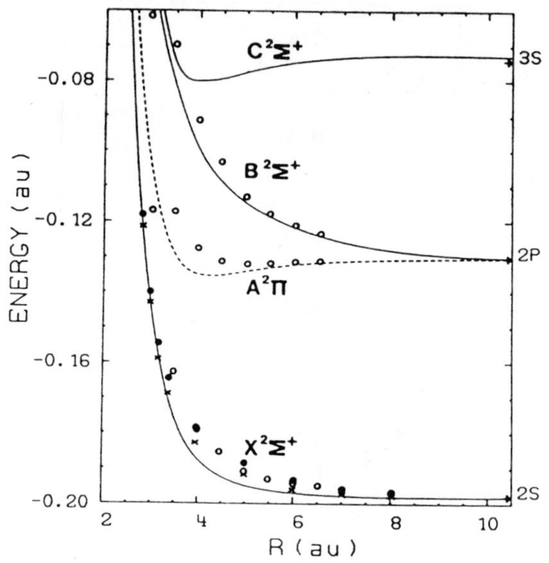

Fig. 10 - Adiabatic potential energies of the lowest states of LiH_2 in $C_{\infty v}$ symmetry, as in Fig. 9.

state. Second, extensive calculations concerning the transition from the ground-state to the first n^2P level of the M-atoms, in the keV energy range, have been performed allowing us to complete the test of the PP molecular-structure calculations against experimental scattering data.

A - Intra-n^2P transitions in the M-atom, in thermal or suprathermal collisions with He

Quantum-mechanical calculations of cross-sections for $n^2P_{1/2} \rightleftarrows n^2P_{3/2}$ transitions in M-atom collisions with He in its ground state have been performed in the energy range from threshold of the fine-structure transition up to 0.5 eV. More generally, the cross-sections for the $n^2P_{jm} \rightleftarrows n^2P_{j'm'}$, Zeeman transitions have been calculated, which allows us to determine any observable. The first n^2P level of all M-atoms, and the second and third n^2P levels of Rb and Cs have been investigated (Pascale,

1983c). The calculated total cross sections are compared with various experimental data : cell-type experimental data, requiring the average of the calculated cross sections over the Maxwell-Boltzman distribution of velocities ; crossed-beam experimental data, sometimes with the evaluation of polarization effects.

A.1 - Theory. In the quantum-mechanical formulation of the scattering problem, one has to solve the Schrödinger equation,

$$H\psi = E\psi \tag{33}$$

where E is the total energy of the system and :

$$H = -\frac{1}{2\mu}\nabla^2_{\vec{R}} + H_A + V(\vec{r},\vec{R}) \ , \tag{34}$$

and to consider the asymptotic behavior of the wave function Ψ which determines the T-matrix, allowing us to calculate the cross-sections of the intra-n^2P transitions. In Eq. (34), H_A is the one-electron Hamiltonian for the M-atom, including the spin-orbit interaction ; $V(\vec{r},\vec{R})$ is the M-atom-rare gas interaction (equal to $V_B + V_{AB} + V_{CT}$ defined above) ; and μ is the reduced mass of the system.

For intra-n^2P transitions induced in M-atom collisions with a rare gas, the quantum-mechanical formulation (see Reid, 1973) follows the theory of Arthurs and Dalgarno (1960) for the collisional excitation of a rigid-rotator, where the rigid-rotator wave function is replaced by the valence electron wave function. It takes advantage of the fact that $V(\vec{r},\vec{R})$ can be expanded in Legendre polynomials,

$$V(\vec{r},\vec{R}) = \sum_{\lambda=0}^{\infty} v_\lambda\ (r,R)\ P_\lambda\ (\hat{r}.\hat{R}) \tag{35}$$

Then, assuming the n^2P level is well isolated in the energy diagram of the M-atom, so that its wave function is little perturbed during the collision, only the $\lambda = 0,\ 2$ terms are involved in the calculations. This assumption is well justified for the first n^2P level, and numerous calculations have been performed in that case. Recently (Pascale, 1983c) the theory has been applied to the second and third n^2P levels of Rb and Cs in collisions with He because their wave functions were found also little perturbed in the region R > 8 a.u. which mainly contributes to the intra-n^2P transitions.

The scattering problem is formulated in a space-fixed axis system, and the scattering wave function ψ is taken to be an eigenfunction of J^2 and J_z since H commutes with the total angular momentum of the system,

$$\vec{J} = \vec{j} + \vec{l} \tag{36}$$

where $\vec{j}$ is the total angular momentum of the M-atom and $\vec{l}$ is the angular momentum for the relative motion of the system. Next, by expanding ψ in terms of eigenfunctions ϕ_{jl}^{JM} of J^2, J_z, j^2 and l^2,

$$\psi^{JM}_{j_1 l_1}(\vec{r}, \sigma, \vec{R}\,;\, E) = \sum_{jl} \frac{1}{R} F^{JE}_{jl\, j_1 l_1}(R)\, \phi^{JM}_{jl}(\vec{r}, \sigma, \hat{R}) \tag{37}$$

two sets of, at most, three coupled equations result :

$$\left[\frac{d^2}{dR^2} + k_2^2 - \frac{l_2(l_2+1)}{R^2}\right] F^{JE}_{j_2 l_2\, j_1 l_1} = \frac{2\mu}{\hbar^2} \sum_{jl} V^{J}_{j_2 l_2\, jl} F^{JE}_{j_2 l_2\, j_1 l_1} \tag{38}$$

where k_2 is the wavenumber of the relative motion in the j_2-channel. In Eq. (37), σ is a spin variable, and j_1, l_1 specify the initial conditions. The matrix elements $V^J_{j_1 l_1 jl}$ are non-zero for $|l-l_1|$ even, and can be explitily written in terms of 3-j and 6-j coefficients, and in terms of $\hat{v}_{\lambda=0,2}(R)$ (the expectation value of $v_{\lambda=0,2}(r,R)$ over the radial wave function of the n^2P level. The terms $\hat{v}_{\lambda=0,2}(R)$ may be evaluated from the $^2\Sigma$ and $^2\Pi$ PP adiabatic potential energies relative to the n^2P level :

$$v_0(R) = \frac{1}{3}\left(E_\Sigma(R) + 2\, E_\Pi(R)\right) \tag{39}$$

and

$$\hat{v}_2(R) = \frac{5}{3}\left(E_\Sigma(R) - E_\Pi(R)\right) \tag{40}$$

For a given value of E (relative to the $n^2P_{1/2}$ level), the two coupled-equation sets (Eq. (38)) are solved under standard boundary conditions and the T^J $(j_2l_2 \; ; \; j_1l_1)$ matrix and then the total cross section for the $j_1m_1 \to j_2m_2$ transition are obtained :

$$\sigma_{j_1m_1 \to j_2m_2}(E;\hat{k}_1) = \frac{4\pi^2}{k_1^2} \sum_{l_1l_2lM} (i)^{l_1-l} \, Y^*_{l_1,M-m_1}(\hat{k}_1) \, Y_{l,M-m_1}(\hat{k}_1) G_{l_1l_2M} G^*_{ll_2M} \tag{41}$$

where G is defined as (Mies, 1973)

$$G_{l_1l_2M} = \sum_J \langle j_1l_1M\text{-}m_1 \, | JM\rangle \, T^J(j_1l_2;j_1l_1) \, \langle j_2l_2m_2M\text{-}m_2|JM) \tag{42}$$

Two types of cross-sections are then defined for purpose of comparisons with experimental data.

For cell-type experiments, one has to average over all orientations of $\hat{k}_1$ with respect to the space-fixed z-axis, and one defines the cross-section $\bar{\sigma}$,

$$\bar{\sigma}\,(j_1m_1 \to j_2m_2) = \frac{\pi}{k_1^2} \sum_{l_1l_2M} G_{l_1l_2M} \, G^*_{l_1l_2M} \tag{43}$$

which obeys both detailed balance and the relation $\bar{\sigma}\,(j_1m_1 \to j_2m_2) = \bar{\sigma}\,(j_1\text{-}m_1 \to j_2\text{-}m_2)$ in absence of a magnetic field. However, the observables which are more usually measured in cell-type experiments are the multipole relaxation cross sections $\sigma_j^{(x)}$ which are derived from the general matrix theory of collisional relaxation (see, for example, Baylis, 1979)

$$\sigma_j^{(x)} = \Lambda_j^{(x)} + \sigma\,(j \to j' \neq j) \tag{44}$$

where $\Lambda_j^{(o)} = 0$ and the $\Lambda_j^{(x \neq o)}$ are related to the Zeeman cross sections $\bar{\sigma}\,(j_1m_1 \to j_2m_2)$. The second term in Eq. (44) is the fine-structure transition cross-section. For purpose of comparison with experiment, all these cross-sections have to be thermally averaged.

For crossed-beam type experiments, the orientation of $\hat{k}_1$ with respect to the z-axis has to be specified ; usually, one takes $\hat{k}_1$ along the z-axis and one defines the cross-section σ_o,

$$\sigma_o\,(j_1m_1 \rightarrow j_2m_2) = \frac{\sigma}{k_1^2} \sum_{l_1 l_2 l} i^{l_1 - l}\,(2l_1+1)^{1/2}\,(2l+1)^{1/2}\,G_{l_1 l_2 m_1}\,G^*_{l l_2 m_1} \tag{45}$$

This cross section does not generally obey detailed balance but one has always the relation $\sigma_o(j_1m_1 \rightarrow j_2m_2) = \sigma_o\,(j_1\text{-}m_1 \rightarrow j_2\text{-}m_2)$ in absence of a magnetic field. Finally, from Eq. (43) or Eq. (45) the conventional fine-structure transition cross-section may be defined :

$$\sigma\,(j_1 \rightarrow j_2) = \frac{1}{2j_1+1} \sum_{m_1,m_2} \sigma_o\,(j_1m_1 \rightarrow j_2m_2) \tag{46a}$$

$$= \frac{1}{2j_1+1} \sum_{m_1,m_2} \bar{\sigma}\,(j_1m_1 \rightarrow j_2m_2) \tag{47a}$$

However, in a crossed-beam experiment, polarization effects may be important (see, Pascale and Perrin, 1980 ; Pascale et al., 1984) and they have to be considered for comparison with experimental data. Thus, for the crossed-beam experiment of Mestdagh et al. (1982) concerning the $K(4^2P)$ + He collision, and where the K-beam is excited from the $4^2S_{1/2}$, F = 2 level to the $4^2P_{3/2}$, F = 3 level, the measured apparent fine-structure transition cross-section $\sigma^{(a)}\,(\frac{3}{2} \rightarrow \frac{1}{2})$ was found (for the particular geometry of this experiment) to be :

$$\sigma^{(a)}(\tfrac{3}{2} \rightarrow \tfrac{1}{2}) = \frac{1}{4} \sum_{m_2} \left[\sigma_o(\tfrac{3}{2}\tfrac{3}{2} \rightarrow \tfrac{1}{2}\,m_2) + 3\,\sigma_o(\tfrac{3}{2}\tfrac{1}{2} \rightarrow \tfrac{1}{2}\,m_2)\right] \tag{48}$$

in place of the conventional cross-section $\sigma^{(c)}\,(\frac{3}{2} \rightarrow \frac{1}{2})$,

$$\sigma^{(c)}\,(\tfrac{3}{2} \rightarrow \tfrac{1}{2}) = \frac{1}{2} \sum_{m_2} \left[\sigma_o(\tfrac{3}{2}\tfrac{3}{2} \rightarrow \tfrac{1}{2}\,m_2) + \sigma_o(\tfrac{3}{2}\tfrac{1}{2} \rightarrow \tfrac{1}{2}\,m_2)\right] \tag{49}$$

In the following, some results that we judge particularly illustrative of these quantum mechanical calculations are reported.

A.2 - First n^2P levels. Calculations have been carried out for the first n^2P level of all the M-atom, for which the spin-orbit energy splitting $\Delta\varepsilon$ ranges from 0.34 cm^{-1} for $Li(2^2P)$ to 554 cm^{-1} for $Cs(6^2P)$.

Figure 11 shows the $\hat{v}_2(R)$ terms for all the M-atom-He systems calculated from the above reported PP calculations. These terms are essential for determining the intra-n^2P transitions. Indeed, in a semiclassical approach

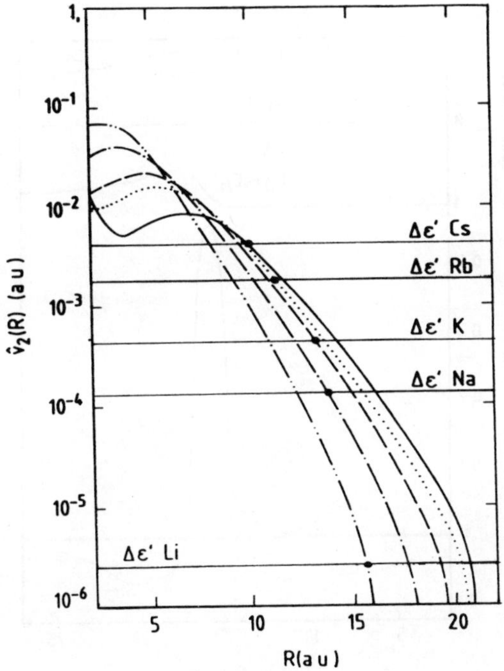

Fig. 11 - $\hat{v}_2(R)$ for all M-atom-He systems (see text). Full curve : Cs, dotted curve : Rb ; dashed curve : K ; dot-long-dashed curve : Na ; double dot-long-dashed curve : Li. The full circles indicate the position of maxima in radial coupling predicted by Eq. (50) and $\Delta\varepsilon' = 5/3\ \Delta\varepsilon$.

of the problem (see, Nikitin, 1975 and references therein), it is shown that fine-structure transitions are mainly induced in the region of a non-adiabatic coupli n g (radial coupling) between the two $\Omega = 1/2$ molecular states (where Ω denotes the absolute value of the projection of the total angular momentum $\vec{j}$ of the M-atom along the internuclear axis) associated with the $n^2P_{1/2 \text{ and } 3/2}$ levels, and determined by the condition

$$| \hat{v}_2(R_1) | = \frac{5}{3} \Delta\varepsilon \tag{50}$$

This non-adiabatic coupling results from the breakdown of the spin-orbit interaction. This is clearly seen in Fig. 12, for the LiHe system as an

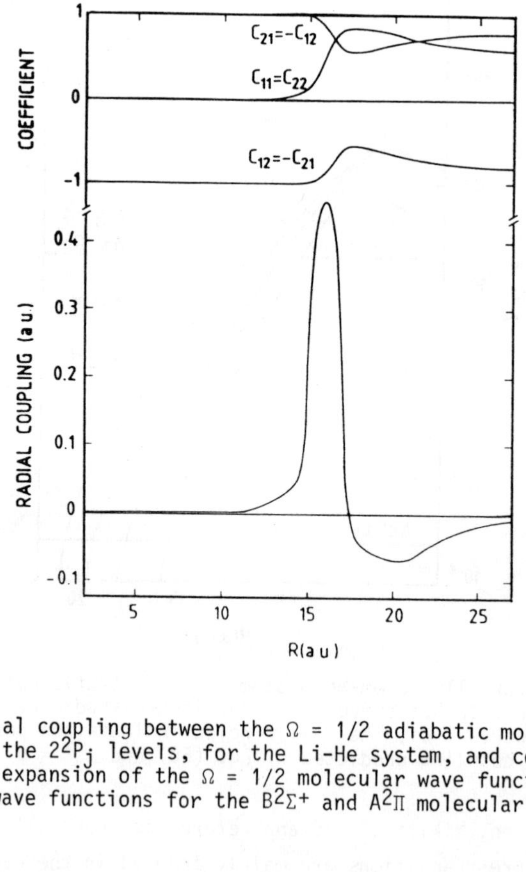

Fig. 12 - Radial coupling between the Ω = 1/2 adiabatic molecular states associated to the 2^2P_j levels, for the Li-He system, and coefficients $C_{ij}(R)$ of the expansion of the Ω = 1/2 molecular wave functions $\psi_i^{1/2}$ in terms of the wave functions for the $B^2\Sigma^+$ and $A^2\Pi$ molecular states (see text).

example, where the radial coupling is seen to have a large maximum in the region R_1 predicted by Eq. (50). Fig. 12 shows also the coefficients of the expansion of the two Ω = 1/2-molecular wave functions $\psi_i^{1/2}$ in terms of appropriate tensorial products of spin functions α or β and the molecular wave function ψ^1 and ψ^0 corresponding to the $A^2\Pi$ and $B^2\Sigma^+$ molecular states :

$$\psi_i^{1/2}(\vec{r},R) = C_{i1}(R)\ \psi^0(\vec{r},R) \otimes \alpha + C_{i2}(R)\ \psi^1(r,R) \otimes \beta \tag{51}$$

where $C_{11} = C_{22}$ and $C_{12} = -C_{21}$ (i = 1 corresponds to the $^2P_{1/2}$ level, and i = 2 to the $^2P_{3/2}$ level). It is clearly seen that for $R < R_1$ one of the two Ω = 1/2 molecular states becomes a pure Π state, while the other one becomes a pure Σ state. Fig. 12 shows also a secondary maximum in the radial coupling, corresponding to the crossing of the $A^2\Pi$ and $B^2\Sigma^+$ poten-

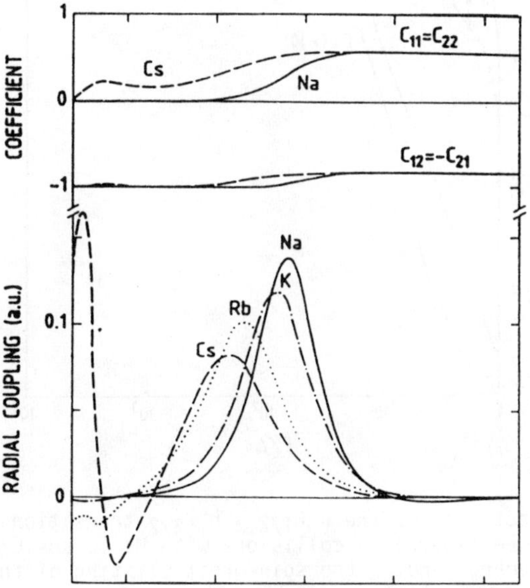

Fig. 13 - Radial coupling and coefficients $C_{ij}(R)$, as in Fig. 12, for the M-atom-He systems as indicated in the figure.

tial curves at large R. When the M-atom changes from Li to Cs, R_1 decreases from about 15.5 a.u. to 10.0 a.u. ; but while the main radial coupling decreases in magnitude, it spreads out more and more around its correspon-ding position R_1 (because the spin-orbit decoupling takes place more and more slowly as seen in Fig. 13). This explains why the cross sections for the $n^2P_{1/2} \to n^2P_{3/2}$ transitions reach more and more rapidly a plateau with increasing energy (see Fig. 14) when going from Li to Cs ; and why this

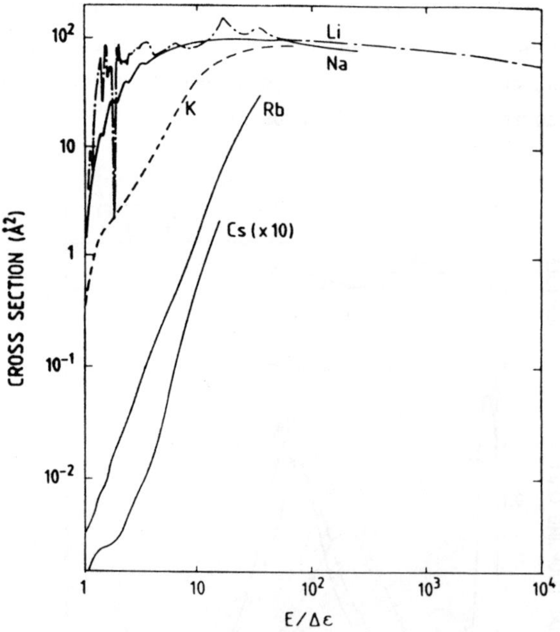

Fig. 14 - Cross sections for the $n^2P_{1/2} \to n^2P_{3/2}$ transition in the M-atom (as indicated in the figure) in collisions with He versus $E/\Delta\varepsilon$, where E is the collision energy and $\Delta\varepsilon$ the spin-orbit plitting of the n^2P level.

plateau does not change too much in magnitude. A pronounced resonance structure is also observed, at low energies, in the fine-structure cross section for Li ; it is primarily due to orbiting in the $A^2\Pi$ adiabatic potential. Finally, it is worthwhile to mention that, in the case of the

M-atom-He collisions, rotational coupling between the 1/2 and 3/2 molecular states (emerging from the $A^2\Pi$ state when spin-orbit is considered) contributes also to the fine structure transition. This rotational coupling, which generally occurs at small R-values where the potential curves are repulsive (Nikitin, 1975) is found to extend to relatively larger R-values for these systems.

Before comparing some results of the quantum-mechanical calculations with experimental data, let us discuss the relative magnitude of the thermal averaged cross sections for disorientation, namely the 1/2 1/2 → 1/2 -1/2 and 3/2 3/2 → 3/2 -3/2 transitions, when going from Li to Cs (see Fig. 15). The 3/2 3/2 → 3/2 -3/2 cross section increases by about three order in

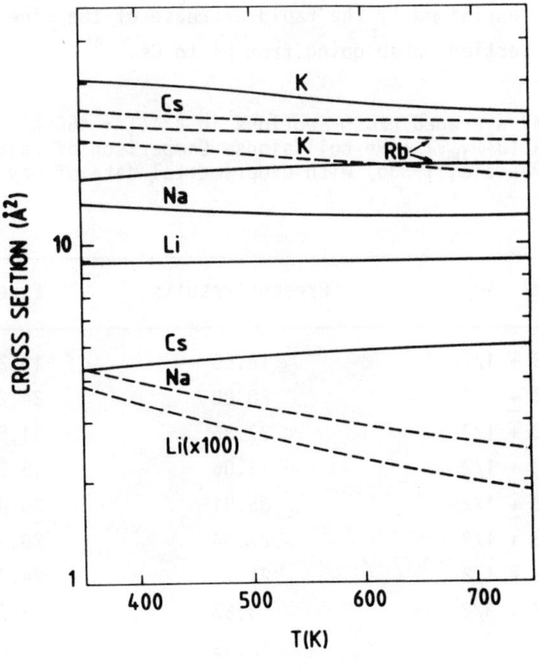

Fig. 15 - Thermal averaged cross sections for 1/2 1/2 → 1/2 -1/2 (full line) and 3/2 3/2 → 3/2 -3/2 (dashed line) transitions in the first n^2P level of the M-atom (as indicated in the figure) in collisions with He, versus temperature.

magnitude from Li to Cs, while the 1/2 1/2 → 1/2 -1/2 cross section goes through a maximum for K and only changes by about one order of magnitude. This can be understood by considering the ratio of the collision time, $\tau_c = \frac{a}{v}$ (where a is the range of the M-atom-He interaction and v the relative velocity of the system), to the spin-orbit coupling time, $\tau_{s.o} = \Delta\varepsilon^{-1}$. This ratio is about 3.10^{-2}, 2, 5, 18 and 39, respectively, for Li, Na, K, Rb and Cs. This means that during the collision with He, the orbital angular momentum $\vec{L}$ and the spin $\vec{S}$ are indeed uncoupled for Li, and strongly coupled for Rb and Cs, while for Na and K the situation is intermediate. So, the disorientation cross section should be the smallest for Li and the largest for Cs. This is indeed observed for the 3/2 3/2 → 3/2 -3/2 transitions. In the case of the 1/2 1/2 → 1/2 -1/2 transitions, which are only allowed through transition to the the $^2P_{3/2}$, the maximum of the cross section for K is explained by the rapid decrease of the fine-structure transition cross section when going from Li to Cs.

Table II - Thermal averaged cross sections in Å^2 (T = 450 K) for Zeeman transitions in Na (3^2P_{jm}) + He collisions. Comparison of calculated cross sections (Pascale, 1983b) with experimental data of Gay and Schneider (1976).

Transitions	Present results	Experiment
$1/2 \pm 1/2 \rightarrow 1/2 \mp 1/2$	12.55	13.2 ± 1.8
$1/2 \pm 1/2 \rightarrow 3/2 \pm 1/2$	20.80	24.2 ± 3.9
$1/2 \pm 1/2 \rightarrow 3/2 \mp 1/2$	27.35	31.5 ± 3.9
$3/2 \pm 3/2 \rightarrow 1/2 \pm 1/2$	15.06	15.8 ± 1.3
$3/2 \mp 3/2 \rightarrow 1/2 \pm 1/2$	35.81	35.4 ± 2.3
$3/2 \pm 3/2 \rightarrow 3/2 \pm 1/2$	27.94	28.3 ± 2.4
$3/2 \pm 3/2 \rightarrow 3/2 \mp 1/2$	21.14	24.3 ± 2.1
$3/2 \pm 3/2 \rightarrow 3/2 \mp 3/2$	3.63	5.6 ± 0.6
$3/2 \pm 1/2 \rightarrow 3/2 \mp 1/2$	12.69	10.8 ± 4.3

Table II illustrates the accuracy of the present quantum-mechanical calculations using PP adiabatic potential energies. Except for the 3/2 $\pm$ 3/2 $\rightarrow$ 3/2 $\mp$ 3/2 transition, all the calculated thermal averaged cross-sections for Zeeman transitions are in excellent agreement with the measurement of Gay and Schneider (1976).

Figure 16 shows comparisons of our results with experiment and previous calculations for the disorientation cross section $\sigma^{(1)}_{1/2} = 2\,\bar{\sigma}\,(1/2\ 1/2 \rightarrow 1/2\ -1/2)$ for Rb $(5^2P_{1/2})$ + He collisions. Our calculations using the PP adiabatic potential energies improve much the calculations of Brouillaud and Gayet (1977) and those of Doebler and Kamke (1977) using liPP adiabatic potential energies. Our results are in excellent agreement with the measu-

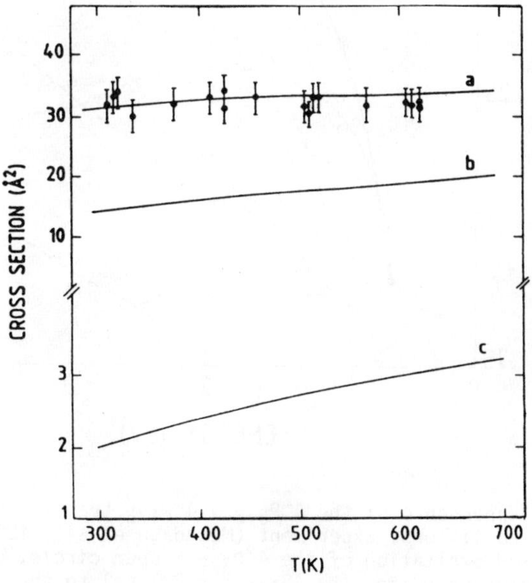

Fig. 16 - Disorientation cross-section $\sigma^{(1)}_{1/2} = 2\,\bar{\sigma}\,(1/2\ 1/2 \rightarrow 1/2\ -1/2)$ for Rb (5^2P) + He collisions, versus temperature. Theory : a) thermal averaged cross sections obtained from PP potential curves (Pascale, 1983 a,b) ; calculations of Brouillaud and Gayet using liPP potential curves (Baylis, 1969) ; calculations of Doebler and Kambe (1979) using liPP potential curves (Pascale and Vandeplanque, 1974). The experimental data, with error bars, are those of Doebler and Kamke (1979).

rements of Doebler and Kamke (1977). The improvement of the PP calculations over previous calculations using liPP adiabatic potential energies is clearly seen also in the comparison of the calculated energy-dependence of the cross sections for fine-structure transitions with crossed-beam experimental data (see Figs. 17 and 18). Figure 17 shows the very good agreement which is observed between experiment (Mestdagh, 1982 ; Mestdagh

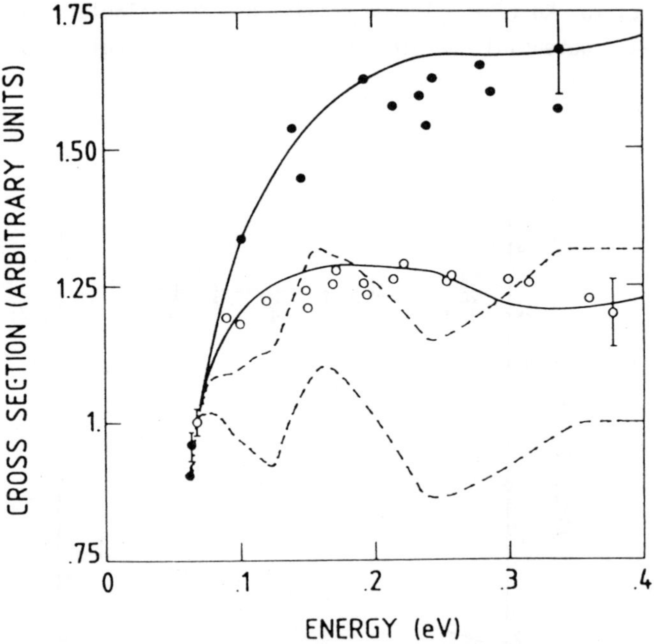

Fig. 17 - Energy-dependence of the $4^2P_{3/2} \rightarrow 4^2P_{1/2}$ transition cross section for K (4^2P) + He collisions. Experiment (Mestdagh et al., 1982) : full circle, unpolarized excitation of the $4^2P_{3/2}$: open circle, circularly polarized excitation of K from the $4S_{1/2}$ F = 2 level to the $4P_{3/2}$ F = 3 level. Theory : dashed line, calculations of Pascale and Perrin (1980) using liPP potential curves (Pascale and Vandeplanque, 1974) ; full line, present results using PP potential curves.

et al., 1982) and theory (when PP adiabatic potential energies are used) concerning polarization effects in the $4^2P_{3/2} \rightarrow 4^2P_{1/2}$ transition in collisions of K with He. We recall that here, polarization effects result from

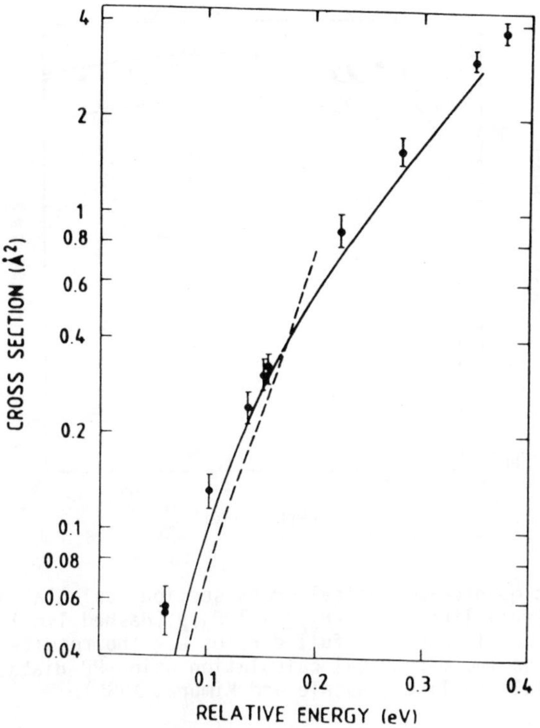

Fig. 18 - Energy dependence of the $5^2P_{1/2} \rightarrow 5^2P_{1/2}$ transition cross section for Rb(5^2P) + He collisions. Theory : full line, present results using PP potential curves ; dashed line, calculations of Olson (1975) using liPP potential curves (Baylis, 1969). The crossed-beam experimental data Mestdagh, 1982), shown with error bars, were normalized to cell measurements (Gallagher, 1968).

the very different energy-dependence of the Zeeman cross sections (Pascale and Perrin, 1980). Figure 18 shows comparisons between crossed-beam experimental data (see, Mestdagh, 1982) and quantum mechanical calculations of the $5^2P_{1/2} \rightarrow 5^2P_{3/2}$ transition cross section in Rb(5^2P) + He collisions. Agreement between experiment and theory (using PP adiabatic potential energy) is seen to be very good both for the absolute values of the cross section, and its energy dependence.

A.3 - Second and third n^2P levels of Rb and Cs. In this case the terms

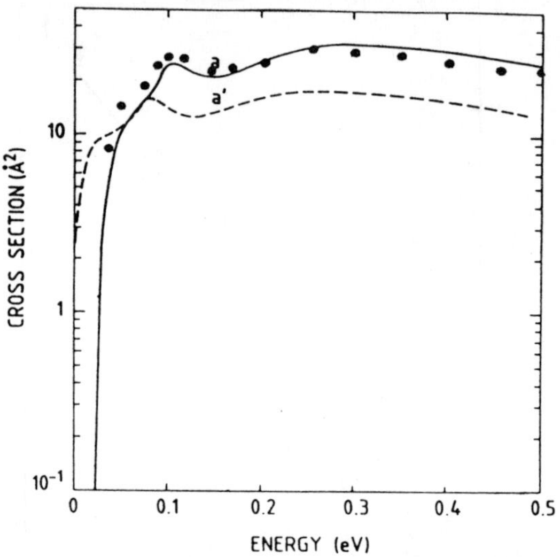

Fig. 19 - Present quantum-mechanical cross section calculations for the $7^2P_{1/2} \rightarrow 7^2P_{3/2}$ (full line) and $7^2P_{3/2} \rightarrow 7^2P_{3/2}$ (dashed line) transitions in Cs (7^2P) + He collisions. The full circles are the results of a four Ω = 1/2 channel quantum-mechanical calculation using PP diabatic potential energies and radial coupling (Pascale and Kimura, 1985).

$\hat{v}_{0,2}(R)$ have been calculated following the method defined above. Because the $\hat{v}_2(R)$ terms may show several maxima, oscillation in the energy dependence of the total cross sections is generally observed. For example, Figure 19 shows the energy-dependence of the $7P_{1/2} \rightleftarrows 7P_{3/2}$ transition cross sections calculated by the quantum-mechanical method developed above and the results of a four Ω = 1/2 channel quantum-mechanical calculation using the diabatic picture (Pascale and Kimura, 1985). In this latter calculation, radial coupling between the four Ω = 1/2 molecular states emerging from the $7^2P_{1/2 \text{ and } 3/2}$ and $6^2D_{3/2 \text{ and } 5/2}$ levels (see Fig. 20) have been calculated ; then, by a unitary transformation the coupled equation set for the scattering problem formulated in the body-fixed frame was transformed from the adiabatic picture to the diabatic picture, and the cross-sections were calculated. Comparison between the two calculations show a good overall agreement ; this indicates that radial coupling is mainly

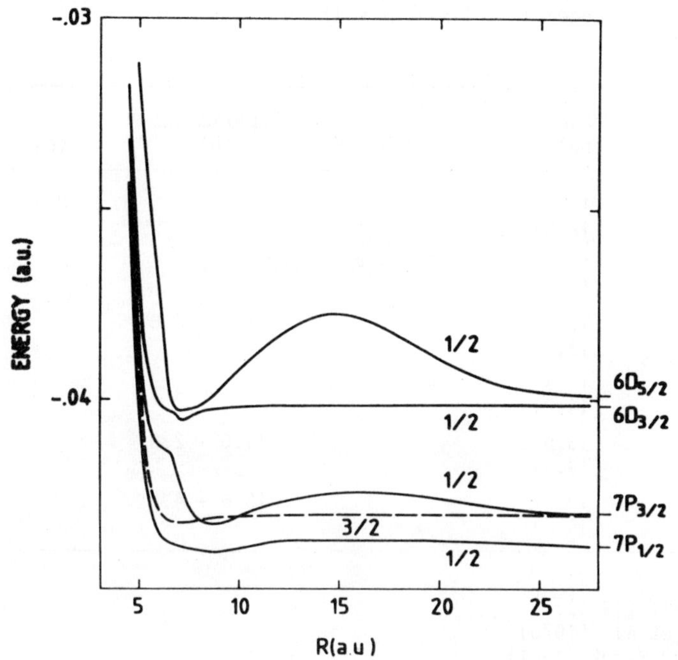

Fig. 20 - Ω = 1/2-PP adiabatic potential energies of the CsHe system, correlated asymptotically to the 7^2P and 6^2D levels of Cs.

responsible for the fine-structure transition and that, moreover, coupling with neighbouring states does not affect the calculation of the cross sections, in the present case. The thermally averaged cross sections calculated for the $7^2P_{3/2} \rightarrow 7^2P_{1/2}$ transition are seen in good agreement with experimental data in Table III. In the case of the $n^2P_{3/2} \rightarrow n^2P_{1/2}$ transitions in Rb (with n = 6, 7) experiment and theory are also found in very good agreement (Pascale, 1983c). In the case of the $8^2P_{1/2} \rightarrow 8^2P_{3/2}$ transition in Cs, the PP quantum-mechanical calculations (Pascale, 1983c) give a substantial improvement over previous liPP semiclassical calculations (Pascale and Stone, 1976), but comparison with experimental data indicates that coupling with the 7^2D level cannot be ignored in that case. Finally, Table IV shows very good agreement between the present results and the recent measurements of Lukaszewski and Jackowska (1984)

Table III - Thermally averaged cross-sections in $\mathring{A}^2$ for the $7^2P_{3/2} \rightarrow 7^2P_{1/2}$ transition in Cs, induced in collisions with He.

T (K)	Present results	Experiment (a)	(b)	(c)
320	11.0 (12.6)*			12.8 ± 2.6
405	11.7 (13.5)	15.2 ± 4.6		
450	12.2 (13.7)		11.0 ± 2.0	
520	12.5 (14.2)	14.9 ± 4.5		
615	13.0 (14.4)		11.0 ± 2.0	
630	13.1 (14.5)		15.6 ± 4.7	

(a) Siara et al. (1974).
(b) Cuvellier et al. (1975).
(c) Munster and Marek (1981).
* four 1/2-channel quantum-mechanical calculation (see text).

Table IV - Thermally averaged cross-sections for depolarization in the second $n^2P_{3/2}$ level of Rb and Cs (in 10^{-14} cm²) for T = 340 K.

	Theory present	Experiment (a)	(b)	(c)	(d)
$Rb(6^2P_{3/2})$	5.11	4.56 (21)	4.3		
$Cs(7^2P_{3/2})$	5.62	4.99 (16)		7.2	7.2

(a) Lukaszewski and Jackowska (1983).
(b) Grosswendt (1969).
(c) Minemoto et al. (1974).
(d) Minemoto and Kakihara (1976).

for depolarization in the second $n^2P_{3/2}$ level of Rb and Cs.

B - Excitation from the ground-state to the first n^2P level of an M-atom in collisions with He, in the keV energy range.

In the following, we consider the process

$$M(n^2S) + He \rightarrow M(n^2P) + He \tag{52}$$

in which an M-atom is excited from the ground state n^2S to the first excited n^2P level in collisions with He. Collisional excitation of an M-atom by rare gases (mainly He and Ne) have been the subject of several investigations, both theoretical and experimental (see, for example, the review article by Andersen and Nielsen, 1982). Two different mechanisms have been proposed for the excitation process : i) for E < 1 keV, where E is the collision energy of the system, the excitation process proceeds through a molecular crossing mechanism involving excited states of the rare gas atom (Courbin-Gaussorgues et al., 1979) ; for E $\gtrsim$ 1 keV the excitation process results from a direct mechanism involving only the M-atom valence electron (Manique et al., 1977 ; Nielsen et al., 1978). For treating the one-electron mechanism, a semiclassical approach was used, in which the wave-function of the system was represented by a linear combination of atomic orbitals ; liPP or MP interactions were used, as well as frozen-core Hartree-Fock interactions. More recently, we have investigated Reaction (52) using a multichannel perturbed-stationary-state (PSS) approach and the PP molecular-structure calculations (Pascale, 1983a).

B.1 - Theory. In the multichannel PSS approach, one has to solve the time-dependent Schrödinger equation,

$$i \frac{\partial}{\partial t} \psi = H_{eff} \psi \tag{53}$$

where H_{eff} is the one-electron effective Hamiltonian used in the PP molecular-structure calculation approach (see Eqs. (7) and (8)) and its t-dependence arises through R(t). The scattering wave function ψ is expanded in terms of Born-Oppenheimer (BO) wave functions,

$$\psi = \sum_i a_i(t)\ \phi_i^{BO}\ (\vec{r},R)\ F_i \tag{54}$$

where

$$H_{eff}\ \phi_i^{BO}\ (\vec{r},R) = E_i(R)\ \phi_i^{BO}\ (\vec{r},R) \tag{55}$$

and $E_i(R)$ is the PP adiabatic potential energy for molecular state i. In Eq. (54), F_i is an electronic translation factor (ETF) (see, Kimura and Thorson, 1981) which can be expanded in terms of the relative velocity $\vec{v}$ of the system, and in terms of f_i, a state-dependent switching function for representing, during the collision, a local propagation of the electron velocity in the quasimolecule. Then, to the first order in $\vec{v}$, the following coupled equations results from Eqs. (53)-(54) :

$$\dot{a}_j = -\ i\ E_i\ a_j + \sum_i \vec{v}\ .\ (\vec{P} + \vec{A})_{ji}\ a_i \tag{56}$$

in which

$$\vec{P}_{ji} = \langle\ j\ |\ -i\ \vec{\nabla}_R\ |\ i\rangle \tag{57}$$

is the non adiabatic coupling, and $\vec{A}_{ji}$ is an ETF correction to the coupling,

$$\vec{A}_{ji} = -\ i\ \langle\ j\ |\ \frac{1}{2}\ f_i\ \vec{\nabla}_r\ |\ i\rangle\ . \tag{58}$$

In the present calculations, because the BO wave functions are expanded over one-center basis functions only, the ETF's are unimportant, but they are crucial in the case of a rearrangement process (see, for example, Kimura et al., 1982). In a rotating coordinate frame (or body-fixed frame), one may separate the coupling term in Eq. (56) into a radial coupling term and a rotational coupling term,

$$\vec{v}\ .\ (\vec{P} + \vec{A}) = \dot{R}\ (P + A)^R + \dot{\theta}\ (P + A)^\theta\ , \tag{59}$$

in which ETF corrections to the usual radial and rotational coupling terms arise. In our approach, a straight-line trajectory has been assumed for describing the relative motion of the system, so that

$$\dot{R} = \frac{v_o}{R}\ z \tag{60}$$

and
$$\dot{\theta} = \frac{b\, v_o}{R^2} \tag{61}$$

where v_o is the initial relative velocity of the system, and b the impact parameter. Recalling that the BO wave function is expanded in terms of STO's centered on the M-ion (center A), for which the non-linear parameters where optimized at $R \to \infty$ (so that no R-dependence arises in the STO's),

$$\phi_i^{BO}(\vec{r},R) = \sum_\alpha C_{i\alpha}^A(R)\, \varphi_\alpha^A(\vec{r}) \quad , \tag{62}$$

the coupling matrix elements can be easily calculated :

$$(P + A)_{ji}^R = -\, i \sum_\alpha \sum_\beta C_{j\beta}^A \frac{d}{dR} C_{i\alpha}^A(R) \langle \varphi_\beta^A | \varphi_\alpha^A \rangle \tag{63}$$

and

$$(P + A)_{ji}^\theta = -\, i \sum_\alpha \sum_\beta C_{j\beta}^A(R)\, C_{i\alpha}^A(R) \langle \varphi_\beta^A |\, i l_y \,| \varphi_\alpha^A \rangle \tag{64}$$

where l_y is the component of the orbital angular momentum along the axis perpendicular to the collision plane. The coupled equation set (Eq. (56)) is then integrated numerically, for an initial relative velocity v_o and impact parameter b, with the usual initial conditions :

$$a_j(-\infty) = \delta_{ij} \, , \tag{65}$$

where i denotes the initial channel, and the probability for the $i \to j$ transition is :

$$P_{ij}(v_o,b) = |\, a_j(+\infty\, ;\, v_o,\, b)\, |^2 \tag{66}$$

Finally, the total cross section for the $i \to j$ transition is :

$$Q_{i \to j}(v_o) = 2\pi \int_0^\infty P_{i \to j}(v_o,b)\, b\, db \tag{67}$$

B.2 - Calculations and results. The multichannel-PSS calculations were performed with a molecular basis set including the first five Σ states and the first three Π states. Radial and rotational coupling terms between all

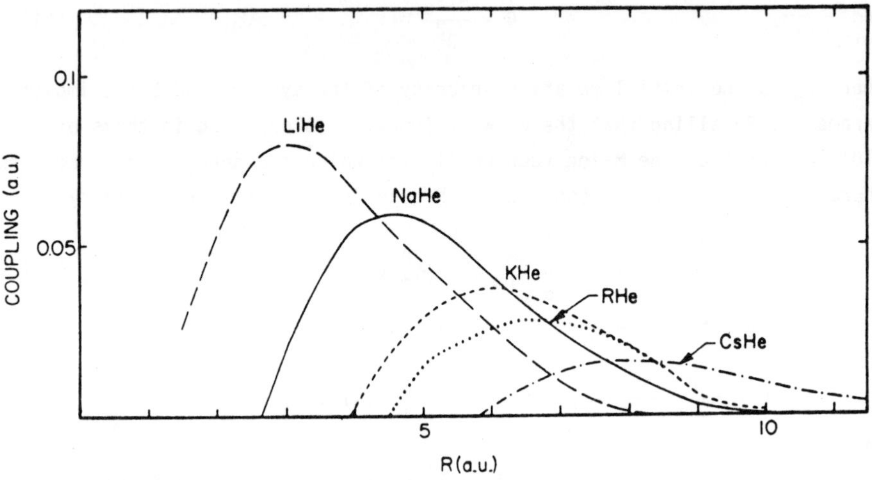

Fig. 21 - Radial coupling between 1Σ and 2Σ molecular states of the M-atom-He systems, as indicated in the figure.

the states were calculated. Figure 21 shows the broad peak of the 1Σ-2Σ radial coupling, which was found the most important radial coupling for these calculations, for all the M-atom-He systems. The position of this peak shifts towards larger R, and its magnitude decreases, as the M-atom changes from Li to Cs. The 1Σ-1Π rotational coupling was also found the most important rotational coupling for these calculations ; it also presents a broad peak which behaves as the 1Σ-2Σ radial coupling when the M-atom changes. However, the decrease of coupling when one goes from Li to Cs will be compensated more or less by the decrease in the energy differences of the molecular states. Radial and rotational coupling between other states involved in the calculations present different characteristics depending on the M-atom ; they may contribute more or less significantly to the collision process, depending upon the collision energy. For example, we have reported in Figs. 22 and 23 the product $P_{i\to j}$ x b, versus b, for excitation from de 1Σ state, i, to excited channel j in Rb(5S) + He collisions, for center of mass energy E_{CM} = 0.5 keV and E_{CM} = 4 keV, respectively. At E_{CM} = 0.5 keV, the dominant channel is found to be the 5pσ state, resulting from the 1Σ-2Σ radial coupling. At E_{CM} = 4 keV, the 5pπ state

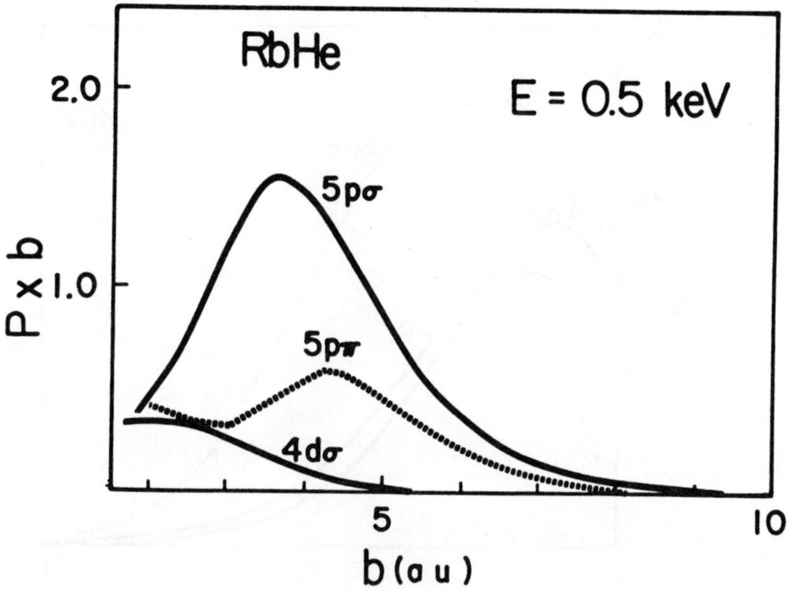

Fig. 22 - Transition probability times the impact parameter b, versus b, for excitation from the 1Σ (5sσ) to the 2Σ (5pσ), 3Σ (4dσ) and 1Π (5pπ) states of Rb He system (as indicated in the figure) at center of mass energy E = 0.5 keV.

becomes the dominant channel for the 5S → 5P excitation in Rb, because of the 1Σ-1Π rotational coupling ; but, also the 4dσ excitation probability becomes comparable with that of the 5pσ excitation (resulting not from a direct excitation through the 1Σ-3Σ radial coupling, but by 1Σ-1Π rotational coupling followed by 1Π-3Σ rotational coupling). The cross sections for excitation of the M-atoms in collisions with He, from the ground-state to the first n^2P level, are shown in Fig. 24, over large range of laboratory energies (note that He is the target). All the cross sections present a broad main peak ; its position shifts towards higher energies when the M-atom changes from Li to Cs and its magnitude decreases in agreement with the behavior of the main rotational and radial coupling and the energy differences discussed above. The position of this main peak is closely related to the Massey parameter $\lambda_M = \frac{\Delta 5\, a}{hv}$ (as discussed, for example, by

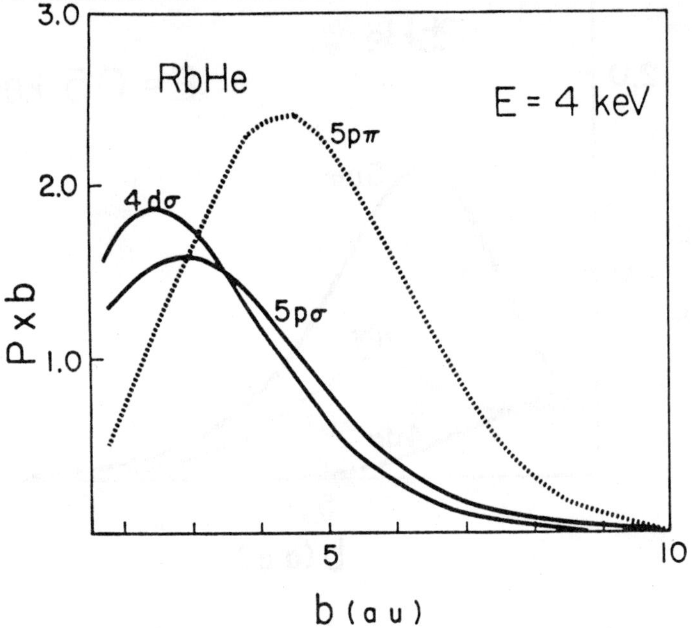

Fig. 23 - Same as Fig. 22 except at E = 4 keV.

Andersen et al., 1979), in which ΔE is the threshold of the n^2S-n^2P transition, v is the relative velocity corresponding to the energy at the peak, and a is an effective range of the interaction. If a is defined, for example, by the position of the maximum in the main radial coupling (see Fig. 21) one finds that λ_M is nearly constant : λ_M = 1.91, 2.15, 2.11, 2.22 and 1.90, respectively, for Li, Na, K, Rb and Cs. The origin of the shoulder observed in the cross sections at lower energies (see Fig. 24) is unclear. However, the structure in the cross sections may be attributed to increassing contribution of the main rotational coupling for the heavier M-atom-He systems, as discussed above, when the energy increases.

Finally, Figures 25 and 26 show comparison between the present results and experimental data for the Na(3S) + He and K(4S) + He collisions, respecti-

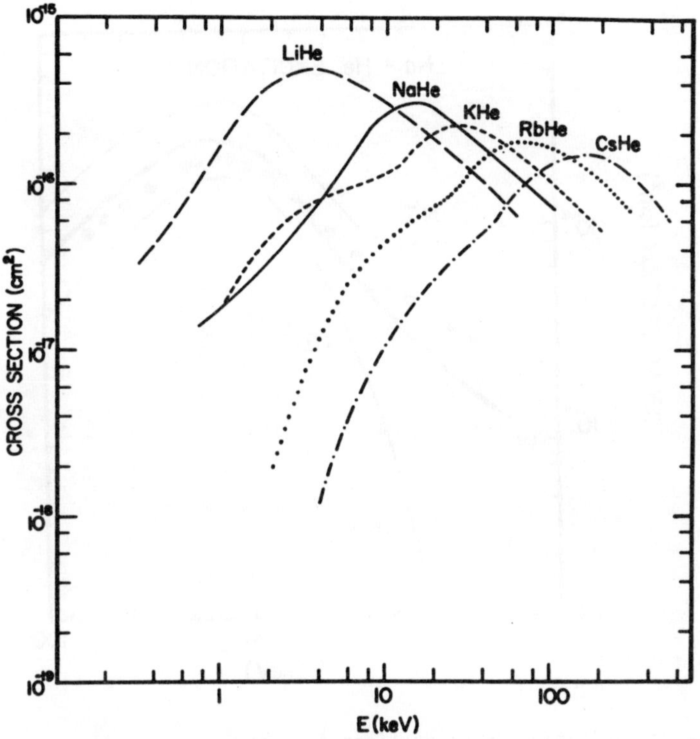

Fig. 24 - Cross-sections for excitation of the M-atom from the ground-state to the first n^2P level, in M-atom-He collisions as indicated in the figure, versus the laboratory energy E.

vely. Agreement between experiment and theory is quite good, both for the position and the absolute value of the main peak in the cross sections. For the Na(3S) + He collision, our multichannel PSS calculations using the PP adiabatic potential energies and molecular wave function above reported, improve significantly, at lower energies, the agreement between experiment and previous calculations based on an atomic orbital expansion and using liPP or HF frozen core interactions. The same observations were made in the case of the Li(2S) + He collision (Kimura and Pascale, 1985).

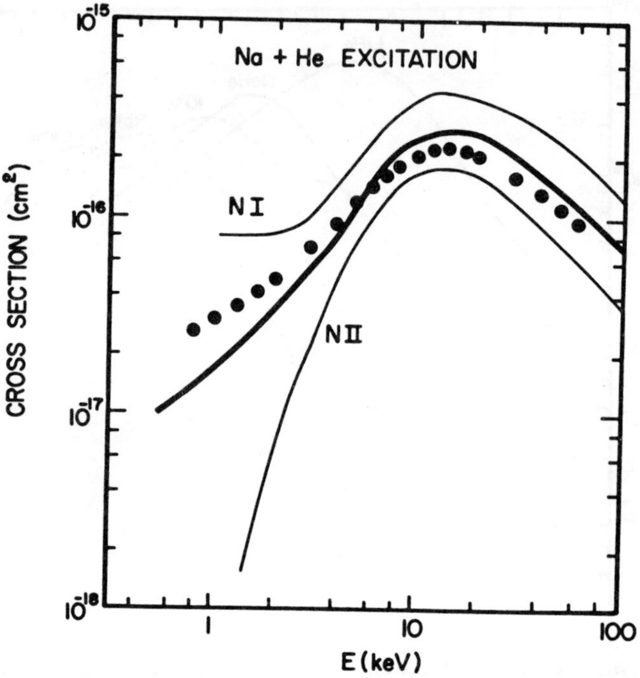

Fig. 25 - 3p excitation cross sections for Na(3S) + He collisions, versus the laboratory energy E. Heavy full curve, present results ; NI, Manique et al. (1977) calculation using liPP interaction (Baylis, 1969) : NII, Manique et al. (1977) calculation using an HF frozen core potential ; full circles, experimental data (Nielsen et al., 1978).

For the K + He collision, our calculations are only in qualitative agreement with the experimental data for the structure observed in the cross-section, at lower energies.

This overall agreement between experimental data and the present results, confirms the reliability of the PP approach for determining quite realistic adiabatic potential energies and wave functions, over a large domain of internuclear distances.

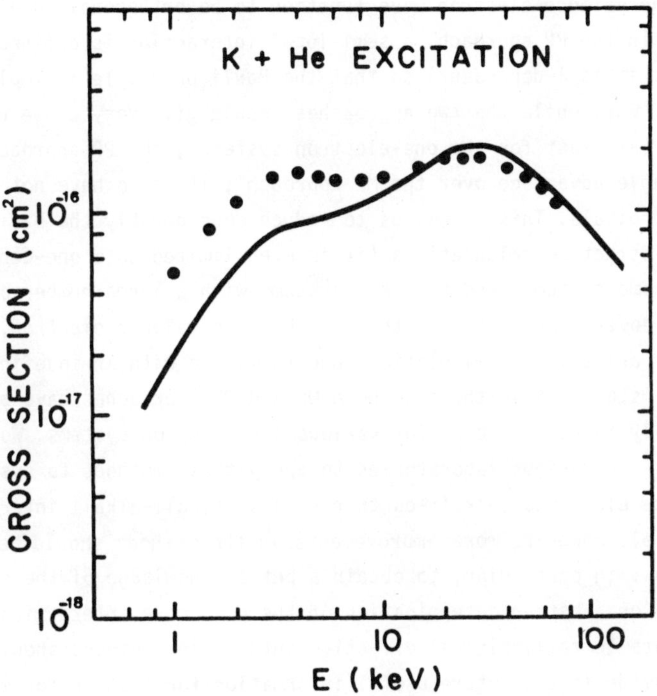

Fig. 26 - 4p excitation cross section for K(4S) + He collisions, versus the laboratory energy E. Full curve, present results ; full circles, experimental data (Andersen et al., 1979).

4. Conclusion

We expect to have shown, from specific but we think quite representative examples, that the use of semiempirical effective interactions (and more particularly, the PP approach) is a quite powerful method for molecular-structure calculations and a detailed study of various collision problems. When the semiempirical effective interactions are carefully determined, their use in molecular-structure calculations or collision problems can lead to quite realistic predictions, which generally compare favorably with experimental data.

In the MP approach, one generally defines a local interaction and one constraints the valence electron wave function to be orthogonal to all the core orbitals. In the PP approach, a semi-local interaction is defined (in the sense that it is l-dependent) so that the Pauli principle is implicitely satisfied. But, while the two approaches should give very close results, in principle (at least for the one-electron systems), the PP approach offers an undeniable advantage over the MP approach ; it is to have not to consider the core orbitals. This allows us to reduce considerably the basis set in molecular-structure calculations (it is even limited to a one-center expansion for some systems) and to treat systems with a large number of electrons. Moreover, for systems with more than one-valence electron, instabilities may arise in MP calculations due to mixing with an indefinite number of virtual states. Nevertheless, both MP and PP approaches have been used successfully in recent years for various one-electron systems. Work is now in progress in various laboratories to apply these methods to systems with two valence electrons (alkali-earth-rare gas, alkali-alkali interactions for example). However, more improvements in the methods should be made in the future : in particular, to obtain a better knowledge of the cross-term everywhere or a better determination of the core-core interaction. Such improvements in semiempirical effective interaction methods should allow one to provide in the future useful information for systems for which *ab initio* approaches are manageable with difficulty. For treatment of reactive collision problems, *ab initio* methods are particularly very difficult to use for calculations of non-adiabatic coupling. Therefore, extension of the PP approach to more than two-center systems may be very useful and should be envisaged in the future.

Acknowledgments

I would like to thank Prof. W.E. Baylis and Dr. F. Rossi for many stimulating and fruitful discussions during the preparation of this paper. I would like to acknowledge also contributions of Prof. R.E. Olson, Drs. M. Kimura and F. Rossi to the work presented here.

References

. Ahmad-Bitar R., Lapatovich R., Pritchard D.E., and Renhorn I. : Phys. Rev. Lett. 39, 1657 (1977).

. Alfaro V. de, and Regge T. : in Potential Scattering, p. 44 (North-Holland Publishing Company, Amsterdam 1965).

. Andersen N., Andersen T., Hedegaard P., and Olsen J.O. : J. Phys. B 12, 3713 (1979).

. Andersen N., and Nielsen S.E. : Adv. At. Mol. Phys. 18, 265 (1982).

. Arthurs A.M., and Dalgarno A. : Proc. R. Soc. A 256, 540 (1960).

. Balling L.C., Wright J.J., and Havey M.D. : Phys. Rev. A 26, 1426 (1982).

. Bardsley J.N. : Case Stud. At. Phys. 4, 299 (1974).

. Barthelat J.C., Durand Ph., and Seraphini A. : Mol. Phys. 33, 159 (1977).

. Baylis W.E. : J. Chem. Phys. 51, 2665 (1969).

. Baylis W.E. : in Progress in Atomic Spectroscopy, ed. Hanle W. and Kleinpoppen H., Part B, Chap. 28 (Plenum Press, 1979).

. Botschwina P., Meyer W., Hertel I.V., and Reiland W. : J. Chem. Phys. 75, 5438 (1981).

. Bottcher C., and Dalgarno A. : Proc. R. Soc. A 340, 187 (1974).

. Brouillaud B. and Gayet R. : J. Phys. B 10, 2143 (1977).

. Buckingham A.C. : Adv. Chem. Phys. 12, 107 (1967).

. Chang E.S. : J. Phys. B 14, 893 (1981).

. Courbin-Gaussorgues C., Wahnon P., and Barat M. : J. Phys. B 12, 3047 (1979).

. Cuvellier J., Fournier P.R., Gounand F., Pascale J., and Berlande J. : Phys. Rev. A 11, 846 (1975).

. Czuchaj E. : Z. Physik A 292, 109 (1979).

. Czuchaj E., and Sienkiewicz J. : Z. Naturforsch A 34, 694 (1979).

. Czuchaj E., and Sienkiewicz J. : J. Phys. B 17, 2251 (1984).

. Czuchaj E., and Sienkiewicz J. : Z. Naturforsch (1985), to appear.

. Dixon R.N., and Robertson I.L. : Mol. Phys. 36, 1099 (1978).

. Doebler H., and Kamke B. : Z. Phys. A 280, 111 (1977).

. Ferray M., Visticot J.P., Lozingot J., and Sayer B. : J. Phys. B 13, 2571 (1980).

. Gadea X.F., Jeung G.H., Pelissier M., and Malrieu J-P., Picqué J.L., Rahmat G., Verges J., and Vetter R. : Laser Chem. 2, 361 (1983).

. Gallagher A. : Phys. Rev. A 172, 88 (1968).

. Gay J-C., and Schneider W.B. : Z. Phys. A 278, 211 (1976).

. Grosswendt B. : Z. Naturforsch 24a, 1424 (1969).

. Hanssen J., Mc Carroll R., Valiron P. : J. Phys. B 12, 899 (1979).

. Hara S. : J. Phys. Soc. Jpn. 22, 710 (1967).

. Havey M.D., Frolking S.E., and Wright J.J. : Phys. Rev. Lett. 45, 1783 (1982).

. Hliva M., Barthelat J.C., and Malrieu J.P. : J. Phys. B 18, 2433 (1985).

. Inouye H., Noda K., and Kita S. : J. Chem. Phys. 71, 2136 (1979) ; and references therein.

. Kahn L.R., Baybutt P., and Truhlar D.G. : J. Chem. Phys. 65, 3826 (1976).

. Karl G., Poll J.D. : J. Chem. Phys. 46, 2944 (1967).

. Kimura M., Olson R.E., and Pascale J. : Phys. Rev. A 26, 3113 (1982).

. Kimura M., and Pascale J. : J. Phys. B 18, 2719 (1985).

. Kimura M., and Thorson W.R. : Phys. Rev. A 24, 1780 (1981).

. Kolos W., and Wolniewicz L. : J. Chem. Phys. 46, 1426 (1967).

. Krauss M. : J. Res. Natl. Bur. Stand., A72, 553 (1968).

. Krauss M., Maldonado P., and Wahl C. : J. Chem. Phys. 54, 4944 (1971).

. Lakowski B.C., Langhoff S.R., and Stallcop J.R. : J. Chem. Phys. 75, 815 (1981).

. Linder F., and Schmidt H. : Z. Naturforsch. A26, 1603 (1971).

. Lukaszewski M., and Jackowska I. : Opt. Commun. 46, 89 (1983).

. Manique J., Nielsen S.E., and Dahler J.S. : J. Phys. B 10, 1703 (1977).

. Masnou-Seeuws F. : J. Phys. B 15, 883 (1982).

. Masnou-Seeuws F., Philippe M., and Valiron P. : Phys. Rev. Lett. 41, 395 (1978).

. Mason C.R., and Peach G. : in Spectral Line Shapes, ed. F. Rostas, Chap. 3, p. 643 (de Gruyter, 1985).

. Mestdagh J.M. : Thèse, Université de Paris-Nord (1982).

. Mestdagh J.M., Berlande J., de Pujo P., Cuvellier J., and Binet A. : Z. Phys. A 304, 3 (1982).

. Mies F.H. : Phys. Rev. A 7, 942 (1973).

. Minemoto T., Goto T., and Kanda T. : J. Phys. Soc. Jpn. 36, 918 (1974).

. Minemoto T., and Kakihara K. : J. Phys. Soc. Jpn. 41, 984 (1976).

. Mó O., Riera A., and Yáñez M. : Phys. Rev. A 31, 3977 (1985).

. Munster P., and Marek J. : J. Phys. B 14, 1009 (1981).

. Nielsen S.E., Andersen N., Andersen T., Olsen J.O., and Dahler J.S. : J. Phys. B 11, 3187 (1978).

. Nikitin E.E. : Adv. Chem. Phys. 28, 317 (1975).

. O'Callagham M., Holstein T., and Gallagher A. (1985): to be published.

. Olson R.E. : Chem. Phys. Lett. 33, 250 (1975).

. O'Malley T.F., Burke P.G., and Berrington K.A. : J. Phys. B 12, 953 (1979).

. Pascale J. : J. Chem. Phys. 67, 204 (1977).

. Pascale J. : Phys. Rev. A 28, 632 (1983a).

. Pascale J. : 13th ICPEAC Berlin (Ed. Eichler J. et al.), 313 (1983b) ; ib., 342 (1983c) ; to be published.

. Pascale J. : in Spectral Line Shape, ed. F. Rostas, Chap. 3, 563-586 (de Gruyter, 1985) ; and references therein.

. Pascale J., and Kimura M. (1985), to be published.

. Pascale J., Mestdagh J.M., Cuvellier J., and de Pujo P. : J. Phys. B 17, 2627 (1984).

. Pascale J., and Perrin M-Y. : J. Phys. B 13, 1839 (1980).

. Pascale J., and Stone P.M. : J. Chem. Phys. 65, 5122 (1976).

. Pascale J., and Vandeplanque J. : J. Chem. Phys. 60, 2278 (1974).

. Peach G. : Comments At. Mol. Phys. 11, 101 (1982).

. Peach G. : in Atoms in Astrophysics, ed. Burke P.G. et al., Chap. 5 (Plenum, 1983).

. Philippe M., Masnou-Seeuws F., and Valiron P. : J. Phys. B 12, 2493 (1979).

. Reid R.H.G. : J. Phys. B 6, 2018 (1973).

. Rossi F., and Pascale J. : Phys. Rev. A 32, 2657 (1985).

. Saxon R., Olson R.E., and Liu B. : J. Chem. Phys. 27, 2692 (1977).

. Siara I.N., Kwong H.S., and Krause L. : Can. J. Phys. 52, 945 (1974).

. Slater J.C., and Kirkwood J.G. : Phys. Rev. A 37, 682 (1931).

. Szudy J., and Baylis W.E. : J. Quant. Spectrosc. Radiat. Transfer 15, 641 (1975).

. Visticot J.P., Pascale J., and Sayer B. : J. Phys. B 18, 2861 (1985).

. Wagner A.F., Wahl A.C., Karo A.M., and Krejci R. : J. Chem. Phys. 69, 3756 (1978).

. Weeks J.D., Hazi A., and Rice S.A. : Adv. Chem. Phys. 16, 283 (1969).

. Williams J.F. : J. Phys. B 12, 265 (1979).

. York G., Scheps R., and Gallagher A. : J. Chem. Phys. 63, 1052 (1975).

Symmetry and Spectra of Neutral and Highly Ionized Many-Electron Atoms

Z.B.Rudzikas

Institute of Physics of the Academy of Sciences
of the Lithuanian SSR
K.Poželos 54, Vilnius, 232600, USSR

1. Introduction

A great variety of optical and spectroscopical quantities of atoms and ions is of interest for many domains of physics and the other branches of science and technology. Often the most complex elements of the periodic table, as well as very highly ionized atoms, are considered. Most of these quantities may be found only with the help of joint efforts of theorists and experimenters. Therefore, the development of powerful methods of theoretical investigation of many-electron systems, elucidating their structure and properties, should be regarded as an urgent issue. This paper presents a brief analysis of presently-accepted or currently-elaborated methods to determine the energy spectra of complex atoms and ions, including multiply charged ones; to identify and classify the energy levels by means of various coupling schemes, and studies of the role of symmetry in many-particle systems. Some new results in this domain, theoretical description of electronic transitions of various multipolarity, as well as regularities in their behaviour along isoelectronic sequences are also considered.

In the next section we shall discuss the methods of calculating energy spectra in non-relativistic and relativistic approximations, as well as the problem of optimizing the coupling scheme (a set of quantum numbers), used for their classification. The third section is devoted to electronic transitions and selection rules for them.

The usefulness of different methods to calculate the energy spectra of atoms and multiply charged ions, as well as the pecularities of their spectra, is discussed in the fourth section. In the last section the role of symmetry, the usage of quasispin and isospin in the theory of many-electron atoms and ions are considered.

2. Obtainment of Energy Spectra of Many-Electron Atom or Ion

Spectra are fundamental characteristic of any atom, the main source to obtain the information of its structure and properties, of many properties of radiating plasma, both laboratory and astrophysical. Multiply charged ions are interesting because of their presence in laser and thermonuclear plasma or in hot astrophysical objects, also because of their possibilities to be used to create the lasers radiating in X-ray regions, and so on.

The most general and effective method to find the spectra of atoms and ions is a variational approach, allowing in various approximations to determine the spectra and the other spectral characteristics of atoms and ions practically for any element of the periodic table. Combining this method with the methods of irreducible tensorial operators and coefficients of fractional parentage opens wide possibilities for investigations of the most complex electronic configurations.

It is well known that a many-particle problem can not be solved exactly. Therefore, normally in atomic spectroscopy there is used the central field approximation, in which the many-particle problem is reduced to a one-particle one. In this approach each electron in a many-electron atom is considered as moving in the field, created by the nuclear charge and the charges of the rest electrons. Practically, this approach is realized in various modifications of the Hartree-Fock self-consisted field method [1].

While obtaining the wave-functions of a many-electron atom, usually the zero-order non-relativistic energy operator (Hamiltonian) H_o is adopted [2] :

$$H_o = T + P + Q \tag{2.1}$$

where

$$T = \sum_{i=1}^{N} \vec{p}_i^{\,2}/2m, \tag{2.2}$$

$$P = -\sum_{i=1}^{N} Ze^2/r_i, \tag{2.3}$$

$$Q = \sum_{i>j=1}^{N} e^2/r_{ij}. \tag{2.4}$$

Here T denotes the kinetic energy of the electrons with respect to the nucleus. P and Q represent the electrostatic interactions of the electrons with the nucleus and between electrons, respectively, whereas e is an absolute value of the electronic change; m is its mass; p denotes its momentum; r_i is the distance of the $i^{\underline{th}}$ electron from the nucleus; Ze represents the nuclear charge; r_{ij} stands for the interelectronic distance; and N is the number of electrons in an atom.

Term splitting is caused by spin-orbit and spin-spin interactions. The spin interaction operators are smaller than the electrostatic interaction operators by the order of magnitude α^2, where α is a fine structure constant. While calculating the oscillator strengths or transition probabilities it is important to define the most accurate energy differences of the levels, between which the transition takes place. Adopting for this purpose a single-configuration approach and the abovementioned zero-order Hamiltonian,one fails to achieve the accuracy desired. Especially this is true for multiply charged ions. To refine the results one has to account for correlation and relativistic effects.

In the central field approximation the Hamiltonian of a many-electron atom, considered with respect to non-relativistic wave-functions and including the terms of the order α^2, may be written in the form (the Hartree-Fock-Pauli(HFP) approximation):

$$H = H_o + W, \qquad (2.5)$$

where W represents all relativistic corrections of the order α^2, including magnetic terms, as well, causing the term splitting. Their explicit form may be found in [2,3]. To evaluate their matrix elements, one has to express all these operators in terms of irreducible tensors. The details of this procedure, as well as the expressions needed may be found in [3].

If relativistic effects are not small or even comparable by the order of magnitude with a non-spherical part of electrostatic interactions, then it is necessary to make use of relativistic wave-functions and relativistic Hamiltonian. The latter usually is chosen in the form of the relativistic Breit operator (the Dirac-Hartree-Fock (DHF) approximation):

$$\mathcal{H} = \sum_{i=1}^{N} (H_i^1 + H_i^2 + H_i^3) + \sum_{i>j=1}^{N} (H_{ij}^4 + H_{ij}^5 + H_{ij}^6), \qquad (2.6)$$

where H^1, H^2 and H^3 are the usual one-particle Dirac operators of an electron, moving in the nuclear charge field. H^1 corresponds to the electron kinetic energy and a one-electron part of the spin-orbit interaction; H^2 denotes the mass effect , whereas H^3 stands for the potential energy. The two-particle operator H^4 describes the Coulomb interaction, H^5 stands for the operator of the magnetic and a part of the retardation interactions, whereas H^6 represents the remaining retardation interactions. Let us point out that the retardation of the interaction is conditioned by the finite value of the light velocity. The explicit expressions for these operators and their matrix elements may be taken from [3].

It ought to be underlined that in a relativistic approach the structure of the electronic shells of an atom changes itself. Due to the dominating role of the spin-orbit interactions each shell of the equivalent electrons $l^N \alpha LSJ$ in a relativistic case splits into a number of subshells $nlj_1^{N_1} j_2^{N_2} \alpha_1 J_1 \alpha_2 J_2 J$, where $j_1 = l - 1/2$, $j_2 = l + 1/2$, $N_1 + N_2 = N$, and α distinguishes degenerate terms, whereas α_i distinguishes the levels.

While calculating the energy spectra of atoms and ions, including those with complex electronic configurations, one has to form the energy matrices, starting with some definite coupling scheme, which ought to be the closest to the real one. For neutral and a few times ionized atoms usually LS coupling takes place. The number of matrices is defined by the allowed values of the resultant angular momentum J. Further it is necessary to calculate all matrix elements E_{ik} (i and k denote the rows and columns of the matrix, respectively) of the energy operator in the approximation chosen (e.g., (2.5) or (2.6)), then to diagonalize each matrix, i.e. to transform it to the form, in which all off-diagonal matrix elements are zero. This can be achieved by the diagonalizing matrix S and the reciprocal one S^{-1} according to the formula

$$(E_{ii}) = S (E_{ik}) S^{-1}. \qquad (2.7)$$

Let us remind that each matrix element E_{ik} is the sum of terms, describing all taken into account interactions (2.5) or (2.6), both inside each electronic shell and between them. Thus, the diagonalization of the energy matrix means that we pass to some intermediate coupling scheme, the wave-functions of which are already eigenfunctions of the energy operator considered. They may be written as follows:

$$\Psi(\beta J) = \sum_{\alpha_i L_i S_i} a(\alpha_i L_i S_i)\, \Psi(\alpha_i L_i S_i J), \qquad (2.8)$$

where a stands for the weights of the wave-functions of the initial, in this case LS, coupling scheme. If numerical value of the weight of one function in (2.8) is much larger than that of the rest functions, then this level may be identified and classified with the help of a set of quantum numbers of this wave-function. If there are several weights of the same order of magnitude, then one has to analyse the fitness of the other coupling schemes and to find the optimal one. The method to optimize the coupling scheme, as well as the usage of various coupling schemes for identification and classification of the energy spectra of complex atoms and ions, including multiply charged ions, has been described in detail in [3]. It is based on a simple relation between the weights c_{jk} and a_{jr} of the wave-functions in two pure coupling schemes according to the formula

$$c_{jk} = \sum_{r} a_{jr} (\varphi_k | \Psi_r). \qquad (2.9)$$

The explicit expressions for the transformation matrices again, may be found in [3].

3. Electronic Transitions

Practically all atomic optics and spectroscopy are based on the radiation or absorption of the quanta of electromagnetic field during electronic transitions. Therefore, it is not sufficient to have the energy spectrum of an atom or ion, but it is necessary to calculate the oscillator strengths or probabilities of various electronic transitions. The latter characteristics usually are devided into electric (Ek) and magnetic (Mk) transitions of the multipolarity k. Of particular importance are electric dipole transitions which are the most popular way for an electron to return from the excited state to the ground one.

In [4] there are obtained the following general expressions for the electric multipole transition probability (in a.u.):

$$W^{Ek}_{1\to2} = \frac{2(k+1)(2k+1)\,\omega^{2k-1}}{k[(2k+1)!!]^2 c^{2k+1}} \left| \left(2 \right\| \left\{ Q'^{(k)}_{-q} + \mathcal{K}\sqrt{\frac{k}{k+1}} \left[Q'^{(k)}_{-q} - \omega Q^{(k)}_{-q} \right] \right\} \left\| 1 \right) \right|^2, \qquad (3.1)$$

$$W^{Ek}_{1\to2} = \frac{2(k+1)(2k+1)\,\omega^{2k+1}}{k[(2k+1)!!]^2 c^{2k+1}} \left| \left(2 \right\| \left\{ Q^{(k)}_{-q} + \mathcal{K}\sqrt{\frac{k}{k+1}} \left[\frac{1}{\omega} Q'^{(k)}_{-q} - Q^{(k)}_{-q} \right] \right\} \left\| 1 \right) \right|^2. \qquad (3.2)$$

In these formulas ω is the transition frequency; K stands for a free parameter, the so called gauge condition of the electromagnetic field potential. The operator of Ek radiation is defined as follows:

$$Q_{-q}^{(k)} = -r^k C_{-q}^{(k)} \quad . \qquad (3.3)$$

It is the first well known form of the Ek radiation operator which, in the case of k = 1, turns into the usual "length" expression for the electric dipole radiation operator. While on the other hand

$$Q'^{(k)}_{-q} = -r^{k-1}\left\{ k\, C_{-q}^{(k)} \frac{\partial}{\partial r} + \frac{i}{r}\sqrt{k(k+1)}\left[C^{(k)} \times L^{(1)}\right]_{-q}^{(k)} \right\} \qquad (3.4)$$

is a new, second form of the Ek radiation operator, which for k = 1 leads to the "velocity" form of the corresponding operator. In these formulas $C^{(k)}$ denotes the spherical function operator, whereas $L^{(1)}$ stands for the angular momentum operator.

The transition probability, being the physical quantity, has not to depend on the non-physical parameter K. This is the case for accurate wave-functions, for which the quantity in square brakets of (3.1) and (3.2) is identically equal to zero. However, in a many-electron case we deal with approximate wave-functions, and, therefore, this dependence takes place. In such cases one can adopt K as a semi-empirical parameter for refined evaluation of the electronic transition characteristics, e.g., in isoelectronic sequences or for the estimation of the accuracy of the wave-functions used. In fact, in general case the dependence of W^{Ek} on K has the form of a parabola, which, with the increase of the accuracy of the wave-function, degenerates into the straight line, parallel to K axis. This can be obviously seen from the results presented in [5], where the dependence of the oscillator strength of electric dipole transitions on K has been considered for the case of the wave-functions of various accuracy, starting with the expressions

$$f^{E1}_{1\to 2}=\frac{2}{3\omega}\left|\left(2\left\|\left\{Q'^{(1)}_{-q}+\frac{\kappa}{\sqrt{2}}\left[Q'^{(1)}_{-q}-\omega Q^{(1)}_{-q}\right]\right\}\right\|1\right)\right|^2, \qquad (3.5)$$

$$f^{E1}_{1\to 2}=\frac{2\omega}{3}\left|\left(2\left\|\left\{Q^{(1)}_{-q}+\frac{\kappa}{\sqrt{2}}\left[\frac{1}{\omega}Q'^{(1)}_{-q}-Q^{(1)}_{-q}\right]\right\}\right\|1\right)\right|^2. \qquad (3.6)$$

The probabilities of magnetic multipole transitions are defined by the formula:

$$W^{Mk}_{1\to 2}=\frac{2(k+1)(2k+1)}{k\left[(2k+1)!!\right]^2}\left(\frac{\omega}{c}\right)^{2k+1}\left|\left(2\left\|{}_mQ^{(k)}_{-q}\right\|1\right)\right|^2, \qquad (3.7)$$

where

$${}_mQ^{(k)}_{-q}=-i^k\frac{r^{k-1}}{c}\sqrt{k(2k-1)}\left\{\frac{1}{k+1}\left[C^{(k-1)}\times L^{(1)}\right]^{(k)}_{-q}+\left[C^{(k-1)}\times S^{(1)}\right]^{(k)}_{-q}\right\}. \qquad (3.8)$$

Here $S^{(1)}$ stands for the spin angular momentum, and $i = \sqrt{-1}$.

Corresponding expressions for the relativistic transition operators, as well as for their matrix elements may be found in [3]. The problem of relativistic corrections to the electron transition operators is discussed in [3], as well.

Concluding this section let us briefly discuss the selection rules for electronic transitions. They are of different degree of accuracy and are caused by the non-zero condition for matrix elements of the transition operators. For exact quantum numbers these conditions hold exactly (e.g., a division of transitions in two groups - electric and magnetic radiation). For approximate quantum numbers (to this group belong, e.g., in the case of complex electronic configurations, all intermediate quantum numbers) the selection rules are rather conditional. This is particularly clear while dealing with intermediate coupling scheme(the wave-functions of the type(2.8)) when many transitions, forbidden in the pure coupling scheme, become allowed and have fairly large values of the oscillator strength. E.g., while considering E1 transition $d^3p - d^3s$ in the intermediate coupling scheme, 2082 lines occur instead of 215, allowed in a pure LS coupling, i.e. the number of allowed transitions increases by the order of magnitude. More or less approximate are the sum rules for electronic transitions, as well.

The selection rules for the transitions strongly depend on the ionization degree of an atom. While considering the characteristics of electronic transitions along the isoelectronic sequence one can see that many transitions, allowed at its beginning, become forbidden for high ionization degrees,and vice versa. This can be explained by the change of the relative role of the nonspherical parts of Coulomb and magnetic interactions in electronic shells and thus, by changes of coupling schemes, which cause disappearance of one selection rules and occurance of others.

For transitions inside the electronic configuration usually the electric quadrupole (E2) and magnetic dipole (M1) radiations are possible. The usage of intermediate coupling schemes for their calculations leads to the substantial increase of the number of allowed E2 transitions. When both the transitions (E2 and M1) are allowed, for a few times ionized atoms the E2 transitions mostly predominate. However, with the increase of the ionization degree the M1 transitions begin to prevail. Moreover, the wavelengths of E2 and M1 transitions then are situated in the visible region of the spectrum, whereas the E1 radiation is displaced to the X-ray region. Therefore, the E2 and M1 radiation becomes fairly useful for diagnostics of the hot temperature plasma.

4. Pecularities of the Spectra of Highly Ionized Atoms

Taking into consideration what was said in the second and third sections, it is possible to calculate the energy spectra practically of any atom or ion of the periodic table. Also it is possible to find oscillator strengths or probabilities of electric and magnetic transitions of any multipolarity and then to use these data for the identification and classification of the experimentally measured or observed astrophysical spectra as well as for plasma diagnostics, and to look for the cases and conditions of inverse population of the levels.

While passing to multiply charged ions the role of the theory increases considerably. Their spectra are in the far ultraviolet or even X-ray spectral region, requiring the special experimental equipment. Therefore, one may use the obtained knowledge, including the spectral region of the ion radiation. Modern theory allows to get such information fairly easily. For multiply charged ions the large deviations from the pure coupling schemes, particularly from LS coupling, are typical. Therefore, in the majority of cases one has to make use of intermediate coupling; to choose, for identification, optimal sets of quantum numbers, the only possibility to find them being theoretical or semiempirical calculations. For extremely highly ionized atoms even the notion of configuration itself loses its normal sense.

Interesting pecularity of the spectra of multiply charged ions is the appearance of the so called sattelite lines. They occure near the resonant lines of the ion given and belong to the ion, having the ionization degree smaller by unity; in this ion two electrons are excited. Sattelite lines at the resonant lines of hydrogen-like ions correspond to the transitions $2l2l' - 1s2l''$ of helium-like ions, etc. The wavelengths of the lines of the ion $A^{(n-1)+}$ are close to those of the resonant lines of the ion A^{n+}. This complicates the identification of the spectra of multiply charged ions. Studies of the intensities of the sattelite lines with respect to resonant ones turned to be an extremely efficient method to diagnose hot-temperature plasma (determination of its density, temperature, etc.).

Further let us consider briefly the accuracy of various methods used for calculations of the spectra of atoms and ions. Analysis of the total energies of 10-20 times ionized atoms shows that accounting for the relativistic effects as corrections (HFP approximation) improves the correspondence between theoretical and experimental results by the order of magnitude. Table 1 presents the numerical values of the energy levels of the excited configuration $1s^2 2p^4$ for Fe^{20+}(in a.u.), found using different approximations. The levels in it are counted from the level $1s^2 2s^2 2p^2\ {}^3P_0$. The data of perturbation theory (PT) are taken from [6], whereas relativistic data (DHF) are from [7]. In the latter paper there has been taken into account, in relativistic approximation, the superposition of the configurations $1s^2 2s^2 2p^2$ and $1s^2 2p^4$.Method SC corresponds to the HFP approach, combined with the superposition of configurations.

Table 1. Energy levels of the configuration $1s^2 2p^4$ of Fe^{20+} (in a.u.)

LSJ	HFP	DHF[7]	PT [6]	SC	Exper. [8]
${}^3P_2$	7,35832	7,55223	7,51919	7,51581	7,50458
${}^3P_0$	7,73837	7,96090	7,92295	7,91596	—
${}^3P_1$	7,78166	7,97666	7,94147	7,92930	7,93963
${}^1D_2$	8,17253	8,35936	8,28585	8,32113	8,29162
${}^1S_0$	9,04921	9,41208	9,34361	9,37746	9,39061
σ	0,2104	0,0466	0,0248	0,0179	

Table 1 illustrates a great role of correlation effects for the configuration consided: single-configuration approximation HFP leads to the root-mean-square deviation, which is larger by the order of

magnitude then that of the rest methods. Combining the HFP approach with the superposition-of-configuration method one gets the best results.

It is possible to judge about the accuracy achieved and about the role of relativistic effects,while calculating in different approaches the wavelengths, looking at Table 2, in which the wavelengths (in Å) of the E1 transitions $1s^2 2s^2 2p^5 3dLSJ - 1s^2 2s^2 2p^6\ {}^1S_0$ for Mo^{32+} are shown.

Table 2. Wavelengths of the transition $2p^5 3d$-$2p^6$ (in Å) for Mo^{32+}

LSJ - ${}^1S_0$	PT [9]	HFP	DHF	Exp. [10]
${}^3P_1$ - ${}^1S_0$	4,869	4,860	4,854	4,847
${}^3D_1$ -	4,812	4,802	4,803	4,804
${}^1P_1$ -	4,643	4,637	4,630	4,630

It is obvious from this table that all the three methods give the results of a high accuracy. However, the best one is the totally relativistic approach (DHF). It ought to be used for even more highly ionized atoms.

The calculations show that accounting for correlation effects already in the frameworks of superposition of quasidegenerated configurations (having the same sets of principal quantum numbers, as the configuration under consideration) allows one to achieve substantial improvement of the energy spectrum, wavelengths, oscillator strengths and probabilities of electronic transitions in multiply charged ions.

There are cases when in the isoelectronic sequence the energies of two configurations cross each other,and then a single-configuration approach is completely inapplicable.

As concerns the structure of a multiply charged ion, it ought to be mentioned that, contrary to the neutral atom, which can be regarded as "diffuse" substance, the multiply charged ion is very compact, "compressed" formation, with near to the nucleus situated electronic shells, split into subshells, the charge density of which is concentrated in a very narrow interval.

Thus, the present status of the theory gives the possibility to carry out the systematic calculations of the spectral characteristics of separate atoms and ions, as well as their isoelectronic sequences, to study regularities in changes of the quantities studied along the isoelectronic sequences, and to use the data obtained for plasma diagnostics, in laser optics, etc.

5. Symmetries in Many-Electron Atom

The use of symmetry properties usually allows one to simplify the problem to describe theoretically the system considered, as well as to get a fairly large number of information about its structure and properties, without solving the equations describing it. The simple symmetry properties, connected with transformations in space and time (inversion, rotations of the three-dimensional space, etc.), are well known and widely used. However, many-particle system may as well possess other symmetry types (rotations in other spaces, transformations, described by the groups of higher ranks, gauge invariance, etc.). Accounting for them may be of great use, too. In conclusion, let us discuss briefly this question. Gauge invariance of the operators of Ek transitions was discussed in the third section. More thoroughly these questions are described in [1].

The angular momentum theory is based on the symmetry of an atom with respect to rotations. Permutational symmetry is reflected in the properties of the spin of a system. Another example of angular momentum theory application in a new space we get when introducing the quasispin Q instead of the seniority quantum number v in a shell $l^N \alpha v LS$ due to the relation $Q = (2l + 1 - v)/2$. The corresponding operator behaves as the spin momentum, its projection being $Q_z = (N - 2l - 1)/2$. Therefore, all the theory may be generalized to cover quasispin space [11].

The spin-angular parts of wave-functions of partially and almost filled shells in a quasispin method differ only in the sign of Q_z. Therefore, the phase relations between various quantities, connecting partially and almost filled shells, may be obtained directly from the Clebsh-Gordan coefficient, which occurs while applying the Wigner-Eckart theorem in quasispin space.

The method of quasispin can be easily generalized to cover the case of complex configurations. Actually, coupling vectorially quasispin momenta, one can as well introduce the total quasispin. In this way one can present a new sort of a multiconfigurational wave-function in the superposition-of-configurations method, in which the weights of superposed configurations, in fact, are usual Clebsh-Gordon coefficients, i.e.,

$$|(l_1+l_2)^N \alpha_1 Q_1 L_1 S_1 \alpha_2 Q_2 L_2 S_2 Q L S M_L M_S) = \sum_{M_{Q_1} M_{Q_2}} \begin{bmatrix} Q_1 & Q_2 & Q \\ M_{Q_1} & M_{Q_2} & M_Q \end{bmatrix} | l_1^{N_1} l_2^{N_2} \alpha_1 Q_1 L_1 S_1 \alpha_2 Q_2 L_2 S_2 L S M_L M_S). \quad (5.1)$$

Calculations show that the basis (5.1) describes the excited states and the superposition of configurations with the same principal quantum numbers, fairly well. In these cases the quantum number of the total quasispin Q is rather exact. In general, the use of theoretical properties of operators and wave-functions in quasispin, orbital and spin spaces leads to the new, very efficient version of the theory of complex atomic spectra [11].

For configurations $n_1 l^{N_1} n_2 l^{N_2}$ the notion of isospin may be successfully used [11], what gives the possibility to define a new basis. It has turned out that the corresponding operator $T^{(1)}_q$ is the irreducible tensor in some new (isospin) space, their components transforming like those of the spin operator. Thus, we can again express all operators in terms of triple tensors acting in the isospin, orbital and spin spaces, correspondingly. The wave-functions, then, may be written as follows:

$$|n_1 n_2 (\ell\ell)^N \alpha T L S M_T M_L M_S) = |n_1 n_2 \ell^{N_1} \ell^{N_2} \alpha T L S M_L M_S), \qquad (5.2)$$

where $T = N/2,\ N/2-1,\ 1/2$ (or 0), $M_T=(N_2-N_1)/2$. We can again apply the Wigner-Eckart theorem to all the three spaces and thus considerably simplify all calculations. In isospin formalism two shells are combined into one, which is considered as partially (almost) filled if $N \leq 4l+2$ ($N > 4l+2$); and completely filled if $N = 8l+4$.

The wave-functions in isospin basis (5.2) may be obtained from the usual one in the following way:

$$|n_1 n_2 (\ell\ell)^N \alpha T L S M_T M_L M_S) = \sum_{\alpha_1 L_1 S_1 \alpha_2 L_2 S_2} |n_1 n_2 \ell^{N_1} \ell^{N_2} \alpha_1 L_1 S_1 \alpha_2 L_2 S_2 L S M_L M_S) \times$$

$$\times (\ell^{N_1} \ell^{N_2} \alpha_1 L_1 S_1 \alpha_2 L_2 S_2 L S | (\ell\ell)^N \alpha T L S M_T), \qquad (5.3)$$

where the last multiplier is a transformation matrix between two bases. Its calculation method is described in [11].

Considering the Coulomb energy operator with respect to the functions (5.3) one sees that in the isospin basis it is possible, already, in an operational form to cover both diagonal and off-diagonal interactions, with respect to configurations, i.e., in this approach one is able at once to account for a part of correlation effects. The calculations show that there are cases, particularly for the configurations of the type $n_1 l n_2 l^N$ ($n_2 \neq n_1+1$), when the new basis is the most preferable.

References

1. Froese-Fischer Ch. The Hartree-Fock Method for Atoms. - N.Y.:John Wiley and Sons, 1977.

2. Bethe H.A., Salpeter E.E. Quantum Mechanics of One- and Two-Electron Atoms. - Berlin, Springer-Verlag, 1957.

3. Nikitin A.A., Rudzikas Z.B.Foundations of the Theory of the Spectra of Atoms and Ions. Nauka Publ., Moscow,1983 (in Russian).

4. Kaniauskas J.M., Merkelis G.V., Rudzikas Z.B. - Lietuvos Fizikos Rinkinys*, v.19, 475 (1979).

5. Rudzikas Z.B., Szulkin M., Martinson I. - Physica Scripta, v.T8, 141 (1984).

6. Bogdanovich P.O., Merkelis G.V., Rudzikas Z.B., Safronova U.I., Šadžiuvienė S.D. - Preprint ISAN , No.4, Moscow, 1977.

7. Desclaux J.P., Cheng K.T., Kim Y.-K. - J.Phys.B: Atom.Molec.Phys., v.12, 3819 (1979).

8. Mori K.,Otsuka M., Kato T. - Nagoya University, Institute of Plasma Physics, Report IPPJ-AM-3, 1977.

9. Safronova U.I., Rudzikas Z.B. - J.Phys.B: Atom.Molec.Phys., v.10, 7 (1977).

10. Boiko V.A., Pikuz S.A., Faenov A.Ya. - Preprint FIAN, No.20, Moscow, 1976.

11. Rudzikas Z.B., Kaniauskas J.M. Quasispin and Isospin in the Theory of Atom. Vilnius, Mokslas Publishers, 1984.

12. Šimonis V.Č., Kaniauskas J.M., Rudzikas Z.B. - Lietuvos Fizikos Rinkinys*, v.22, No.4, 3(1982).

---

* English translation: Soviet Physics Collection

OPTICAL TESTS OF QUANTUM MECHANICS

K. Wódkiewicz

Institute of Theoretical Physics,
Warsaw University, Warsaw, 00-681, Poland

I. INTRODUCTION

Inspired by Einstein's idea of a ghost field Born published in 1926 his famous paper "Quantum Mechanics of Collision Phenomena" [1] in which he proposed the statistical interpretation of the wave function. Discussing a scattering process Born came to the conclusion that the wave function of the electron determines the probability for the scattering of the charged particle into some fixed direction. In his next paper with the same title [2] Born elaborated the probabilistic interpretation of the wave function further. In this paper Born says clearly that he starts from a remark made by Einstein "that the waves are only there to show the way to the corpuscular light-quantum" and that they form " a ghost field which determines the probability of a light-quantum to take a definite path".

Einstein always felt uneasy about the fact that his theory of spontaneous emission leaves time and direction of an elementary process to chance. In order to overcome this difficulty Einstein speculated during the early 1920's about the role of such a ghost field in the dynamics of light quanta and wave fields [3].

It is interesting to note that in the same paper Born pointed out that the possibility of hidden variables cannot be completely ruled out [2]. For Born these hidden variables were equivalent to additional parameters not yet introduced into the theory. This parameters could in principle determine individual events and individual properties of particles. In classical mechanics these coordinates are "phases" of the motion i.e., coordinates of the particle at a certain moment.

According to Heisenberg [4] the probabilistic interpretation of the wave function can be regarded as a quantitative formulation of the concept of potentiality, taken from Aristotle's philosophy. The same attitude can be found in latter writing of Popper [5] when he says that: "...waves are mathematical representations of propensities, or of dispositional properties of the particle to take up certain states".

In 1932 Wigner generalized Born's concept of a q-space probability to a phase-space introducing his famous function [6]:

$$W_\psi(q,p) = \int \frac{dx}{2\pi\hbar} \psi^*(q+x/2)\psi(q-x/2)\exp(ipx/\hbar), \quad (1)$$

where for simlicity of the arguments we choose to work only in one space dimension. The marginal averages of the Wigner function correspond to momentum and position distributions:

$$\int dq W_\psi(q,p) = \frac{|\tilde{\psi}(p)|^2}{2\pi\hbar} \quad \text{and} \quad \int dp W_\psi(q,p) = |\psi(q)|^2. \quad (2)$$

Wigner realized also that W_ψ is not everywhere nonnegative and as such cannot be really interpreted as the simultaneous probability for coordinates and momenta.

In 1935 a criticism of the probabilistic interpretation of the wave function has been raised by Einstein Podolsky and Rosen (EPR) [7]. The EPR argument has raised the important question of the completeness of the quantum mechanical description. This point repeated many time in the literature can be clearly related to the concept of hidden variables. With these variables the probabilistic interpretation of the wave function is gone because each microscopic individual event in q-space can be described by a deterministic function $A(q,\lambda)$ where now due to the introduction of the hidden parameter λ this dynamical variable is no longer a quantum statistical operator. In actual experiments due to our lack of informations about these parameters we can obtain only ensemble averages of $A(q,\lambda)$. As a result a hidden variable theory predicts that the expectation value of this dynamical variable is equal to:

$$\langle A\rangle = \int d\lambda \rho(\lambda)\, A(q,\lambda), \quad (3)$$

where $\rho(\lambda) > 0$ is a positive distribution of these parameters characteristic to the performed experiment.

From these remarks follows a series of natural questions related to the fundamental interpretation of probabilities in quantum mechanics. For example is a hidden variable theory equivalent to quantum mechanics, and if not, what are the physical differences? Can we use the phase-space parameters of the Wigner function as "phases" of the motion, are they related somewhat to hidden variables? What is the physical interpretation of the negative values of the Wigner function and what this has to do with the Heisenberg concept of potentiality or propensity?

II. BELL'S INEQUALITIES

Until 1965 it has not been clear at all if the hidden variable approach can or cannot reproduce quantum mechanical results.

Because of this, most of the published literature up to this date deals mostly with philosophical implications of the completeness argument raised by EPR.

In 1965 Bell published his famous proof that a local (in Bell's sense) hidden variable theory cannot reproduce all quantum mechanical correlations and average expectation values of dynamical variables [8]. In other words a local hidden variable (LHV) theory is constrained and these constrains are given by the Bell's inequalities. Let us illustrate these inequalities in typical optical effects where the polarization of photons is measured.

In "The Principles of Quantum Mechanics" Dirac explains the quantum properties discussing polarization of photons passed through a crystal which acts as a polarizer [9]. According to the quantum mechanical interpretation when we make the photon meet the crystal we are subjecting it to an observation. The effect of making this observation is to face the photon entirely into the state of parallel or entirely into the state of perpendicular polarization. The photon has to make a sudden jump from being partly in each of these two states to being entirely in one or other of them. This means that behind the crystal sometimes one will find a whole photon of energy equal to the energy of the incident photon on the back side and other times one will find nothing. Which of the two states it will jump into, cannot be predicted but is governed only by probability law.

The concept of a random orientation of the photon polarization can be easily related to the idea of hidden variables. In such an approach the polarization of the photon is described by $A(\vec{a},\lambda)$ where $\vec{a}$ is the polarizer direction and λ is some hidden variable. In the simplest model it is enough to assume that the angle of polarization carries a "hidden direction" λ and during a measurement of the polarization the crystal "performs" an average over the hidden angles.

We can carry this remark further discussing the EPR correlations. According to Bohm [10] the EPR wave function can be simplified if we consider a source of light emitting two photons in the following state of polarization (s-wave):

$$|\psi\rangle = \frac{1}{\sqrt{2}} \left(|+\rangle_a |-\rangle_b + |-\rangle_a |+\rangle_b \right), \tag{4}$$

where $|+\rangle$ and $|-\rangle$ are two states of linear polarization. Such a state is analyzed by two polarizers with axes given by $\vec{a}$ and $\vec{b}$ respectively.

According to quantum mechanics the probability to detect photons with polarizations aligned along the axes $\vec{a}$ (given by angle θ_a) and $\vec{b}$ (given by angle θ_b) respectively is given by:

$$P(\vec{a},\vec{b}) = \frac{1}{2}\cos^2(\theta_a - \theta_b). \tag{5}$$

The joint probability $P(\vec{a},\vec{b})$ of detecting photons at the polarizers a and b exhibits quantum mechanical correlations which are at the core of the EPR argument.

In classical physics the polarizer acts on the light beam according to the Malus law. This law predicts an attenuation of light through a linear polarizer by an amount $\cos^2(\theta_a-\lambda)$ where θ_a is the polarizer angle with some given fixed direction and λ is an initial orientation of the electric field polarization. This law can be used as a basis of a hidden variable theory with the initial orientation of the polarization described by a hidden angle λ. During the measurement process the linear polarizer ($\vec{a}$ or $\vec{b}$) averages over the unknown hidden angle λ. As a result we obtain:

$$P(\vec{b}) = P(\vec{a}) = \int d\lambda \ \cos^2(\theta_a-\lambda) = 1/2, \tag{6}$$

where we have assumed a uniform distribution $\rho(\lambda)$ of the hidden angle.

Now we can apply the hidden variable approach and the Malus law to calculate joint correlations of detecting photons by polarizers $\vec{a}$ and $\vec{b}$. Because these events are independent (locality in Bell sense) the probability of such a measurement is simply a product of two $\cos^2$ functions. The correlation is introduced as a result of a hidden variable average over the random angle λ:

$$P(\vec{a},\vec{b}) = \int d\lambda \cos^2(\theta_a-\lambda)\cos^2(\theta_b-\lambda) = \frac{1}{4}(1+\frac{1}{2}\cos 2(\theta_a-\theta_b)) \tag{7}$$

which is clearly different from the quantum mechanical result given by Eq.(7). This example shows the simplest difference between quantum and hidden variable theory.

One can show that for arbitrary $\rho(\lambda)$ the following combination of the hidden variable probabilities:

$$S = P(\vec{a},\vec{b})-P(\vec{a},\vec{b}')+P(\vec{a}',\vec{b})+P(\vec{a}',\vec{b}')- P(\vec{a})-P(\vec{b}), \tag{8}$$

where $\vec{a},\vec{b},\vec{a}'$ and $\vec{b}'$ are four arbitrary directions of the polarizers, has to satisfy the following inequality:

$$-1 \leqslant S \leqslant 0. \tag{9}$$

This inequality is one form of Bell's inequalities, as derived by Clauser and Horn [11].

If we take the optimal and the minimal values for the upper-limit and lower-limit of S, one can simplify this inequality further. For a coplanar geometry of the polarizers such that the angles between $\vec{a},\vec{b},\vec{a}'$ and $\vec{b}'$ are 22.5°. we obtain that the relation (9) reduces to:

$$D= \left|P(22.5^\circ)-P(67.5^\circ)\right| \leqslant 1/4. \tag{10}$$

The inequality (9) depends on the four joint correlations $(P(\vec{a},\vec{b}), P(\vec{a},\vec{b}'), P(\vec{a}',\vec{b}), P(\vec{a}',\vec{b}'))$ and on the two single probability distributions $(P(\vec{a}), P(\vec{b}))$. It is possible to obtain an inequality that involves only joint distributions. If we introduce the following combination of these functions:

$$s = P(\vec{a}',\vec{b}) + P(\vec{a}',\vec{b}') + P(\vec{a},\vec{b}) - P(\vec{a},\vec{b}') \tag{11}$$

we can obtain the followinng inequality:

$$-2 \leqslant s \leqslant 2. \tag{12}$$

Such an equality has been obtained by Bell. The three parameters S, D and s can be verify in realistic photon correlation experiments.

III. OPTICAL TESTS

With the progress of high resolution spectroscopy and with modern optical methods correlations of photon polarizations can be measured with high accuracy. Due to a very high efficiency of photon polarizers optical tests are much better than for example any spin correlation measurement.

The EPR-Bohm wave function (4) can be generated if pairs of low-energy photons are emitted in certain radiative cascade.

Because of this, modern quantum optics can offer the best tests of quantum mechanics. These tests can verify the fundamental properties of quantum mechanical probabilities and correlations versus LHV prediction.

For example the quantum mechanical expression (5) predicts that the ratio r, defined as the minimal joint probability divided by its maximal value is equal to zero ($r=0$). The hidden variable theory based on the uniform average of the hidden or "unknown" polarization angle predicts for this ratio $r=1/3$. In 1967 Kocher and Commins made us of the 6S-4P-4S cascade decay of Ca atoms to measure this ratio [12]. The experimental result was:

$$r = 0.15 \pm 0.02 .$$

Recent experiments of photon correlations performed by Aspect et al. [13] in the same Ca photon cascade leads to the following results:

$D = 0.3072 \pm 0.0043$.

$S = 0.101 \pm 0.020$ (with time-varying analyzers, Einstein locality).

$s = 2.697 \pm 0.015$.

The polarization correlation of the two photons emitted simultaneously by metastable atomic deuterium has been measured by Perrie et al.[14]. The result of this experiment leads to:

$$D = 0.268 \pm 0.010 .$$

All these experiments agree with quantum mechanics and violate Bell's inequalities given by Eqs.(9,10,12) by several standard deviations.

IV. QUANTUM PROPENSITIES

At this point one can asks a natural question. In Dirac's description on photon polarization there are some obvious elements of a random average over the polarization when one says that the photon has to make a sudden jump into the state of parallel or perpendicular polarization. What went wrong with the LHV approach in photon polarization processes?
The answer is quite obvious if one remembers that in quantum mechanics one deals with <u>probability amplitudes</u> rather than probabilities and one should remember to sum these amplitude first, before squaring the result. The Malus law can then be used for quantum mechanical amplitudes i.e., the amplitude for passing through the linear polarizer is attenuated by an amount cos(−) where is a random and evenly distributed initial orientation of the photon polarization.

Summing over all polarizations, squaring and multiplying the result by a factor of two (two photons are involved) one obtains the quantum result (5). Moreover such a result indicates that the quantum mechanical probability cannot be written in the form (7). Indeed we can write the quantum mechanical probability in the following form:

$$P(\vec{a},\vec{b}) = \iint d\lambda_a d\lambda_b \, \rho(\vec{a},\vec{b},\lambda_a,\lambda_b)\cos^2(\theta_a-\lambda_a)\cos^2(\theta_b-\lambda_b), \qquad (13)$$

where we have applied the Malus law to quantum mechanical amplitudes [15]. Because in this equation we have singled out the $\cos^2$ terms typical for a LHV theory (see Eq.(7)), the definition $\rho(\vec{a},\vec{b},\lambda_a,\lambda_b)$ contains all the remaining terms of the quantum mechanical expression. Now this expression looks like an ensemble average over some Born's "phases" of the polarization characterized by the random angle λ. A simple inspection of this formula indicates that indeed quantum mechanics looks like a hidden variable theory but with a distribution of the unknown angles λ_a and λ_b which depends on the state of the polarizers $\vec{a}$ and $\vec{b}$, moreover this function can be negative. The hidden-like form of the quantum mechanical probability (13) shows that the Bell locality assumption is violated and that the distribution of the quantum hidden variables can be negative.

So far we have discussed the case of photon polarization correlation i.e., a very simple quantum mechanical world confined to a two-dimensional Hilbert space. From now we shall study a nonrelativistic particle described in an infinite dimensional Hilbert space. The phase-space motion of such a particle is described by the Wigner function W . This brings us closer to the properties of the Wigner function which, as we have pointed it out can be negative.

What is the relation of these negative values with the hidden-look-like form of quantum mechanics that we have obtained for photon polarizations? In what sense one can speak of q and p as quantum "phases" of motion?

Let us start with a measurement of a particle (which is in a pure state $\hat{\rho} = |\psi\rangle\langle\psi|$ projecting it on a state $|\phi\rangle$ that characterizes the "polarizer": $Tr(\hat{P}_\phi \hat{\rho})$. In this expression $\hat{P}_\phi = |\phi\rangle\langle\phi|$ is the projector on the state $|\phi\rangle$ that we shall call from now the projector on some filter state or simply a filtering device. Physically in order to compare states in a realistic laboratory set-up we have to "bring" or "displace" the filter towards the measured object. In order to compare states in phase-space, we need to "shift" one with respect to the other by amounts q and p respectively. In the q-coordinate, this shift can be done by the operator $\exp(iq\hat{p})$ where $\hat{p}$ is the space translation generator. In the p-space, the operator $\exp(ip\hat{q})$ can be used because $\hat{q}$ is the Galilean boost. Combining these two operation in the unitary operator $U(q,p) = \exp(iq\hat{p} + ip\hat{q})$ we can "displace" the filter projection operator $\hat{P}(q,p) = U^{+}(q,p)\hat{P}_\phi U(q,p)$. Now we will define a positive definite probability [16]:

$$Pr(q,p) = kTr(\hat{P}(q,p)\hat{\rho}). \tag{14}$$

This probability is parametrized by the phase space (q,p) and the coefficient k is chosen in such a way that if we integrate over all possible points in phase-space we obtain:

$$\iint dqdpPr(q,p) = 1. \tag{15}$$

The expression (14) can be regarded as a phase-space <u>propensity</u> of the state $|\psi\rangle$ to have a momentum and position different from the filter state $|\phi\rangle$ by amount q and p. This propensity reflects also the tendency of the state to take up certain states prescribed by the displaced filter. This tendency or propensity can be regarded also as a phase-space probability distribution (everywhere positive) that involves both the measured object and the filtering state. The normalization (15) leads in this case to $k^{-1} = 2\pi\hbar$.

The relation of the propensity (14) to the phase-space Wigner functions of the states $|\phi\rangle$ and $|\psi\rangle$ can obtained in a straightforward way. It reads [17]:

$$Pr(q,p) = W_\phi * q * p W_\psi \tag{16}$$

where $*q$ ($*p$) denotes a space (momentum) convolution. This formula has a very simple geometrical meaning. The quantum mechanical propensity is a phase-space overlap of two Wigner functions. The first function describes the state of the object to be measured. The other Wigner distribution describes the filtering device. The (q,p) integration can be regarded as an average over the "phases" of the motion of the particle $|\psi>$ with respect to a filter distribution function W_ϕ. Even if W_ψ and W_ϕ can be negative the overall result is always positive for all possible states $|\psi>$ and $|\phi>$.

From the formal point of view the quantum mechanical propensity has the hidden-look-like form if we associate (q,p) with and the Wigner distribution of the filter with $\varrho(\lambda)$. Any restriction of $\varrho(\lambda)$ to be positive results in a LHV form of quantum mechanics. It shows also that in this case not all quantum propensities can be obtained in such a way. This is a different form of the Bell's constrains of LHV theories. Propensities that need to be described with negative $\varrho(\lambda)$ leads to correlations which are outside of the Bell's constrains.

This construction shows that a hidden variable look-like theory with negative distributions with averages described by Eq.(16) is equivalent to quantum mechanics. Such theories are local in Einstein's sense but are nonlocal in Bell's sense.

ACKNOWLEDGMENTS

This research was partially supported by the Polish program MR-I-7.

REFERENCES

[1] M. Born Z. Phys.37,863(1926)
[2] M. Born Z. Phys.38,803(1926)
[3] A. Pais Science 218,1193(1982)
[4] W. Heisenberg in "On Modern Physics" (Orion Press, New York, 1961) pp.9-10.
[5] K. R. Popper "Quantum Theory and the Schism in Physics" (Hutchinson, London, 1982) pp.125-130.
[6] E. Wigner Phys. Rev. 40,749(1932)
[7] A. Einstein, B. Podolsky and N. Rosen Phys. Rev. 47,777(1935)
[8] J. S. Bell Physics 1195(1965)
[9] P. A. M. Dirac "Quantum Mechanics" (Oxford Univ. Press, London, 1935) ch.1.
[10] D. Bohm "Quantum Theory" (Prentice-Hall, Englewood Cliffs, 1951) p.614.
[11] J. F. Clauser and M. A. Horn Phys. Rev.D10,526(1974)
[12] C. A. Kocher and E. D. Commins Phys. Rev. Lett. 18,575(1967)
[13] A. Aspect, P. Grangier and G. Roger Phys. Rev. Lett.

47,460,(1981); 49,491,(1982);A. Aspect, J. Dalibard and G. Roger Phys. Rev. Lett. 49,1804,(1982)
[14] W. Perrie, A. J. Duncan, H. J. Beyer and H. Kleinpoppen Phys. Rev. Lett. 54,1790,(1985)
[15] K. Wódkiewicz Phys. Lett. 112A,276(1985)
[16] K. Wódkiewicz Phys. Rev. Lett. 52,1064(1984)
[17] K. Wódkiewicz in "Quantum Electrodynamics and Quantum Optics" ed. A. O. Barut (Plenum, New York, 1984) p.305.

EXPERIMENTS ON LASER INDUCED FLUORESCENCE AND COLLISIONAL QUENCHING

José Campos

Cátedra de Física Atomica. Facultad de Física
Universidad Complutense de Madrid.
Ciudad Universitaria. 28040. Madrid. Spain

ABSTRACT

Basic methods for quenching cross sections measurement are outlined and experiments on lifetimes and quenching cross sections of Na and I_2 are described. The extension of this kind of studies to laser excitation from levels previously populated by electron impact is shown. Examples of experiments on Ne, Xe and N_2 are given.

1. INTRODUCTION

The aim of this contribution is to describe some experiments on depopulation measurement of atomic and molecular levels and to discuss their results. The knowledge of the radiative and collisional processes leading to level depopulation and the measurement of radiative lifetimes and collision cross sections have a great interest. Besides the relevance of these studies for basic research in atomic and molecular physics a continuous need for these data exists, regarding its applications to plasma and laser physics and to astrophysics.

In the present studies atoms or molecules were confined in a resonance cell or in an electron collision chamber depending on the particular experiment. In order to follow the population evolution of levels,laser induced fluorescence methods and single photon coincidence techniques were used. The collision experiments that are described were carried out under gas-kinetic conditions. As it is well known, under this conditions only deactivating superelastic collisions, excitation transfer and collisions with change of total angular momentum can occur.

To study collisions in which a level of an atom or a molecule deexcites by collision with another atom or molecule,a suitable method is the observation of the radiation emitted from the level. The intensity of this emission will be reduced by the deactivating collisions. Through this processes of collisional quenching of radiation the corresponding cross sections can be found. An alternative method for determination of quenching cross sections is the measurement of the collisional shortening of the level lifetime. Generally, lifetimes are obtained by emission methods but if the levels are metastable the depopulation can be studied by measuring the absorption of light from the level to a suitable upper level. Absorption can be monitored through the intensity of the fluorescent light coming from radiative decay of the upper level.

If it is assumed that the only effect of collisions with foreign atoms or molecules is level depopulation (i.e. collisional population can be neglected) the well known Stern-Volmer expression relating quenching cross section and level lifetime is

$$1/\tau_{eff} = 1/\tau_0 + \langle v_r \rangle \cdot \sigma \cdot N$$

where τ_{eff} is the effective lifetime in presence of a density N of colliding particles, τ_0 the zero pressure lifetime, σ the cross section and $\langle v_r \rangle$ the mean relative velocity of the colliding particles.

In gas-kinetic conditions $\langle v_r \rangle$ is given by the expression $\langle v_r \rangle = \sqrt{8kT/\pi\mu}$, μ being the reduced mass of the colliding particles, k the Boltzmann constant and T the absolute temperature.

According to the above expression a plot of $1/\tau_{eff}$ versus pressure of quenching particles is linear. If the experimental correlation of $1/\tau_{eff}$ and pressure can be fitted by a straight line, at least in a reasonable pressure range, it can be assumed that the colliding system follows a Stern-Volmer kinetics and the above hypothesis are valid in the experimental range. The corresponding cross section is deduced from the slope of the line and extrapolation to zero pressure gives the unperturbed lifetime.

For the intensity quenching a similar expression can be deduced in a easy way

$$I_0 / I = \tau_0 / \tau_{eff} = 1 + \tau_0 . N . \sigma . \sqrt{8kT/\pi \mu}$$

being I_0 the intensity without collisional deactivation and I the intensity with collisional depopulation. Evidently, if the level lifetime is known, the cross section is deduced from the slope of the linear plot of I_0/I versus pressure.

Most of the experiments described in this work are on atom-molecule and molecule-molecule collisions.

From the point of view of efficiency as quenching colliders it can be affirmed that, in general, molecules are more effective quenchers than atoms. The molecular vibration plays a fundamental role as a sink of the excess energy released in the quenching process. On the other hand, in these collisions the amount of energy transfer to translational motion is small.

Notwithstanding the great theoretical and experimental efforts made in this field the detailed mechanisms of these interactions are not yet fully understood. The most important test for the existing theoretical models are the experimental vibrational population after quenching and the temperature dependence of the cross sections. Thus these measuremets are of special interest.

In the present work we describe recent experiments carried out by our group of Madrid on quenching of resonance radiation, molecular fluorescence quenching, two-step electron-photon excitation and level population and depopulation measurements by absorption laser spectroscopy.

In every section references to the existing literature are given. For general information the classical text of Massey (Ref 1) and the work of Hertel (Ref 2) shall be useful.

2. DIRECT EXCITATION EXPERIMENTS

In this section we describe experiments on atomic and

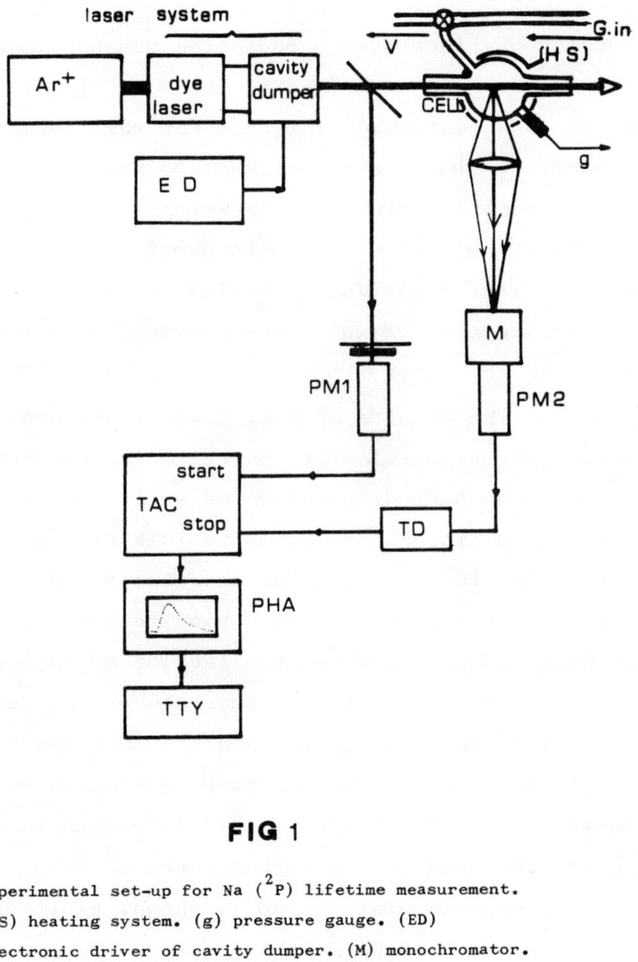

FIG 1

Experimental set-up for Na (^2P) lifetime measurement. (HS) heating system. (g) pressure gauge. (ED) electronic driver of cavity dumper. (M) monochromator. (TD) time delay. (PHA) pulse height analyzer. (TAC) time to amplitude converter.

molecular fluorescence quenching performed by laser excitation from the ground state.

2.1 Na (^2P) fluorescence quenching by diatomic molecules.

The quenching of fluorescence radiation has been studied from the theoretical and experimental point of view for a long time and there is a continuous interest for this theme (Refs. 3-7). The object of the experiment (Ref.8) reported here was to obtain cross section values for deexcitation of Na (^2P) atoms by N_2 and CO molecules at low temperature (i.e. low kinetic energy of the colliding particles). Previous experiments of fluorescence in flames provided values for the cross sections in the lower temperature range 400-600 K (Refs 11, 13-15). The results of Ref. 5 were obtained with Na atoms of relatively high velocity (energies in the range 0.07-0.35 eV) produced by photodisociation of Na I. Theoretical models for these collisions are describe in Refs. (2,5-7,16-19). As said in the introduction the experimental dependence of cross sections on temperature is a good test of the theory. The present work was prompted for the lack of measurements at temperatures below 400 K. At these temperatures the density in a resonance cell of Na atoms is low and the corresponding fluorescence light is weak. Pulsed laser excitation and single photon detection allow to obtain Na (^2P) lifetimes in strong quenching conditions and at low temperature.

In Fig. 1 the experimental set-up is shown. Excitation of ^2P resonant lines has been performed by a dye laser pumped by Ar^+ laser. The tunable laser consisted of a dye laser joined to a cavity dumper (Spectra Physics 580 A and 344 S, respectively). Cavity dumping is made with an acousto-optic modulator driven at high frecuency by an rf oscillator. The repetition frecuency was 40 kHz and the spectral resolution 0.05 Å.

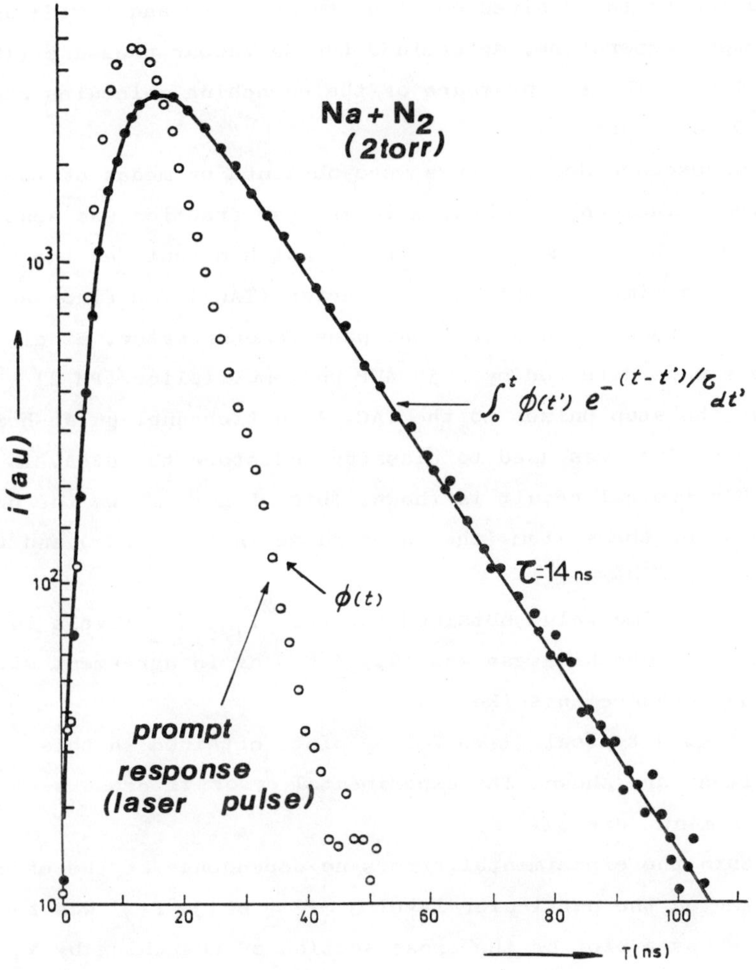

FIG 2

A typical result of Na (^2P) deexcitation in presence of N_2. O Laser. ● Experimental curve. — result of convolution of laser pulse and a 14 ns exponential decay.

The resonance cell was electrically heated and the temperature controlled by several termocouples. The cell was mantained at the desired constant temperature and a side arm, at lower temperature, determined the Na vapour pressure (10^{-5}-10^{-6} Torr). The gas pressure of the quenching molecules ranged from 0 to 6 Torr.

Fluorescence decay curves were obtained by means of the delayed coincidence method. A laser beam fraction was sent to a EMI 9781 B photomultiplier (PM 1) which output fed the start input of a time to amplitude converter (TAC). The fluorescence light was focused on a 50 Å bandpass monochromator. Single photons were detected by a 56 AVP photomultiplier (PM 2) giving the stop pulses to the TAC. A multichannel pulse-height analyzer (PHA) was used to classify and store the data. In Fig. 2 a typical result is shown. This figure shows the prompt response of the system (the laser pulse in this case) and the fluorescence decay.

The lifetime value obtained for the $^2P_{1/2,3/2}$ levels in absence of quenching gas was 16.3 ± 0.3 ns in agreement with previous measurements (Ref 3).

In Fig. 3 typical Stern-Volmer plots obtained in this experiment are shown. The experimental errors in cross sections measurement are 8%.

Within the experimental errors no dependence of the effective lifetime on the particular level $3\ ^2P_{1/2}$ or $3\ ^2P_{3/2}$ was found. Fig 4 shows a plot of the cross section of quenching by N_2 versus the reciprocal of the mean kinetic energy $\langle E \rangle^{-1}$. We include in the figure previous results from other authors, Refs (5,9,10-15). From previous measurements a linear relation between the cross section and $\langle E \rangle^{-1}$ (i.e. T^{-1}) can be inferred. The values of this experiment obtained at the lowest kinetic energy add support to this dependence. As it is known,

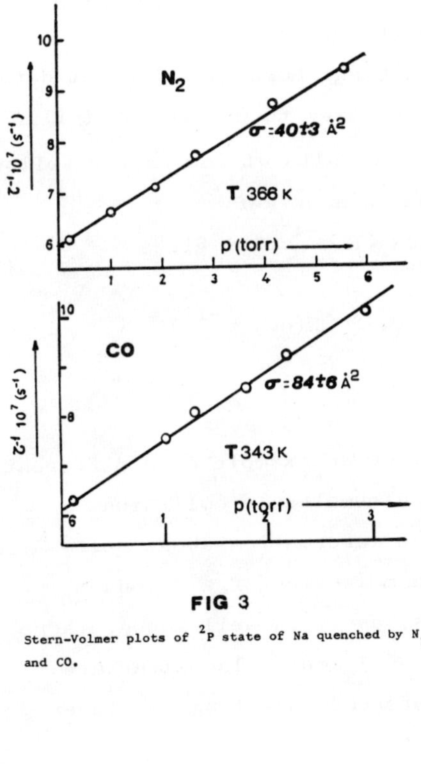

FIG 3

Stern-Volmer plots of ^2P state of Na quenched by N_2 and CO.

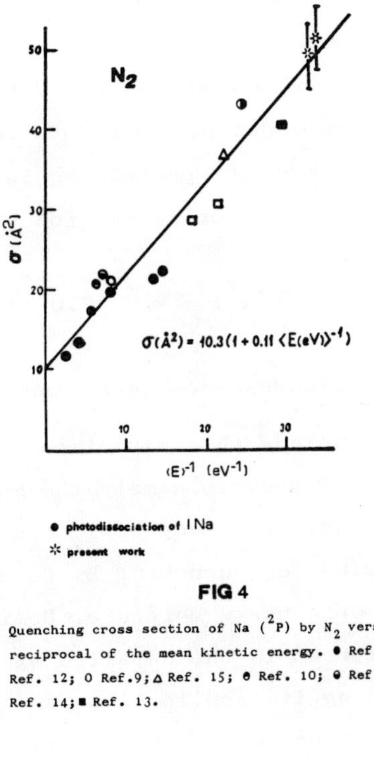

FIG 4

Quenching cross section of Na (^2P) by N_2 versus the reciprocal of the mean kinetic energy. ● Ref.5 ; □ Ref. 12; O Ref.9; △ Ref. 15; ◐ Ref. 10; ◒ Ref. 11; ◑ Ref. 14; ■ Ref. 13.

the aforesaid linear dependence is predicted by the "absorbing-sphere" model and the ionic intermediate state model. Refs (2,5,6,16-18).

The result obtained was

$$\sigma(\text{Å}^2) = 10.3\ (1 + 0.11 \langle E(ev) \rangle^{-1})$$

being σ the cross section for Na (^2P) quenching by N_2. This result agrees with the prediction of Ref. 6.

There are less data for CO quenching. More experimental data at intermediate energy would be necessary to confirm a similar linear dependence. Nevertheless our results at low energy follow the trend of the photodissociation data of Ref. 6. A linear fitting to the cross section versus $\langle E \rangle^{-1}$ plot gives

$$\sigma(\text{Å}^2) = 17.0\ (1 + 0.12 \langle E(ev) \rangle^{-1})$$

2.2 I_2 fluorescence quenching

We include this experiment to show an example of a different quenching process, namely the molecule-molecule collision quenching. The study of I_2 can give information about self-quenching and quenching by foreign molecules of a resonant state of a heavy molecule. Besides, as it is well known, there is interest in the detailed study of I_2 molecular structure regarding its applications as a reference spectrum for laser spectroscopy works (Ref. 19).

2.2.1 Collisional and radiative depopulation of vibrational levels of the state B $^3\Pi 0^+_u$ of I_2.

Previous works about deexcitation processes of I_2 states can be found in Refs (20-30). Brown and Klemperer (Ref 20) studied the fluorescence of the B $^3\Pi\ 0^+u$,v =15 level and the quenching for foreign molecules. Steinfeld and Klemperer (Ref 21) studied the vibrational level v = 25 and Capelle and Broida (Ref 23) several levels with v in the range 5-70. The results of these previous experiments showed the change of radiative lifetimes for the different vibrational levels and the dependence of the foreign gas quenching cross sections on the colliding molecules and vibrational state of I_2. Also a correlation exists of quenching efficiency with the mass and polarizability of the colliding foreign particle. In a semi-empirical way it was found that cross sections are proportional to the product of polarizability of quenching molecules (α) and the reciprocal of the relative velocity of the colliding partners. This last quantity is proportional to $\sqrt{\mu}$ (μ reduced mass) if the temperature is constant. Besides from the experiments it appears that there is not remarkable contribution to quenching efficiency from permanent electric dipole moment and that there are a complex set of interactions including electrostatic polarization and specific chemical effects able to give rise to repulsive states contributing to induced predissociation of I_2.

In order to measure the lifetime and quenching cross sections for a new I_2 level (v = 14) and for molecules not previously studied (CO , CH_4) we performed an experiment (Ref 31) to measure quenching by H_2, CO and CH_4 . On the other hand,the B $^3\Pi\ 0_u^+$, v = 14 level is close to a minimum in the natural predissociation of this electronic state of I_2, so the lifetime of low rotational levels does not depend on J (Ref 25),leading to a more easy interpretation of results.

In Fig 1 the experimental set-up is shown. The B $^3\Pi\ 0_u^+$, v = 14 , J = 0-8 levels were excited by a N_2 pumped tunable dye laser. The laser resolution was 0.15 Å and the excitation

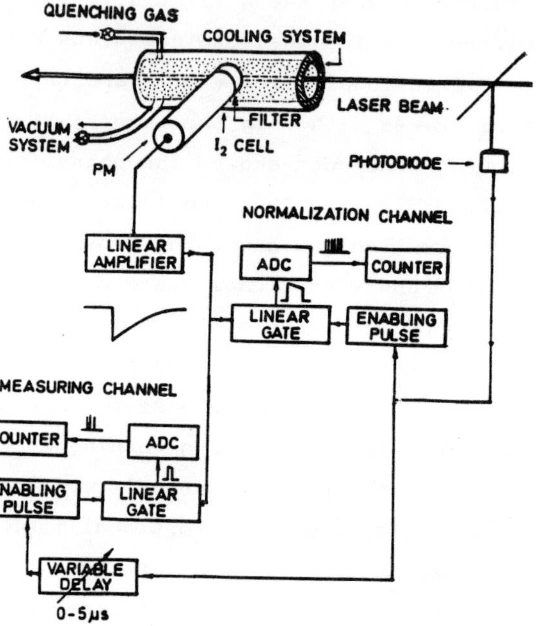

FIG 5

Experimental set-up for lifetime measurement of I_2 levels.

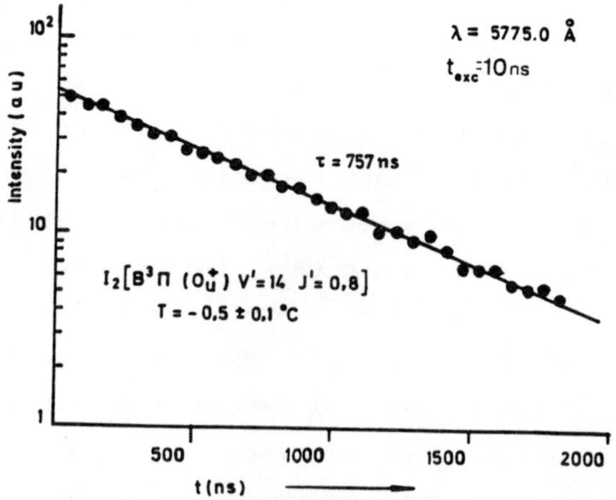

FIG 6

Deexcitation of the level B $^3\Pi$ 0_u^+ (v = 14)of I_2 at 25 mTorr.

wavelength was 5775 Å. The laser pulse peak power and duration were 10 nsec and 6 kW, respectively. The pulsation frecuency was 7 Hz.

The iodine cell was connected to a vacuum system through a liquid nitrogen trap and to a gas inlet. The I_2 pressure was controlled by cooling the cell and mantaining a constant temperature. The I_2 pressure could be varied in the range 30-40 mTorr. The pressure of the foreign molecules was varied in the range 0.05-5 Torr in the collision deexcitation experiments.

Fluorescent light from I_2 was detected by a EMI 981 6 B photomultiplier cooled to -25 ºC. The wavelength was selected by an interferential filter at 6510 Å. This emission corresponds to transitions to the X $^1\Sigma_g^+$ $v''= 9,10$ vibrational levels. The fluorescent light decay was measured by analog sampling (Refs. 32,33) of the photomultiplier signal. The start signal for the electronic system is given by the laser light pulse through a photodiode. The photodiode signal is shaped in square pulses of adjustable width (sampling interval) and delayed a selectable time. This pulse constitutes the enabling pulse for a linear gate. This gate transmits the photomultiplier signal during the sampling interval. The sampled signal by means of an analog to digital converter is stored in a counter. Repeating the measurements for different delays, the decay curve of the level is obtained. The time range for decay measurements was 3 μs . There is a normalization channel that measures the total fluorescence for every laser pulse and corrects for laser power or photomultiplier gain fluctuations.

In Fig 6 a typical decay curve of the mentioned level is shown. A strong self-quenching exists for I_2. The corresponding Stern-Volmer plot is shown in Fig 7. The lifetime value from this experiment is 1.25 ± 0.05 μs in agreement with the result of Ref 25 for the J = 53 level. The self -

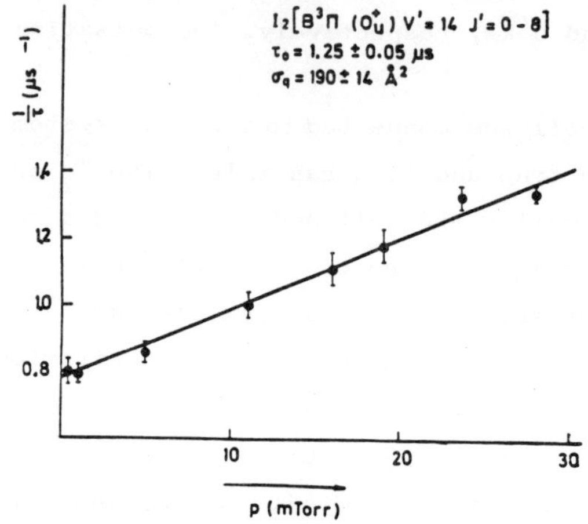

FIG 7

Stern-Volmer plot for the B $^3\Pi$ 0_u^+ (v =14) level of I_2.

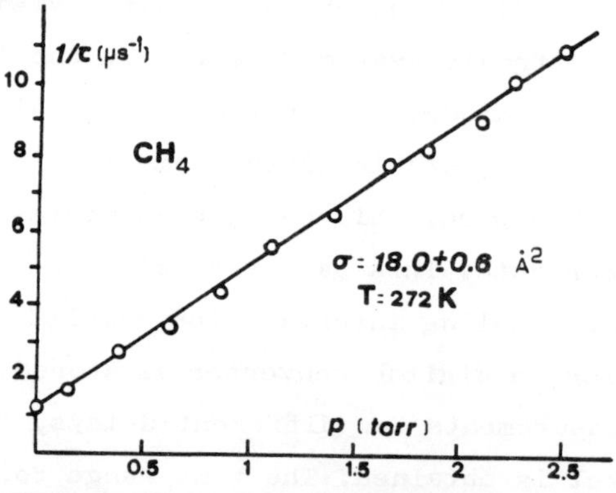

Quenching cross sections ($Å^2$) for I_2(B $^3\Pi$(0_u^+), v =14)

H_2	CO	CH_4
2.5 ± 0.3	15.1 ± 0.4	18.0 ± 0.5

Cross section ratio

	H_2 : CO : CH_4	CO/CH_4
Experiment:	1 : 5.0 : 7.2	0.84
Semiempirical:	1 : 9.3 : 10	0.93

FIG 8

Stern-Volmer plot for quenching of B $^3\Pi$ 0_u^+ (v =14) level of I_2 by CH_4. Summary of results.

quenching cross section is $\sigma = 190 \pm 14$ Å^2 in agreement with the result of Ref 26.

In Fig 8 a typical Stern-Volmer plot of quenching of I_2 by CH_4 is shown. The values for the quenching cross sections for H_2, CO and CH_4 were 2.5 ± 0.3, 15.1 ± 0.4 and 18.0 ± 0.6 Å^2, respectively. To extend the pressure range (0-20 Torr) of these measurements, intensity quenching measurements were made with results in agreement.

The semiempirical proportionality of quenching cross sections with the product of polarizability and square root of reduced mass was compared with the results of this experiment. Polarizability values have been obtained from Ref 34. The ratio of cross sections are in reasonable agreement with the semiempirical law. Discrepancy is about 35% for CO/H_2 and CH_4/H_2 ratios and 1% for the CH_4/CO ratio. Fig 8 includes a table with these values.

In summary, it seems interesting to continue the research in this molecule-molecule quenching collisions that can be very useful to test elaborate theoretical calculations.

3. ELECTRON-PHOTON TWO-STEP EXCITATION

Direct laser excitation techniques have a limited range of application with regard to the different atoms and the number of levels of every atom that can be excited. To extend this range selective excitation methods are based on two-photon excitation and stepwise excitation. In Fig 9 some examples of this kind of laser excitation are shown.

To reach high excited levels we employ the method of double excitation electron-laser. Firstly, a level is populated by electron impact and when the electron beam is interrupted a pulsed laser excites a higher level (Fig 9). The atomic fluorescence decay gives directly the lifetime of the upper

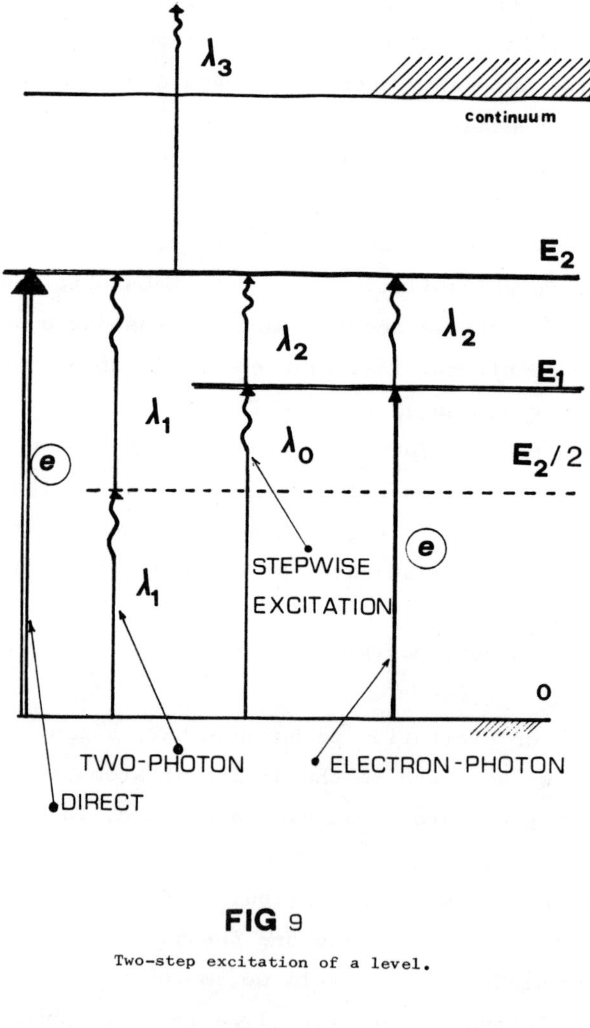

FIG 9

Two-step excitation of a level.

level. The experiments described correspond to electron impact excitation of relatively long lifetime ($\approx 1 \mu s$) levels in order to get a great population, nevertheless with a high electron current the method can be applied to any level. As we shall show in the following examples on N_2 and Ne experiments, valuable information about population evolution of the intermediate state can be obtained from the upper level fluorescence.

3.1 Fluorescence from the upper level.

We describe two different experiments based on electron-photon stepwise excitation. The first one is about lifetime measurements on Xe I by pulsed and continuous excitation of the intermediate state. The second on Ne I lifetime measurements. These examples show that selective excitation from metastable and from resonant levels can be achieved.

3.1.1 Lifetimes of 7p levels of Xe I

The object of the experiment (Ref 35) was to study the lifetime of levels belonging to the $5p^5 7p$ configuration of Xe by a cascade-free method and to compare the results with elaborate theoretical calculations performed in intermediate coupling scheme of Refs (36,37) and with the laser experiments of Ref 38 carried out by using a flowing afterglow of Xe diluted in Ar. The lifetime value of Ref (38) was obtained by extrapolating the results to zero Ar pressure.

Excitation was made from the metastable $6s\,(3/2)_2$ level that was previously populated by a pulsed electron beam of 1.75 ms duration, 20 mA peak current and 7 Hz repetition frequency.

The experimental system is shown in Fig 10. A pulse generator

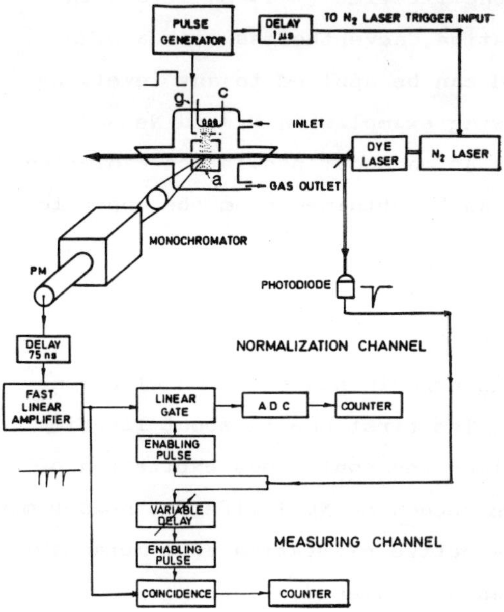

FIG 10

Experimental set-up for the measurement of lifetimes of 7p levels of Xe.
(g) grid. (c) cathode.

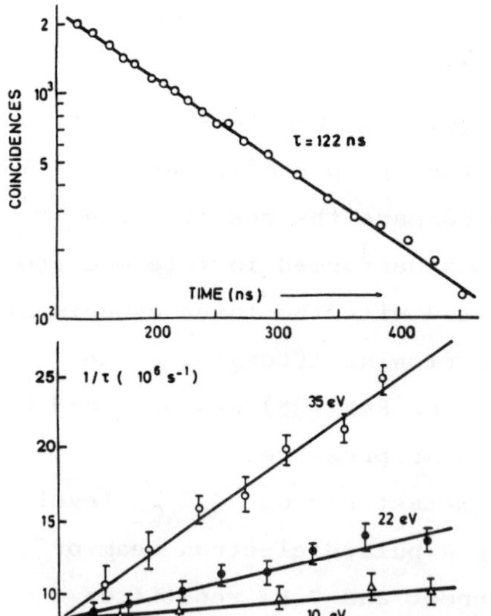

FIG 11

Upper part: typical decay result for the $7p(3/2)_2$ level of Xe.
Lower part: Continuous current results for the same level.
Plots of the reciprocal value of lifetime versus current density.

gives a positive pulse to a control grid of an electron excitation chamber producing the pulsed electron beam. The electron energy depends on the anode potential.

A tunable dye laser pulsed by a nitrogen laser was used. With the dye used (of coumarin type) the spectral range was 4400-4480 Å. The peak power was 1 kW and the resolution 0.15 Å. One μs after the end of the electron pulse the laser is triggered.

In this experiment single photon detection was used. It was measured the variation in the number of single photon fluorescence pulses as the time interval from the end of the laser excitation increases. The light from the chamber was selected by a 3 Å monochromator and detected by a Philips 56 AVP photomultiplier. The electronic apparatus is similar to the system used for the I_2 experiments described in part 2.2 but in this case time resolution was 10 ns and time range 600 ns. Measurements can be performed in 1 ns steps. The function of the normalization channel is to measure the total fluorescence signal emitted in the first 100 ns in order to make corrections for fluctuations of the laser power and for other accidental changes in the experimental conditions, such as gas pressure or electron current. In Fig 11 a typical result is shown. Measurement were made at 4 mTorr pressure.

In this experiment lifetimes for $7p(3/2)_2$ and $7p(5/2)_3$ levels were measured. The laser exciting wavelengths were 4624.2 Å and 4671.2 Å, respectively. For the first level the decay curve was obtained through the 4843 Å line (transition to the state $6s(3/2)_1$) and for the second level deexcitation was studied through the same line as excitation. The lifetime results are 121 $\pm$ 3 ns and 128 $\pm$ 7 ns. Both results agree with the values of Ref 38 (120 $\pm$ 12 and 120 $\pm$ 30 ns), with the jK-Coulomb calculation of this work (111 and 116 ns) and the dipole velocity calculations of Ref (36,37).

We also made lifetime measurement with a continuous electron current. The decay curves were also singly exponential but the corresponding lifetimes decrease as the electron current increases. This effect is shown in Fig 11. The plot of reciprocal lifetime versus current density is linear, the slope depending on the energy of the impinging electrons. From these experimental facts inference can be done of quenching of the excited atoms by charged particles present in the discharge. By extrapolating to zero density of current the resulting lifetime agrees with the value obtained with pulsed electron excitation.

These results indicate that in similar cases the lifetime values obtained by using laser excitation from metastable atoms produced in a low energy continuous arc can be extrapolated to give the unperturbed lifetime. This result may be useful for measurements in discharges when pulsed electron excitation is not possible.

3.1.2 Lifetimes of 3p levels of Ne I.

In this experiment levels of the $2p^5 3p$ configuration of Ne were populated by laser excitation from the metastable and resonant levels of the $2p^5 3s$ configuration and the fluorescent decay measured. The object of this work (Ref 39) was to improve the accuracy of lifetime data by using the laser-electron stepwise excitation method and to compare the results with those obtained with the afterglow experiments performed at higher pressure, Ref (38,40)

Lifetime measurements were made by using the delayed coincidence method in a similar way as described in sections 2.1 and 3.1.1. The experimental set-up is shown in Fig 12. The

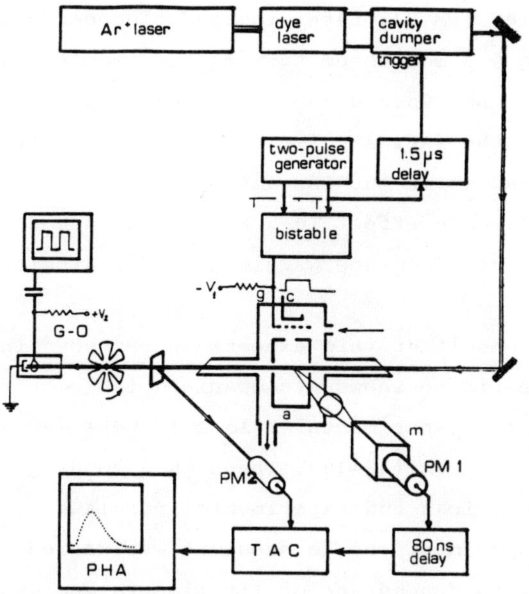

FIG 12

Experimental set-up for lifetime measurement in Ne by two-step electron-laser method. (G-O) optogalvanic resonance monitoring system. (a) anode. (g) grid. (c) cathode. (m) monochromator.

laser repetition frequency was 40 kHz and the resolution 0.5 Å. The energy of the colliding electrons was in the range 20-70 eV. The electron beam was pulsed at a repetition rate of 30 kHz and the pulse duration was 6 μs . Laser pulses were produced 1.5 μs after the end of the electron pulse. The start signal for the TAC was given by an EMI 9781 B photomultiplier (PM 2). Laser induced fluorescence light was selected by a 5 Å monochromator (m) and single photons were detected by an EMI 9816 B cooled photomultiplier that fed the stop input of the TAC. A pulse height analyzer (PHA) was used to classify and store the TAC output.

The optogalvanic effect in a Ne hollow-cathode lamp was used to find the resonance wavelengths for atomic fluorescence excitation.

A typical result of this experiment is shown in Fig 13. Also the same figure shows a comparison table of the present results with the experimental values of Refs (40-44) and the calculations of Refs (45-46). There is a good agreement with previous theoretical and experimental results.

In this experiment the Ne pressure was varied in the range 40-150 mTorr. No dependence of lifetime values with gas pressure was found within the experimental error (1.5%). This fact can be attributed to the relatively fast radiative decay of levels. Information about quenching cross sections can be found in Ref 40 where experiments at higher pressure are described. For lifetime measurement the present method enables excitation from resonant states. This is made by taking advantage of lifetime lengthening by resonance trapping.

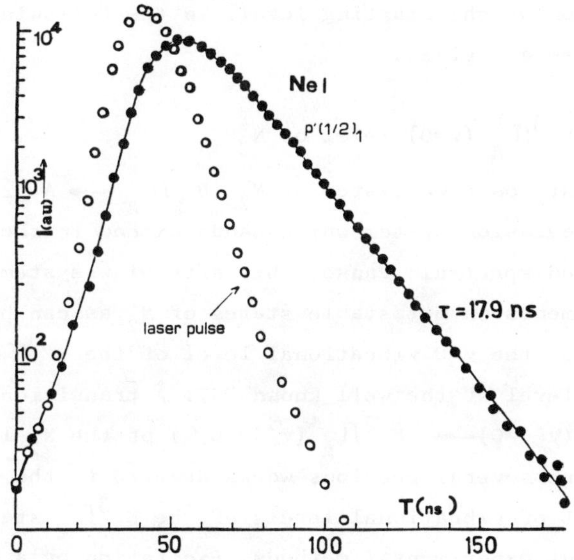

Level	Experimental (ns) This work	Chang and Setser	Kandela and Schmoranzer	Bennett and Kindlmann	Klose	Martín and Campos	Theoretical (ns) Feneuille	Mehlhorn
$2p_1$	14.0±0.1			14.4±0.3	14.7±1.3	14.7±0.8	13.7	15.1
$2p_2$	17.9±0.1 17.6±0.3	19.8±0.4	18.28±0.07	18.8±0.3	16.3±1.4	19.9±2.5	17.8	18.9
$2p_3$	16.9±0.2		17.38±0.08	17.6±0.2	23 ±6.6	21.5±2	16.7	16.9
$2p_4$	20.2±0.1 20.2±0.3	19.8±0.5	18.69±0.04	19.1±0.3	22 ±5.4	20.9±2.5	18.2	19.1
$2p_5$	19.0±0.2	19.2±0.4	19.05±0.05	19.9±0.4	18.9±2.8		18.8	20.2
$2p_6$	20.6±0.2	19.0±0.3	19.38±0.03	19.7±0.2	22 ±3.2	20.5±2	18.8	20.2

FIG 13

A typical result of population evolution after pulsed excitation of a $2p^5 3p$ level of Ne. Lifetime values for these levels.

3.2 Measurement of intermediate level population

It was said in preceding sections that fluorescence arising from a level previously excited by laser absorption from other level can give valuable information about population and depopulation of the starting level. In the following sections two examples are given.

3.2.1 The B $^3\Pi_g$ (v=0) level of N_2

The first positive system of N_2 (B $^3\Pi_g \longrightarrow$ A $^3\Sigma_u^+$) is a prominent emission system which bands extend in the visible and infrared spectral range. This afterglow system is relevant for experiments on metastable states of N_2, as can be seen in Ref 47. Also the v=0 vibrational level of the B $^3\Pi_g$ state is the lower level of the well known 3371 Å transition (C $^3\Pi_u$ (v′=0) $\longrightarrow$ B $^3\Pi_g$ (v″= 0)) of the N_2 laser.

There are several previous works devoted to the lifetime measurement of vibrational levels of the B $^3\Pi_g$ state by means of different experimental methods. Excitation by a pulsed rf discharge was used in the experiment of Ref 48 and pulsed electron excitation was employed in Refs (49-51). Beam time of flight and laser-beam interaction were the methods of Refs 52 and 53, respectively. Most of these experiments are in disagreement. The disaccord is greater for low vibrational levels. In particular for the first vibrational level there are, to our knowledge, only two previous results : $8.0 \pm 1\ \mu s$ (Ref 48) and $4.9 \pm 0.5\ \mu s$ (Ref 51). This situation can be attributed to the low intensity of the emission bands arising from the B $^3\Pi_g$ (v =0) level. Therefore we considered interesting to perform lifetime measurements by a different technique based on absorption from this level of direct N_2 laser radiation. The amount of absorption measured through fluorescence

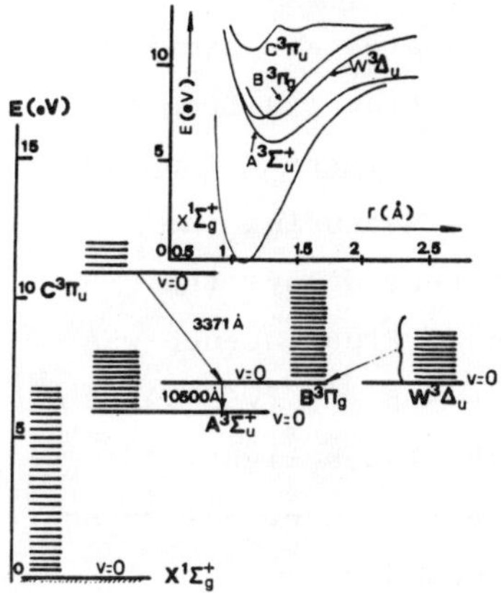

FIG 14

Energy levels of N_2. Potential curves for electronic states.

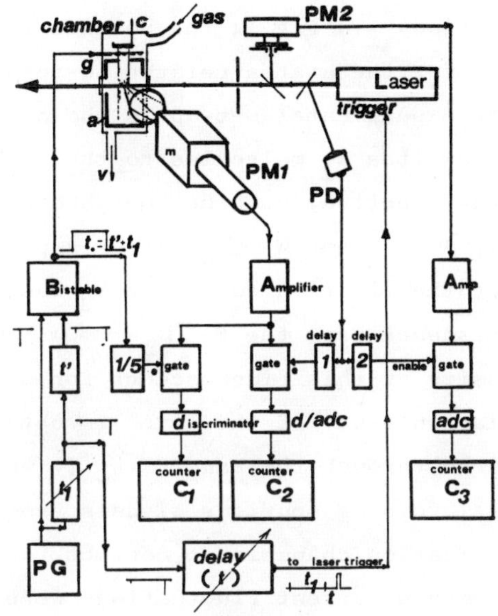

FIG 15

Experimental set-up for lifetime measurement of B $^3\Pi_g$ (v =0) vibrational level of N_2 by laser absorption. (PD) photodiode. (m) monochromator. (PG) pulse generator.

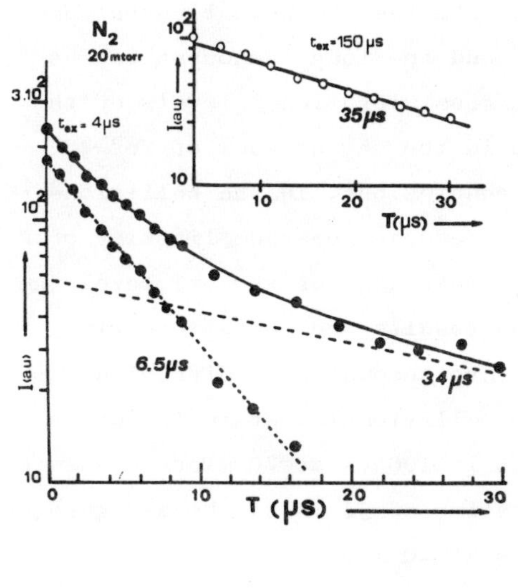

FIG 16

Typical results for the decay of B $^3\Pi_g$(v = 0) vibrational level of N_2. (t_{ex}) is the excitation time for each case.

intensity of any band arising from the C $^3\Pi_u$ (v=0) level is proportional to the B $^3\Pi_g$ (v=0) population if there is not saturation. In Fig 14 the potential energy curves and energy levels for the states related to this experiment are shown.

The experimental set-up is shown in Fig 15. An electron pulse excites N_2 molecules to the B $^3\Pi_g$ state. This excitation is non-selective, so other neighbouring states are excited, namely A $^3\Sigma_u^+$ and W $^3\Delta_u$. After a preset delay (t) from the end of the electron current, a N_2 laser is triggered inducing fluorescence from the C $^3\Pi_u$ (v=0) level. Measuring the intensity of this fluorescence for different delays the population decay of the level is obtained. Fluorescence was measured through the band $C^3\Pi_u$ (v'=0) $\longrightarrow$ B $^3\Pi_g$(v''=2) at 3805 Å, avoiding spurious signals from the laser light. Normalization channels to correct for laser power, gas pressure and electron current fluctuations were included.

Typical results are shown in Fig 16. The decay curves are composed of a long decay component and a faster component. Short electron excitation pulses favour enhancement of the fast component as depicted in Fig 16. The fast component corresponds to the B $^3\Pi_g$ (v=0) level decay and the long component can be attributed to radiative cascade from vibrational levels of the W $^3\Delta$ state. In Ref 51 and also in the recent work of Ref 47 the authors arrived at similar conclusions. In the collision-free beam experiment of Ref 47 an average cascade lifetime of 145 μs was found for the B $^3\Pi_g$ state and for the v=1 level the lifetime found was 255 μs. These results are compatible with the present one, because taking into account only diffusion processes the time constant for collisional lost in the walls of our cell of excited molecules is 100 μs at 20 mTorr pressure Correcting for this effect a lifetime of 54 μs can be assigned to the cascade populating levels at 20 mTorr.

Measurements made in the pressure range 10-40 mTorr give a result of 6.5 $\pm$ 0.2 μs for the B $^3\Pi_g$(v=0) level lifetime. The diffusion corrected value is 7.0 $\pm$ 0.5 μs. These values are in good agreement with the trend of recent laser measurements of Ref 53 where lifetimes from v=5 to 12 were measured and it was confirmed that the level lifetime decreases as the vibrational number v increases. The lifetime result from the aforementioned reference is 5.87 $\pm$ 0.21 μs for the v=5 level.

In summary, the present result confirms the strong population of B $^3\Pi_g$ levels from neighbour metastable states and add a new value relevant for comparison with precise theoretical calculations of transition moments for the B $\longrightarrow$ A system of N_2.

3.2.2 Metastable levels in Ne

Lifetime measurement of metastable levels is one of the most useful applications of the double excitation electron-laser method.

In Fig 17 we show the set-up employed in this experiment. The experimental arrangement is similar to that used in section 3.1.2 but with several changes. As in the previously described experiments, metastable excitation is made by impact of a pulsed electron beam, but in this case fluorescence is induced by means of a continuous dye laser. A typical result is shown in Fig 18.

In this example the metastable level was the 3s $(3/2)_2$ level of Ne, laser radiation at 5975 A excites the 3p$'(3/2)_1$ level and fluorescence is observed through the 6266 Å transition. Single photon were recorded by a multiscaling started syncronously with the electron pulse. The multiscaling dwell time was 10 μs and the number of channels 256. The electron beam was pulsed at 100 Hz frequency and the gas pressure was

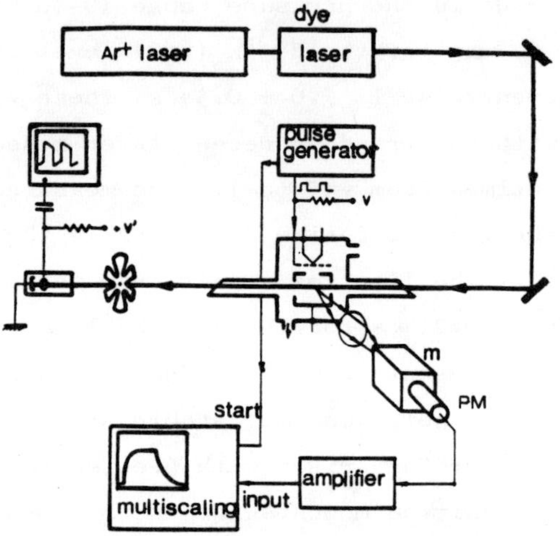

FIG 17

Experimental set-up for the measurement of decay time of metastable levels.

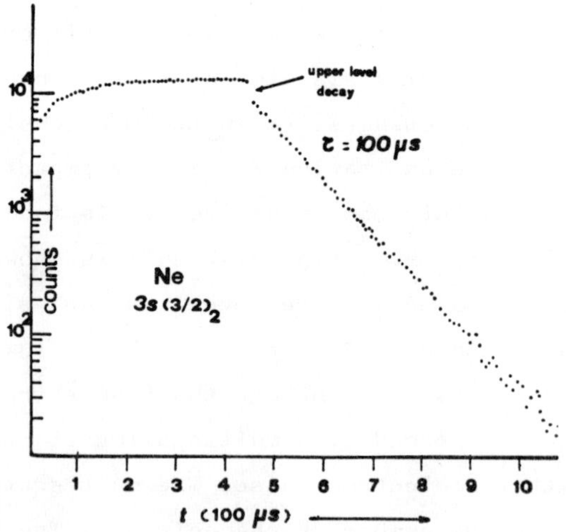

FIG 18

Typical result for the $3s(3/2)_2$ level of Ne at 100 mTorr.

100 mTorr.

As can be seen in Fig 18 the decay curve is singly exponential. The 100 μs lifetime value results from quenching and diffusion processes. The time constant for collisional lost in the chamber walls (lowest-order diffusion mode) was 200 μs in this experiment.

This experimental set-up is useful to perform measurements of quenching of metastable atoms by foreign atoms or molecules maintaining a constant Ne partial pressure and adding deactivating agents in known amounts.

REFERENCES

1.- H.S.W. Massey, Electronic and ionic impact phenomena, Vol III, Slow collisions of heavy particles. Clarendon Press. Oxford (1971)

2.- I.V. Hertel, Adv. Chem. Phys. 45, 341 (1981)

3.- P.L. Lijnse, Report i 398, Fysisch Laboratorium, Rijksumivesiteit Utrecht, The Netherlands (1972)

4.- L. Krause, Adv. Chem. Phys. 28, 267 (1975)

5.- J.R. Barker and R.E. Weston, Jr., J. Chem. Phys. 65, 1427 (1976)

6.- J.R. Barker, Chem. Phys. 18, 175 (1976)

7.- C. Bottcher and C.V. Sukumar, J. Phys. B 10, 2853 (1977)

8.- I. Tanarro, F. Arqueros and J. Campos, J. Chem. Phys. 77, 1826 (1982)

9.- D. R. Jenkins, Proc. Roy. Soc. A 293, 493 (1966): ibid A 306, 413 (1968): ibid A 303, 453 (1968)

10.- H.P. Hooymayers and P.L. Lijnse, JQSRT 9, 995 (1969)

11.- P.L. Lijnse and R.E. Elsenaar, JQSRT 12, 1115 (1971)

12.- C. Bastlein, G. Baumgartner and B. Brosa, Z. Phys. 218, 319 (1969)

13.- B.P. Kibble, G. Copley and L. Krause, Phys. Rev. 159 , 11 (1967)

14.- W. Demtröder, Z. Phys 166, 42 (1962)

15.- E.E. Hulpke, E. Paul and W. Paul, Z. Phys 218, 319 (1969)

16.- E. Bauer, E.R. Fisher and F.R. Gilmore, J. Chem. Phys. 51, 4173 (1969)

17.- E.R. Fisher and G.K. Smith, Appl. Opt. 10, 1803 (1971)

18.- P. L. Lijnse, Chem. Phys. Lett. 18, 73 (1973)

19.- W. Demtröder, Laser Spectroscopy, Springer Ser. Chem. Phys. (Springer, Berlin 1981)

20.- R. L. Brown and W. Klemperer, J. Chem. Phys. 41, 3072 (1964)

21.- J.I. Steinfeld and W. Klemperer, J. Chem. Phys. 42, 3475 (1965)

22.- R.B. Kurzel and J.I. Steinfeld, J. Chem. Phys. 53, 3293 (1970)

23.- G.A. Capelle and H.P. Broida, J. Chem. Phys. 58, 4212 (1973)

24.- J.A. Paisner and R. Wallenstein, J. Chem. Phys. 61, 4317 (1974)

25.- M. Broyer, J. Vigué and J.C. Lehmann, J. Chem. Phys. 63, 5429 (1975)

26.- J. Derouard and N. Sadeghi, Chem. phys. Lett. 102, 324 (1983)

27.- S. Gerstenkorn and P. Luc, "Atlas du Spectre de la molecule d'Iode", CNRS, Paris (1978)

28.- S.L. Dexheimer, M. Durand, T.A. Brunner and D.E. Pritchard, J. Chem. Phys. 76, 4996 (1982)

29.- J. Vigué, M. Broyer and J.C. Lehmann, J.Phys 42, 949 (1981)

30.- J. Derouard and N. Sadeghi, Chem. Phys. 88, 171 (1984)

31.- C. Colón, M. Ortiz and J. Campos, J. Mol. Spectrosc, 112 (1985)

32.- M. Ortiz and J. Campos, JQSRT 26, 107 (1981)

33.- P. Martin, M. Ortiz and J. Campos, Anales Fís. B 79, 246 (1983)

34.- H.H. Landolt and R. Bornstein, Zehlenwerte and Functionen Vol 1, part 3, p 514 Springer, Berlin 1951.

35.- M. Ortiz, P. Martin and J. Campos, Physica 124 C, 416 (1984)

36.- A.V. Loginov and P.F. Gruzdev, Opt. Spectrosc. 41, 104 (1976)

37.- M. Aymar and M. Coulombe, At. Data Nucl. Data Tables 21, 537 (1978)

38.- H. Horiguchi, R.S.F. Chang and D.W. Setser, J. Chem. Phys. 75, 1207 (1981)

39.- I. Tanarro, F. Arqueros and J. Campos, Phys. Rev. A 27, 2533 (1983)

40.- R.S.F. Chang and D.W. Setser, J. Chem. Phys. 69, 3885 (1978): ibid 72, 4099 (1980)

41.- S.A. Kandela and H. Schmoranzer, Phys. Lett. 86 A, 101 (1981)

42.- W.R. Bennet,Jr and P.J. Kindlmann, Phys. Rev. 149, 38 (1966)

43.- P. Martin and J. Campos, Anales Fís. 73, 276 (1977)

44.- J.Z. Klose, Phys. Rev. 141, 181 (1966)

45.- S. Feneuille, M. Klapisch, E. Koening and S. Liberman, Physica 48, 571 (1970)

46.- E. Mehlhorn, J. Opt. Soc. Am. 59, 1453 (1969)

47.- H. Geisen, D. Neuschäfer, Ch. Ottinger and A. Sharma, Acta Phys Pol., A 66, 289 (1984)

48.- M. Jeunehomme, J. Chem. Phys. 45, 1805 (1966)

49.- A.W. Johnson and R.G. Fowler, J. Chem. Phys. 53, 65 (1970)

50.- D.E. Shemansky and A.L. Bradfoot, JQSRT 11, 1385 (1971)

51.- S.T. Chen and R.J. Anderson, Phys. Rev. A 12, 468 (1975)

52.- M. Hollstein, D.C. Lorents, J.R. Peterson and J.R. Sheridan, Can. J. Chem., 47, 1858 (1969)

53.- E.E. Eyler and F.M. Pipkin, J. Chem. Phys 79, 3654 (1983)

POLARIZATION OF COLLISIONALLY REDISTRIBUTED LIGHT

Nils Andersen

Institute of Physics, University of Aarhus, DK-8000 Aarhus C, and
Physics Laboratory, H.C. Ørsted Institute, DK-2100 Copenhagen Ø
Denmark

ABSTRACT

A brief introduction is given to the physics of far-wing collisional redistribution of light. The relation between the quasimolecular potential curves of the collision complex and the intensity and polarization of the redistributed light is analyzed within the reorientation model. In particular, the dependence of polarization properties of the redistributed photons on the polarization of the absorbed photons, quasimolecular quantum numbers, geometrical properties, and concepts such as excitation and decoupling radii, is discussed for the simple case of a $^1P \rightarrow {}^1S$ transition, and illustrated by recent data for barium - rare-gas systems.

1 INTRODUCTION

The redistribution in frequency and polarization of photons scattered off collision complexes is interesting from many points of wiew [1,2] and is currently a very active field of atomic physics [3]. Allard and Kielkopf [4] may be consulted for a broad review and historical introduction, while Burnett [5] reviews the most recent progress within this area. The point of view adopted here is to investigate what can be learned about atomic interactions from such studies. The brief account below is to a large extent based upon two recent papers [6,7], which may be consulted for further details, and extensive lists of references. The basic model is closely related (by time reversal) to the quasistatic theory of line-broadening, as reviewed by Gallagher at an earlier summer school in this series [8].

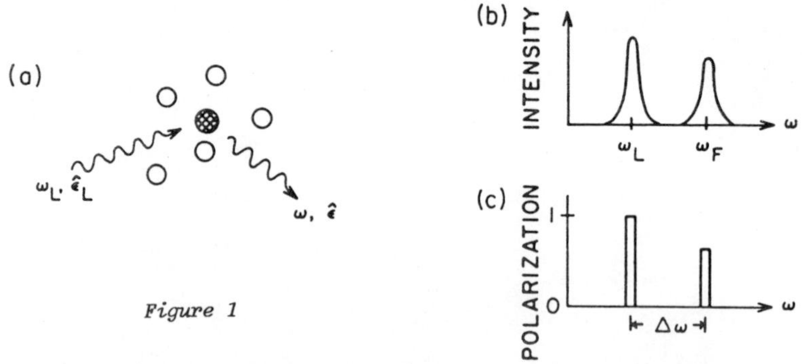

Figure 1

Figure 1(a) shows shematically the geometry of a redistribution experiment. An atom A sourrounded by perturbers is irradiated by photons from, e.g., a tunable laser with frequency ω_L and polarization $\hat{\varepsilon}_L$. The intensity and polarization $\hat{\varepsilon}$ of frequency-analyzed photons scattered in some direction are studied. Below we shall in particular study the cases where the incident light is linearly or circularly polarized, with a direction of observation either perpendicular or along the direction of the incident light beam.

Assume that the incident frequency ω_L is not far from the frequency ω_0 of a resonance transition of the free atom. For reasonably low intensity of the incident light, the scattered light will then, as shown in Fig. 1(b) and (c), consist of two components. One component is due to Rayleigh scattering off the atom and has the same frequency ω_L as the incident photons. This component is also present without perturbers and is of no interest in the present context. The other component has the frequency ω_0 and an intensity proportional to the perturber density. Furthermore, the polarization of this component will in general be different from that of the Rayleigh peak. We shall now discuss the physical origin of the fluorescence peak.

2 THE ABSORPTION-REEMISSION PROCESS

The absorption of a photon with frequency ω_L and the subsequent emission of a photon with frequency ω_0 may be understood within the simple scheme outlined in Figure 2.

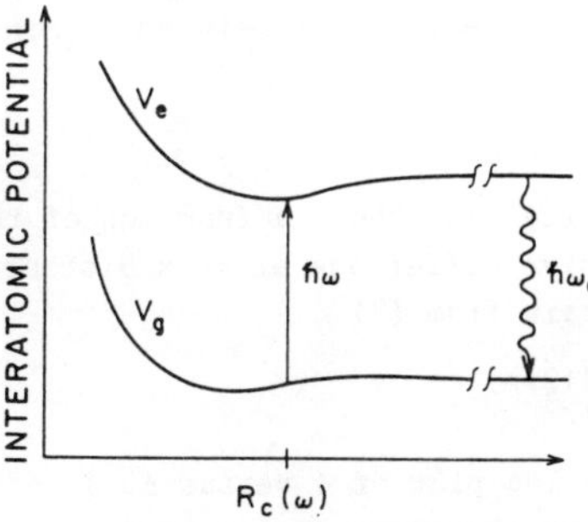

Figure 2

The atom A undergoes a collision with one of the surrounding perturber atoms, thereby forming a quasimolecule with a lifetime τ_C of typically a picosecond at thermal energies. The collision complex may now evolve in the following steps:

i) Absorption:; In the quasistatic region, where $\Delta\omega=|\omega_L-\omega_0|>>\tau_C^{-1}$, a photon is absorbed at a well-localized internuclear distance $R_C=R_C(\omega_L)$, the so-called Condon point, where the energy difference between the ground-state potential curve V_g and an excited state V_e matches the photon energy,

$$\Delta V = V_e(R_C) - V_g(R_C) = \hbar\omega_L \quad (1)$$

ii) Reorientation: The collision complex then evolves along the excited-state potential curve V_e. In this region, the electronic motion is locked to the rotating internuclear axis.

iii) Decoupling: Further out, at some (more or less well-defined) distance R_D, the decoupling radius, the quasimolecule separates into two noninteracting atoms.

iv) Decay: The excited atom A decays back to the ground state by emission of a flourescence photon, typically after some nanoseconds.

The transition from region ii) to iii), and in particular the validity of the concept of a decoupling or 'locking' radius, has been studied extensively in a recent paper by Hertel et al. [9].

We shall now analyze in more detail how information on the atom-perturber interaction may be extracted from studying the changes in intensity and polarization of this fluorescence when ω_L, and thereby the Condon point, varies.

2.1 Intensity of the redistributed light

The fluorescence intensity $I=I(\omega_L)$ corresponding to an incident photon frequency ω_L is propotional to the number of atoms at the corresponding distance $R_C(\omega_L)$,

$$I(\omega_L)d\omega_L \propto 4\pi R_C^2 dR \tag{2}$$

or, using Eq. (1) to obtain the relation between $d\omega_L$ and dR,

$$I(\omega_L) \propto 4\pi R_C^2 \left|\frac{d}{dR}\Delta V\right|^{-1}_{R_C} . \tag{3}$$

The fluorescence intensity is thus a function of the Condon radius and the slope of the potential-curve difference at this distance. For a simple power potential $\Delta V \propto R^{-n}$, we obtain from (3)

$$I(\omega_L) \propto (\Delta\omega)^{-(1+3/n)} . \tag{4}$$

Thus, in this case, a log-log plot of I versus $\Delta\omega$ directly yields the effective power of the potential from the slope of the curve.

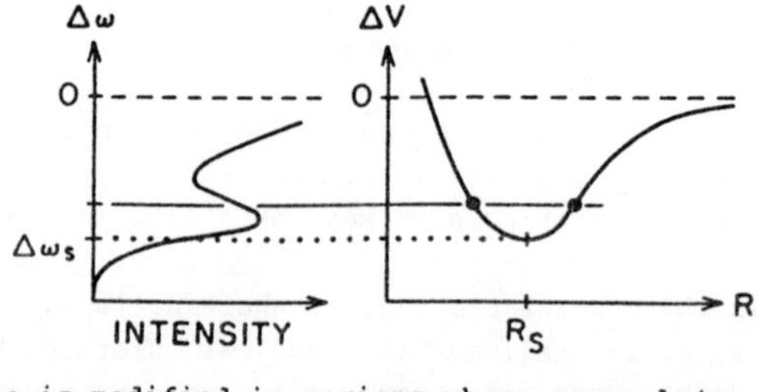

Figure 3

This simple picture is modified in regions where several (or no) Condon points contribute to the fluorescence. As indicated in Fig. 3, this is especially so near a maximum or minimum of ΔV, which gives rise to a so-called satellite, followed by an exponential drop of the intensity at even larger detunings.

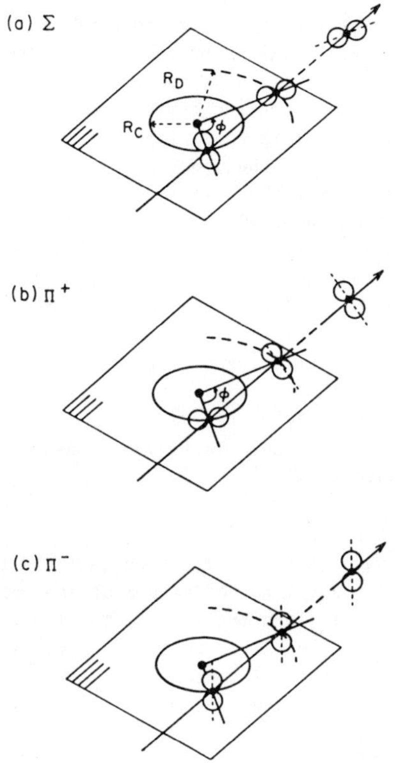

Figure 4

2.2 Polarization of the redistributed light

Turning the attention now to the polarization, we first analyze the qualitative aspects and then derive formulas for specific cases. Consider Fig. 4(a) which shows an atom on a (straight-line) trajectory colliding with a perturber. Assume that excitation takes place at an internuclear distance R_C to a Σ state, i.e., the dumbbell-shaped charge cloud is aligned along an internuclear axis, as shown. (Note that in general, this distance is obtained twice along the trajectory, once on the way in and once on the way out.) The charge cloud is locked to the rotating internuclear axis out to some distance R_D, where decoupling takes place, from whereon it stays fixed in space until the optical decay. (We here neglect effects due to subsequent collisions and possible hyperfine structure which can be measured or evaluated and corrected for.) The excitation amplitude is proportional to the exciting electric field component along the internuclear direction. Depending on the size of the rotation angle, the resulting radiation pattern will to a varying extent preserve memory of the initial polarization direction of the incident photons, the resulting light becoming more and more depolarized, the further out R_D is located.

Before calculating, it is instructive also to consider excitation to a Π state, i.e., the case where the p orbital is oriented perpendicular to the internuclear axis. Figure 4(b) and (c) show the two cases of a Π^+ and a Π^- orbital, labelled according to their reflection symmetry with respect to the scattering plane, positive for an orbital in this plane and negative for the orbital perpendicular to the plane. Keeping the geometry fixed, it is seen that a Π^+ orbital will experience exactly the same fate as the Σ orbital just considered, except for an initial rotation of 90^0. The Π^- orbital, however, stays fixed in space all the way from excitation to decay and thus does not contribute to depolarization. Therefore, on the average, excitation to a Π orbital will experience smaller depolarization than will excitation to a Σ orbital.

We shall evaluate this effect for some important observation geometries. Most of the resulting formulas can be found in the literature, rigorously derived from the general theory [6,7,10-14]. Here, a more heuristic approach will be attempted, in the hope to bring out clearly the essential features of the underlying physical mechanisms.

2.2.1 Absorption of linearly polarized light

We first analyze the case of a single internuclear axis fixed in space, having a polar angle (θ,ϕ) with respect to the axis z of the polarization direction of the light, which is incident along the y axis. The light emitted along the x axis, perpendicular to these two directions, is polarization-analyzed. With the assumption above, we then get by elementary geometry for excitation of a Σ orbital and a Π orbital, respectively.

$$\begin{cases} I_{\parallel}^{L}(\Sigma) \propto \cos^2\theta \cdot \cos^2\theta & \text{(5a)} \\ I_{\perp}^{L}(\Sigma) \propto \cos^2\theta \cdot \cos^2\phi \cdot \sin^2\theta & \text{(5b)} \end{cases}$$

$$\begin{cases} I_{\parallel}^{L}(\Pi) \propto \sin^2\theta \cdot \sin^2\theta & \text{(5c)} \\ I_{\perp}^{L}(\Pi) \propto \sin^2\theta \cdot \cos^2\phi \cdot \cos^2\theta & \text{(5d)} \end{cases}$$

Here, $I_{\parallel}$ and $I_{\perp}$ are the intensities observed through a linear polarizer with transmission axis parallel and perpendicular to the z axis. The upper index L=Linear refers to the polarization of the incident light.

If, instead of staying fixed in space, we allow the internuclear axis to rotate an angle Ω before the optical decay, we get, again by elementary spherical geometry, by averaging over all directions of rotation of the internuclear axis the somewhat complicated results,

$$\begin{cases} I_{\parallel}^{L}(\Sigma) \propto \cos^2\theta(\cos^2\theta\cos^2\Omega + \frac{1}{2}\sin^2\theta\sin^2\Omega) & \text{(6a)} \\ I_{\perp}^{L}(\Sigma) \propto \cos^2\theta(\cos^2\phi[\frac{1}{2}\cos^2\theta\sin^2\Omega+\sin^2\theta\cos^2\Omega] + \frac{1}{2}\sin^2\phi\sin^2\Omega) & \text{(6b)} \end{cases}$$

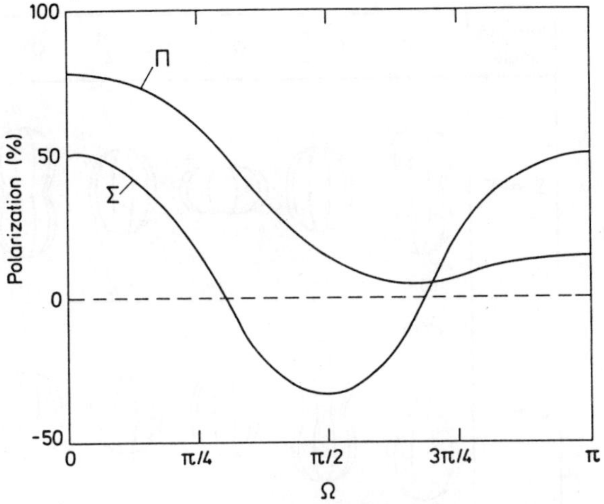

Figure 5

$$\begin{cases} I^L_{\parallel}(\Pi) \propto \sin^2\theta(\frac{1}{2}\cos^2\theta\sin^2\Omega+\sin^2\theta[\frac{3}{8}(1-\cos\Omega)^2+\cos\Omega)]) & (6c) \\ I^L_{\perp}(\Pi) \propto \sin^2\theta(\cos^2\phi[\cos^2\theta\{\cos\Omega + \frac{3}{8}(1-\cos\Omega)^2\} \\ \quad + \frac{1}{2}\sin^2\theta\sin^2\Omega] + \sin^2\phi\frac{1}{8}(1-\cos\Omega)^2) & (6d) \end{cases}$$

However, an actual experiment is mostly performed with a cell, thereby averaging over all orientations (θ,ϕ) of the internuclear axis, thus leading to a radiation pattern having rotational symmetry with respect to the polarization direction of the exciting light. Averaging in addition over (θ,ϕ) yields simply

$$\begin{cases} I^L_{\parallel}(\Sigma) \propto 1 + 2\cos^2\Omega & (7a) \\ I^L_{\perp}(\Sigma) \propto 2 - \cos^2\Omega & (7b) \end{cases}$$

$$\begin{cases} I^L_{\parallel}(\Pi) \propto 4 + 2\cos\Omega + 2\cos^2\Omega & (7c) \\ I^L_{\perp}(\Pi) \propto 3 - \cos\Omega - \cos^2\Omega & (7d) \end{cases}$$

With the usual definition of linear polarization $P_L=(I_{\parallel}-I_{\perp}/(I_{\parallel}+I_{\perp})$, we thus get (cf. [6,11,12])

$$\begin{cases} P^L_L(\Sigma) = (-1+3\cos^2\Omega)/(3+\cos^2\Omega) & (8a) \\ P^L_L(\Pi) = (1+3\cos\Omega+3\cos^2\Omega)/(7+\cos\Omega+\cos^2\Omega) & (8b) \end{cases}$$

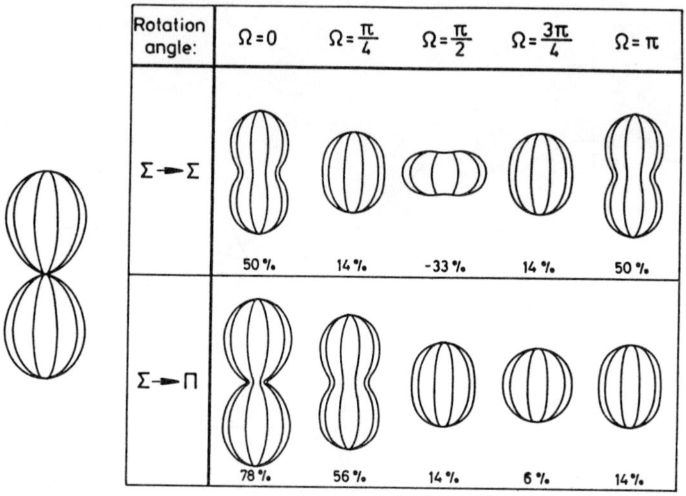

Figure 6

The variation of P_L^L with Ω is shown in Fig. 5. The Σ curve is symmetric with respect to $\Omega=90^0$, as it obviously should be, while the asymmetry of the Π curve is due to the Π^- contribution. The shapes of the corresponding (angular part of the average) charge cloud for the excited state is shown in Fig. 6.

In the special case of $\Omega=0$, i.e., no rotation of the internuclear axis before decay, one obtains $P_L^L(\Sigma)=1/2=50\%$ and $P_L^L(\Pi)=7/9=78\%$, in agreement with the results of Vigué et al. [14].

The final step is the estimation of the average rotation angle Ω. This angle will depend on the shapes of the trajectories. For not too close encounters, the assumption of a straight-line trajectory (SLT) is a reasonable one. Using furthermore that the excitation probability is inversely proportional to the derivative of ΔV along the trajectory, yields the following expressions for the excitation probability p, averaged over all impact parameters b:

$$\langle p\rangle_b \propto 4\pi R_C^2\, \Delta V'(R_C)^{-1} \tag{9a}$$

in agreement with Eq. (3). Furthermore,

$$\langle p \cos\Omega\rangle_b \propto 4\pi R_C^2\, \Delta V'(R_C)^{-1}\, \frac{2}{3}\cdot\left(\frac{R_C}{R_D}\right) \tag{9b}$$

$$\langle p \cos^2\Omega\rangle_b \propto 4\pi R_C^2\, \Delta V'(R_C)^{-1}\left[\frac{1}{3} + \frac{2}{5}\left(\frac{R_C}{R_D}\right)^2\right] . \tag{9c}$$

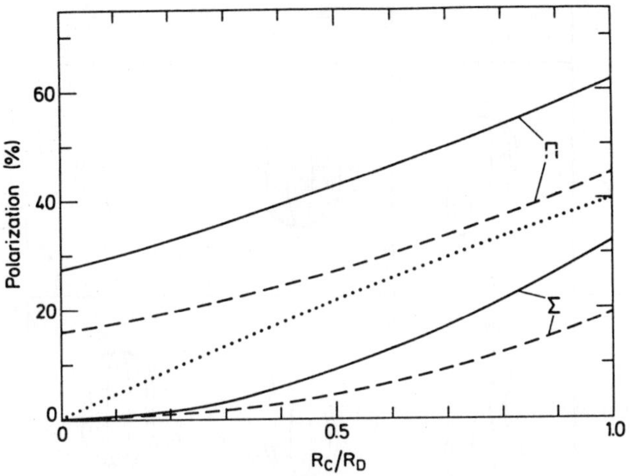

Figure 7

Using the abbreviation $x=R_C/R_D$, we then get

$$\begin{cases} P_L^L(\Sigma) = 9x^2/(25+3x^2) & (10a) \\ P_L^L(\Pi) = (15+15x+9x^2)/(55+5x+3x^2)\ . & (10b) \end{cases}$$

The variation of P_L^L with x ($0 \leqslant x \leqslant 1$) is shown in Fig. 7 by the solid lines.

We notice that the two ranges are hardly overlapping,

$$0\% \;=\; 0 \leqslant P_L^L(\Sigma) \leqslant 9/28 \;=\; 32\% \qquad (11a)$$

$$27\% \;=\; 3/11 \leqslant P_L^L(\Pi) \leqslant 13/21 \;=\; 62\%\ . \qquad (11b)$$

Thus, if the excitation takes place to only one potential curve, the fluorescence polarization will label it uniquely as a Σ or Π curve.

The angular parts of the average charge cloud density for x=0, 1/2, 1 are shown in Fig. 8 for $\Sigma \to \Sigma$ and $\Sigma \to \Pi$, respectively.

Equations (10) were derived assuming SLT. Closer inspection of the details of the derivation shows that any deviation from SLT will cause the polarization to increase. (In general, this analysis is not trivial, but consider the extreme case where excitation can only take place in head-on collisions. Then $\Omega \simeq 0$, giving a large polarization.)

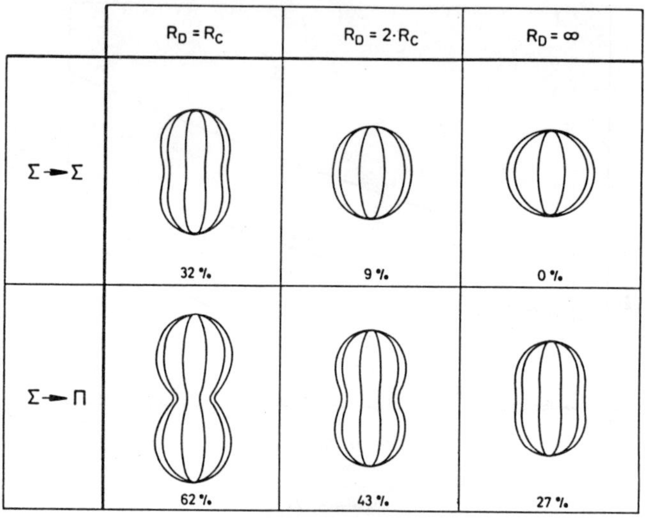

Figure 8

2.2.2 Absorption of circularly polarized light

We shall now analyze the same geometry as above, however using circularly polarized incident light. In addition, this excitation mode allows nonzero circular polarization to be observed, and we shall calculate the circular polarization of the light emitted along the incident beam direction too.

Starting with the linear polarization measured in a direction perpendicular to the incident photon beam, one can go through a similar analysis as that above. However, it is easier to note that for this purpose, the incident beam can be considered as a sum of two beams of equal intensity with linear polarization perpendicular to one another, thereby,

$$\begin{cases} I_{\parallel}^{C} = \frac{1}{2}(I_{\parallel}^{L}+I_{\perp}^{L}) & \text{(12a)} \\ I_{\perp}^{C} = \frac{1}{2}(I_{\perp}^{L}+I_{\perp}^{L}) = I_{\perp}^{L}\,. & \text{(12b)} \end{cases}$$

The signs ∥ and ⊥ still refer to the z axis. The upper index means C=Circular. Insertion of Eq. (7) in (12) yields

$$\begin{cases} P_{L}^{C}(\Sigma) = (-1+\cos^{2}\Omega)/(7-\cos^{2}\Omega) & \text{(13a)} \\ P_{L}^{C}(\Pi) = (1+3\cos\Omega+3\cos^{2}\Omega)/(13-\cos\Omega-\cos^{2}\Omega)\,, & \text{(13b)} \end{cases}$$

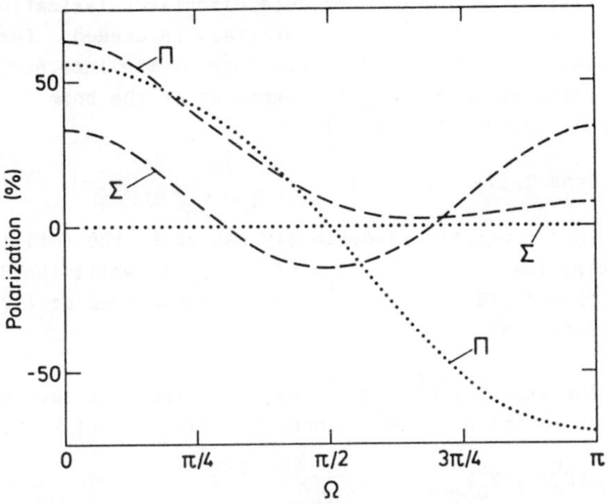

Figure 9

analogous to (8). The variation of P_L^C is shown by dashed lines in Fig. 9, which is analogous to Fig. 5. The Ω dependence is Fig. 9 is somewhat weaker than in Fig. 5, but the curve shapes are similar, as expected from (12). Assuming SLT insertion of Eq. (9) in (13) yields

$$\begin{cases} P_L^C(\Sigma) = 9x^2/(50-3x^2) & (14a) \\ P_L^C(\Pi) = (15+15x+9x^2)/(95-5x-5x^2)\,. & (14b) \end{cases}$$

These curves are indicated in Fig. 7 by dashed lines. We notice from (14) that

$$0\% = 0 \leq P_L^C(\Sigma) \leq 9/47 = 19\% \quad (15a)$$

$$16\% = 3/19 \leq P_L^C(\Pi) \leq 13/30 = 45\%\,. \quad (15b)$$

Again, the behaviour of the P_L^C curves is similar to the corresponding ones for P_L^L, though yielding smaller values.

A general relation between the two polarizations can be derived from Eq. (12),

$$P_L^C = P_L^L/(2-P_L^L) \quad (16)$$

Finally, we shall consider the circular polarization of the light emitted in the same direction as the incident beam.

The first thing to note is that excitation to a Σ state only can result in circular polarization. Thus, provided that no Π mixing takes place,

$$P_C^C(\Sigma) = 0\,. \quad (17a)$$

For Π-state excitation, in order to produce circular polarization, simultaneous, coherent excitation of the π^+ and π^- orbitals is needed. For reasons of symmetry, when averaging over all orientations of the internuclear axis, the resulting angular momentum has a nonzero component in the beam direction only. For incident RHC photons, one obtains [7],

$$P_C^C(\Pi) = 5\cos\Omega/(7+\cos\Omega+\cos^2\Omega) . \tag{17b}$$

The factor $\cos\Omega$ in the numerator is obviously caused by the rotation of the angular momentum during the collision ($\langle L_y\rangle \propto I_{RHC}-I_{LHC}$), while the denominator is the sum of Eqs. (7c) and (7d), cf. (12a). The Ω dependences of (17) are shown by dotted curves in Fig. 9.

A measurement of, for example, P_L^L and P_L^C for a Π state thus determines the (cosine of the) average rotation angle Ω since, according to (13b) and (17c),

$$\cos\Omega_\Pi = 4P_C^C/(3-P_L^L) . \tag{18}$$

In the special case of SLT, insertion of (9) in (17b) yields

$$P_C^C(\Pi) = 25x/(55+5x+3x^2) . \tag{19}$$

The variation with $x=R_C/R_D$ is shown by dotted curves in Fig. 7.

3 RESULTS FOR BARIUM - RARE-GAS SYSTEMS

In this section, we shall illustrate how the concepts developed above allow interpretation of the redistribution, as observed in the neighbourhood of the BaI 6^1S-6^1P resonance line (λ=5535 Å) for barium atoms perturbed by the heavy rare gases argon and xenon, in terms of characteristics of the corresponding quasi-molecular potential curves. This atomic transition is particularly convenient since it is not complicated by fine or hyperfine structure, and the wavelength region of interest, ±300 Å, is well within the reach of tunable dye lasers.

Extensive linear polarization data exist also for the strontium - rare-gas systems [15], while linear and circular polarization results for the sodium D lines have recently been published by Behmenburg et al. [16].

Figure 10 shows results for the red (a) and blue (b) wings of the Ba-Xe system. The lower panels present the intensities (or redistribution coefficients k_r) measured per barium and per perturber atom and has been multiplied by $\Delta\omega^2$ in order to better display the structures. (In this plot, a Lorentzian curve shape, varying as $(\Gamma^2+\Delta\omega^2)^{-1}$, will yield a constant value in the far-wing region.) The absolute value, accurate to within 20%, has been determined by normalization to known Rayleigh cross sections. The upper panels show the corresponding linear polarizations measured perpendicular to the polarization vector of the linearly polarized incident light.

For the red wing, we notice a satellite structure around 40 cm^{-1}. Here, the polarization drops to about 10%, indicating predominantly Σ excitation, cf. Fig.

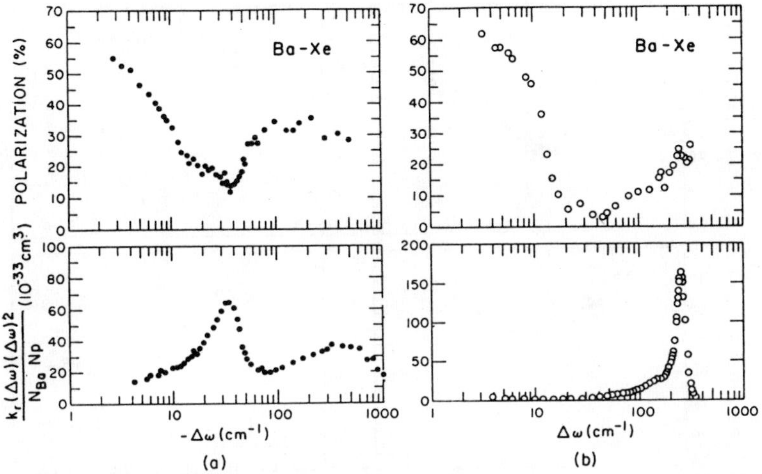

Figure 10

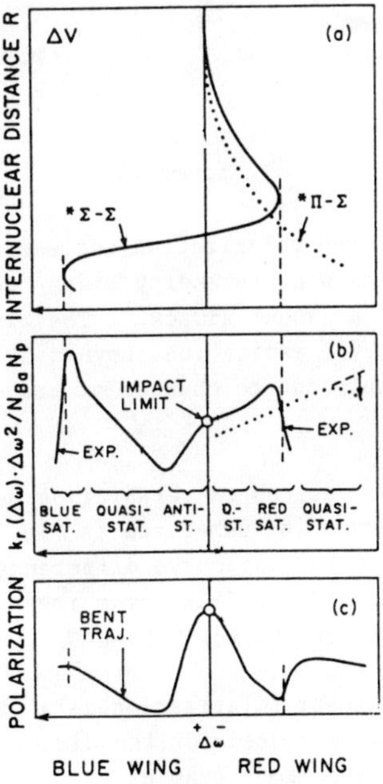

Figure 11

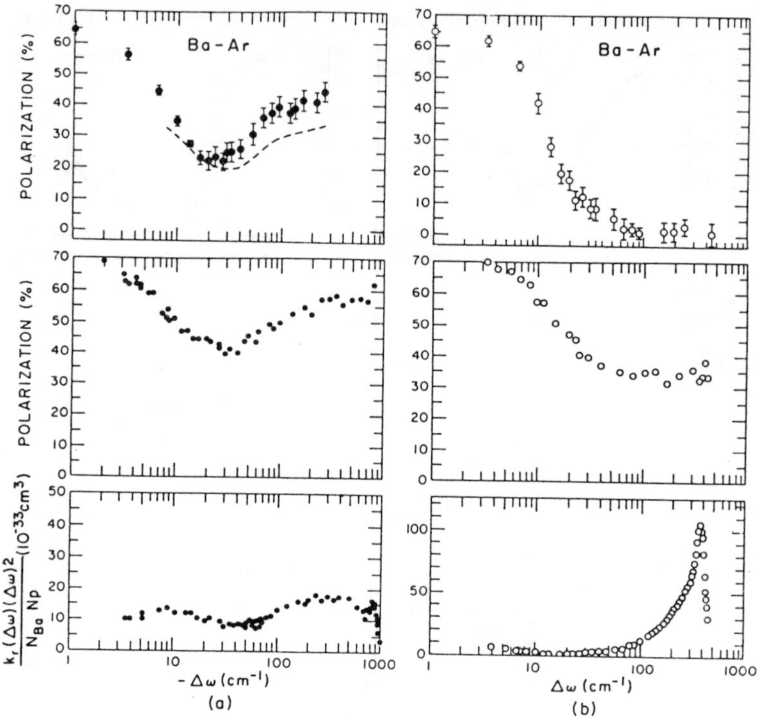

Figure 12

8. Beyond the satellite, the polarization increases to about 30%, apparently due to Π excitation. The blue-wing intensity stays very small until a pronounced satellite develops at about 250 cm^{-1}. The blue-wing polarization drops rapidly to almost zero, i.e., Σ excitation. Beyond 50 cm^{-1}, the polarization increases steadily, probably due to changes of trajectory, deviating more and more from SLT.

Interpretation of these observations in terms of shapes and quantum numbers of corresponding potential curves is summarized in Fig. 11 which shows the qualitative behaviour of (a) the potential-curve differences ΔV, and the intensities (b) and linear polarizations (c) resulting from these. A detailed discussion of the various regions is presented in [6].

For the system Ba-Ar, both linear polarization data, measured as for Ba-Xe, and circular polarization data, measured for the direction along the incident, in this case circularly polarized, light beam exist. The circular polarization is determined in a two-step excitation scheme, involving a probe laser in the second step to circumvent absorption problems for the resonance light in the barium vapour [6].

The lower panel of Fig. 12 shows the red (a) and blue (b) wing intensities, and above that, the linear polarizations (middle panels). The qualitative behaviour is much the same as that seen for Ba-Xe in Fig. 10, although with less pronounced structures, and the qualitative interpretation is thus also as shown in Fig. 11. The circular polarizations (upper panels) show for the blue wing a decrease to almost zero, confirming the Σ labelling of the potential curve responsible. It is interesting, however, that the polarization stays significant out to about a hundred wave numbers, which is due to the polarization being very sensitive to even a tiny admixture of a Π component.

For the red wing, the dashed line indicates the circular polarization predicted from the measured linear polarization, using the formalism developed in the previous section and assuming SLT. The agreement with the measured data is seen to be good, with increasing difference with increasing detuning, most probably due to a gradual break-down of the SLT assumption. Further details may be found in [7].

4 CONCLUSIONS

Although brief and somewhat simplified, it is hoped that the presentation above has served to illustrate how careful analysis of the redistribution in intensity and polarization of photons scattered off collision complexes may serve as a rich source of information about interatomic forces and molecular quantum numbers and guide the development of simple models that allow us to visualize the time evolution of the excited electron charge cloud during the collision.

5 ACKNOWLEDGEMENTS

My involvement in the experiments described above took place during stays in 1983 and 1984 as a Visiting Fellow at the Joint Institute for Laboratory Astrophysics of the National Bureau of Standards and the University of Colorado in Boulder, Colorado. I am much indebted to my colleagues at JILA, and in particular to Joe Alford, Keith Burnett, and Jinx Cooper, for introducing me to this fascinating topic. Helpful discussions on the reorientation model and polarization properties with Ingolf Hertel, Paul Julienne, and Gerard Nienhuis are gratefully acknowledged.

6 REFERENCES

1) J. Cooper, in: Spectral Line Shapes, edited by K. Burnett (de Gruyter, Berlin, 1983), Vol. 2, pp. 649-60
2) K. Burnett, contribution to this volume
3) See, for example, the series of proceedings of the International Conference on Spectral Line Shapes, the latest of which is edited by F. Rostas (de Gruyter, Berlin, 1985), Vol. 3
4) N. Allard and J. Kielkopf, Rev.Mod.Phys. 54(1982)1103-82
5) K. Burnett, Phys.Rep.C 118(1985)339-401

6) W.J. Alford, N. Andersen, K. Burnett, and J. Cooper, Phys.Rev.A 30(1984) 2366-80

7) W.J. Alford, N. Andersen, M. Belsley, J. Cooper, D.M. Warrington, and K. Burnett, Phys.Rev.A 31(1985)3012-6

8) A. Gallagher, Acta Phys.Pol.A 54(1978)761-5

9) I.V. Hertel, H. Schmidt, A. Bähring, and E. Meyer, Rep.Prog.Phys. 48(1985) 375-414

10) G. Nienhuis, J.Phys.B 16(1983)1-24

11) E.L. Lewis, M. Harris, W.J. Alford, J. Cooper, and K. Burnett, J.Phys.B 16 (1983)553-625

12) P.S. Julienne and F.H. Mies, in: Proc.10ème Coll. sur la Physique des Coll. Atomiques et Electroniques (Aussois, France, 1984) Conférences Invitées, pp. 103-14

13) P.S. Julienne and F.H. Mies, Phys.Rev.A 30(1984)831-43

14) J. Vigué, P. Grangier, G. Roger, and A. Aspect, J.Physique 42(1981)L531-5

15) W.J. Alford, K. Burnett, and J. Cooper, Phys.Rev.A 27(1983)1310-3

16) W. Behmenburg, V. Kroop, and F. Rebentrost, J.Phys.B 18(1985)2693-2704

MULTIPHOTON SPECTROSCOPY OF COLLISION AND REACTION DYNAMICS

Keith Burnett

Spectroscopy Group
Blackett Laboratory
Imperial College
London SW7 2BZ
UNITED KINGDOM

ABSTRACT

I shall discuss the prospects for using multiphoton excitation to study collision and reaction dynamics. The theory of multiphoton absorption by collision complexes is now undergoing careful scrutiny. We want to know how much one can learn in practice by studying multiphoton absorption during collisions and reactions.

For some simple cases of atomic collisions the tale to be told is relatively simple: a lot can be learned ! I'll discuss the experimental problems as well as some possible resolutions. My analysis will be mostly qualitative although some quite sophisticated techniques are needed for detailed analysis of specific cases.

COHERENT TRANSIENTS FOR COLLISIONAL STUDIES

J.-L. Le Gouët, J.-C. Keller, M. Defour, M. Saïdi*

Laboratoire Aimé Cotton, Centre National de la Recherche Scientifique II, Bâtiment 505, F-91405 Orsay

I. Introduction

Numerous high resolution spectroscopic techniques have emerged during the last fifteen years in connection with the development of tunable lasers. They have been used for the precise mapping of the atomic energy level diagrams and for the determination of the natural width and of the relaxation broadening of spectral lines. In addition some of these techniques have proved efficient for acting on the external degrees of freedom of the atoms through the way of selective excitation. Such a capability is demonstrated in the example of the optical piston which has been presented during this Summer School [1,2]. Driving atomic external parameters out of equilibrium and monitoring their rethermalization brings informations on relaxation processes. In a vapor, laser excitation drives the atomic velocity distribution out of thermal equilibrium and the monitoring of equilibrium restoration brings information on collisional processes. The two aspects of the problem - namely the formation of a non equilibrium structure and the return to equilibrium - have been actively investigated. This paper intends to present them in the specific case of the time resolved coherent transient spectroscopic techniques in an atomic vapor. Following an elementary procedure we derive the shape of the structure which is built by the light pulses inside the atomic velocity distribution and we describe its collisional relaxation. Attention is focussed on the influence of the temporal coherence of the excitation. Experimental results are discussed.

The section II is devoted to the mathematical description of the system and of its evolution in the frameof the density matrix formalism. In section III, monochromatic excitation is assumed and the density matrix equation is solved in a perturbative way for various excitation pulse sequences. This provides us with the shape of the non equilibrium structure which is prepared by the light pulses. Relaxation by elastic scattering is analyzed in section IV. Some recent stimulated photon echo experiments are examined in section V. Finally in section VI we assume that the sample is excited by incoherent light pulses. After solving the density matrix equation we compare the effects of monochromatic and broad band excitation. In the light of new experimental results we foresee some developments for the coherent transient spectroscopy with incoherent light as applied to the monitoring of the atomic translational motion.

II. Description of the Physical System

II.1. The sample

We consider a pure atomic vapor which is illuminated by light pulses close to resonance with a transition between two non degenerate levels a and b. The transition-frequency is ω_{ab}. The transition spectrum is characterized by its homogeneous width γ_{ab} and by its inhomogeneous width Ω_D which, in a vapor, identifies with the Doppler width. Thus $\Omega_D = k\bar{v}$ where $\bar{v}$ is the atomic mean velocity and $\vec{k}$ is the radiation wavevector. We assume that no transition of any type can occur between level a or b and any other level inside the atoms. These atoms may be regarded as two level systems. When excited in level b, atoms decay to level a with a spontaneous decay rate γ_b. The optical thickness of the sample is supposed to be smaller than unity. Therefore no propagation effect such as self-induced transparency [3] or self-focussing [4] can occur. More generally atoms are supposed to be driven by the incident field alone. Coupling with the induced field is neglected so that no superradiant pulse [5] can develop.

II.2. The external field

The field is described by its classical expression $\vec{E}(\vec{r},t) = \vec{E}_0(t)\cos(\omega t - k\vec{r} + \varphi(t))$ where $\vec{E}_0(t)$ and $\varphi(t)$ are slowly varying functions with regard to optical oscillations. The temporal width of the field envelope is τ_L. Polarization is linear along Ox and propagation is along Oz. Describing the field as a plane wave we assert that it is *spatially coherent*. However the time varying phase $\varphi(t)$ indicates that the field is not yet assumed to the monochromatic, that is to say we do not yet impose any assumption concerning the *temporal coherence* of the field.

II.3. The equation of motion

The atomic density matrix ρ satisfies the equation

$$i\hbar\dot{\rho} = [H,\rho] + \text{relaxation terms} \qquad (1)$$

where the hamiltonian operator H includes the atomic hamiltonian H_0, the translation motion operator $-\frac{\hbar^2}{2m}\Delta$ and the interaction operator $-\vec{\mu}.\vec{E}(\vec{r},t)$ between the atomic dipole moment $\vec{\mu}$ and the external field $\vec{E}(r,t)$. Collisional processes are discarded so that only spontaneous emission enter the relaxation terms. In the classical limit when the radiation pressure effect is neglected, the density matrix may be regarded as a function of position, velocity and time [6]. Then the translation term $-i\left[\frac{\hbar}{2m}\Delta,\rho\right]_{\alpha\alpha}$, reduces to $-\vec{v}\vec{\nabla}\rho_{\alpha\alpha}$, and eq.1 may be expanded as :

$$\dot{\rho}_{aa} + \vec{v}\vec{\nabla}\rho_{aa} = i\chi(\rho_{ba} - \rho_{ab})\cos(\omega t - kz + \varphi) + \gamma_b\rho_{bb}$$

$$\dot{\rho}_{bb} + \vec{v}\vec{\nabla}\rho_{bb} = i\chi(\rho_{ab} - \rho_{ba})\cos(\omega t - kz + \varphi) - \gamma_b\rho_{bb} \qquad (2)$$

$$\dot{\rho}_{ab}+\vec{v}\vec{\nabla}\rho_{ab} = i\chi(\rho_{bb}-\rho_{aa})\cos(\omega t-kz+\varphi)+(i\omega_{ab}-\gamma_{ab})\rho_{ab} \qquad (2)$$

where $\chi = \mu_{ab}E_o/\hbar$

The rapidly oscillating factors of the off diagonal elements can be removed with the substitution

$$\rho_{ab} = \tilde{\rho}_{ab}\exp i(\omega t-kz) \qquad (3)$$

and by neglecting non resonant high frequency terms that oscillate as $\exp 2i(\omega t-\vec{k}\vec{r})$. This procedure, which is known as the rotating wave approximation, reduces eq.2 to :

$$\dot{\rho}_{aa} = \frac{i}{2}\chi(e^{i\varphi}\tilde{\rho}_{ba}-e^{-i\varphi}\tilde{\rho}_{ab})+\gamma_b\rho_{bb} \qquad (a)$$

$$\dot{\rho}_{bb} = \frac{i}{2}\chi(e^{-i\varphi}\tilde{\rho}_{ab}-e^{i\varphi}\tilde{\rho}_{ba})-\gamma_b\rho_{bb} \qquad (b) \qquad (4)$$

$$\dot{\tilde{\rho}}_{ab} = \frac{i}{2}\chi e^{i\varphi}(\rho_{bb}-\rho_{aa})+(i\Delta-\gamma_{ab})\tilde{\rho}_{ab} \qquad (c)$$

where $\Delta = \omega_{ab}-\omega+\vec{k}\vec{v}$

II.4. Solutions to the equation of motion

There is no general solution to eq.4. Exact solutions may be obtained only in some limiting cases. For instance, eq.4 may be solved when :

i) $\gamma_b\tau_L \ll 1$, $\Omega_D\tau_L \ll 1$ and φ is constant

ii) $\gamma_b\tau_L \ll 1$, φ is constant and χ is a plateau function of time with steep edges [7]

iii) $\gamma_b\tau_L \ll 1$ and $\theta^2 \ll 1$, where $\theta = \int\chi(t)dt$ is the pulse area.

Since the main features of the coherent transient formation do not depend on the field strength, we restrict our discussion to the small pulse area limit (case iii)). In the next section, the formation of coherent transient is detailed. The additional condition of temporal coherence (φ = constant) is fulfilled by the incident light pulses. This condition is withdrawn in section VI.

III. Coherent Transients after Coherent Excitation

III.1. Single pulse excitation

The sample is illuminated at time t, by a temporally coherent pulse. Its amplitude $E_o(t)$ is a smoothly varying function of time and its phase φ is constant. Such a pulse is said to be Fourier transform limited. Its spectral width Ω_L and its duration τ_L verify $\Omega_L\tau_L \sim 2\pi$. The pulse area θ_1, verifies the condition $\theta_1^2 << 1$. For sake of simplicity we assume in addition that $\Omega_D\tau_L << 1$. Then $\Omega_L >> \Omega_D$ and the entire axial velocity distribution is coherently excited. In other words, during the illumination period, Doppler effect can not build any phase shift between the field and the atomic dipoles. We assume that relaxation proceeds from spontaneous emission only and that $\gamma_b\tau_L << 1$. Solving eq.4 to first order in θ_1 one obtains :

$$\tilde{\rho}_{ab}^{(1)}(v_z,t_1^+) = -\frac{i}{2}\theta_1 n_o(v_z) \tag{5}$$

where $n_o(v_z) = \rho_{aa}^o(v_z)-\rho_{bb}^o(v_z)$ is the thermal equilibrium population difference, and where a + (-) indicates a time just after (before) a pulse. After the pulse is switched off, the coherence ρ_{ab} evolves and decays according to eq.4c. At time t one obtains :

$$\tilde{\rho}_{ab}^{(1)}(v_z,t) = \tilde{\rho}_{ab}^{(1)}(v_z,t_1)\exp(i\Delta-\gamma_{ab})(t-t_1) \tag{6}$$

This coherence originates an oscillating dipole moment density P_x along Ox :

$$\begin{aligned} P_x = \mathrm{Tr}\mu_x\rho &= \mu_{ab}(\tilde{\rho}_{ab}e^{i(\omega t-\vec{k}\vec{r})}+c.c.) \\ &= n_o(v_z)\mu_{ab}\theta_1\sin(\omega t-\vec{k}\vec{r}+\Delta(t-t_1))\exp-\gamma_{ab}(t-t_1) \end{aligned} \tag{7}$$

This oscillating dipole density emits an electromagnetic field, in quadrature with the exciting pulse. In the radiation zone at $\vec{r}$, the electric field density which issues from the dipole density located at $\vec{r}_o$ is [8]

$$E(\vec{r}_o,\vec{v};\vec{r},t) = i\mu_{ab}\theta_1 k^2(\hat{R}\Lambda\hat{x})\Lambda\hat{R}\exp(i(kR+\vec{k}\vec{r}_o-\omega t-\Delta(t-t_1))n_o(\vec{v})/R + c.c. \tag{8}$$

where $\vec{R} = \vec{r}-\vec{r}_0$, $\hat{R} = \vec{R}/R$ and $\hat{x}$ is a unit vector along Ox. The total field radiated by the sample is :

$$E(\vec{r},t) = \int d\vec{r}_0 d\vec{v} E(\vec{r}_0,\vec{v};\vec{r},t) \tag{9}$$

The sample may be divided into slices orthogonal to Oz. Summation over $\vec{r}_0$ is a sum over contributions from all the slices. The elementary fields which radiate from each point of a slice located at z_0 behave like the Huyghens wavelets of diffraction theory. These wavelets build the pattern of the diffraction of a plane wave by an aperture positioned at z_0 and sized to the radial dimensions of the laser beam. Therefore, the field which radiates from each slice of the sample, after the exciting pulse extinction, is a diffraction limited plane wave which propagates along Oz. Its divergence is λ/a where a is the incident beam radius. Finally, contributions from all the slices interfere constructively in forward direction. Summation over $\vec{v}$ makes different velocity classes interfere. The velocity integral in eq.9 reduces to :

$$\hat{n}_0(k(t-t_1)) = \int d\vec{v} e^{ik\ v_z(t-t_1)} n_0(\vec{v}) \tag{10}$$

It expresses that the Doppler phase shift which develops between velocity classes turns them out of phase and turns off their electromagnetic emission at a time $\sim (ku)^{-1}$ after the exciting pulse extinction. This burst of light which is emitted by the sample in forward direction and which dies out a time $(ku)^{-1}$ after the exciting pulse is named a free induction decay signal.

Despite of the signal extinction, the dipoles are still present in the sample and they keep on oscillating with a damping-time-constant γ_{ab}^{-1}. Besides, the phase shift between velocity classes $[v_z]$ and $[v_z']$ is not arbitrary. It equals $k(v_z-v_z')(t-t_1)$ provided no collisional velocity change disturbs its building. Thus, there is still a "hidden order" inside the sample. Illumination by additional light pulses is needed to reveal this order.

III.2. Excitation by a pulse sequence

A second pulse illuminates the sample at time t_2. We assume that this second pulse is in phase with the first one. From eqs.4 one obtains :

$$\tilde{\rho}_{ab}^{(3)}(t_2^+) = \frac{\theta_2^2}{2}(\tilde{\rho}_{ba}^{(1)}(t_2^-)-\tilde{\rho}_{ab}^{(1)}(t_2^-)) \tag{11}$$

This expression indicates that the second pulse originates a coherence component $\tilde{\rho}_{ab}^{(3)}$ from a component $\tilde{\rho}_{ba}^{(1)}$. After the pulse, $\tilde{\rho}_{ab}^{(3)}$ evolves with a phase factor $\exp i\Delta(t-t_2)$ while $\tilde{\rho}_{ba}^{(1)}$, from which it issues, has accumulated a phase factor $\exp -i\Delta t_{12}$ during the time interval between the pulses. The resulting total phase shift $\Delta(t-t_2)-\Delta t_{12}$ cancels at time $t = t_2 + t_{12}$ and the components in the different velocity classes fall again in phase, as they were at time t_1^+. The corresponding dipole moment density is :

$$P_{ab}^{(3)} = -n_o(\vec{v})\mu_{ab}\theta_1\frac{\theta_2^2}{2}\sin(\omega t-\vec{k}\vec{r}+\Delta(t-t_2-t_{12}))\exp -\gamma_{ab}(t_{12}+t-t_2) \tag{12}$$

Thus, at time $t = t_2 + t_{12}$ a burst of light is emitted with the same properties of directivity and of duration as the free induction decay signal. This is the conventional photon echo signal [9,10]. The signal formation relies on the storage of the excitation memory between pulses. It appears that this memory is retained by coherences alone. Level populations are not involved except at the moments of excitation. Therefore the signal intensity reflects the relaxation of the atomic dipoles. This is revealed by the relaxation factor $\exp -\gamma_{ab}(t_{12}+t-t_2)$.

However excitation memory is stored in populations too. According to eq.4, the following perturbation in the level populations results from the coupling of the second pulse with the coherences built by the first one :

$$n_{ab}^{(2)}(v_z,t) = \rho_{aa}^{(2)}(v_z,t)-\rho_{bb}^{(2)}(v_z,t) = -\theta_1\theta_2 n_o(v_z)\cos\Delta t_{12}\, e^{-\gamma_{ab}t_{12}-\gamma_b(t-t_2)} \tag{13}$$

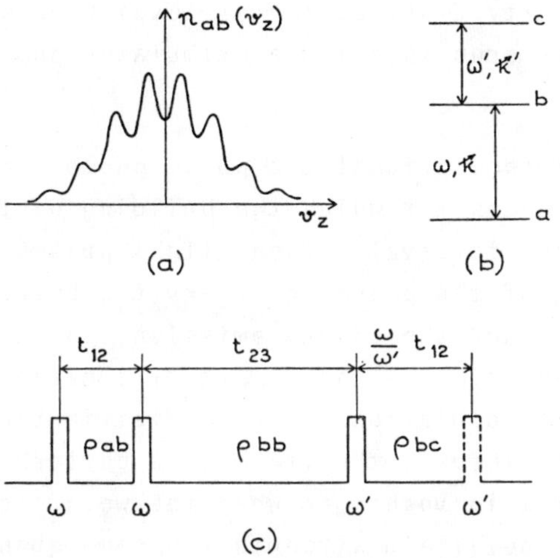

Figure 1 : a) modulation in the velocity distribution of the level population difference ; b) three level scheme ; c) time sequence of light pulses in a stimulated photon echo experiment.

The phase evolution of atomic coherences between the two pulses is thus revealed by a sinusoïdal modulation of the population difference in velocity space (fig.1a). Probing of this modulation is conveniently achieved by a third pulse which is turned to resonance with a transition b-c and which illuminates the sample at time t_3 (fig.1b,c). Then the dipole moment component which is built from $n_{ab}^{(2)}(v_z,t)$ on transition b-c is :

$$P_{bc}^{(3)} = n_o(v_z)\mu_{bc}\frac{\theta_1\theta_2\theta_3}{4}\sin(\omega' t-\vec{k}'\vec{r}+\Delta'(t-t_3)-\Delta t_{12})\times$$

$$\times\exp-\gamma_{ab}t_{12}-\gamma_b t_{23}-\gamma_{bc}(t-t_3) \qquad (14)$$

where $\vec{k}'$ and ω' are respectively the wave vector and the frequency of the third pulse and $\Delta' = \omega_{bc}-\omega'+\vec{k}'\vec{v}$. Rephasing occurs at time $t_2 = t_3+\frac{k}{k'}t_{12}$. Then a photon echo is emitted which reflects the evolution of coherence ρ_{ab} between t_1 and t_2, of coherence ρ_{bc} between t_3 and t_e and of the modulation inside the velocity distribution of level b between t_2 and t_3 [11,12,13]. This echo is named a stimulated photon echo (S.P.E.).

In summary, the conventional two pulse photon echo is produced by atomic coherences for which the building of the phase shift kv_zt_{12} during the interval between light pulses is followed by the building of the phase shift $-kv_zt_{12}$ between the second pulse excitation and the signal emission. In addition, the three pulse echo *involves* the carving of a periodic structure inside the level population velocity distribution. These features are established in the frame of a perturbative calculation but they remain unchanged when the weak field approximation fails. For any field strength the same quantities are involved and they evolve in the same way during intervals between light pulses. More quantitatively, the relaxation terms may be factorized out of the signal expression as in eqs.12 and 14. This property holds at any field strength and for any relaxation process provided relaxation during light pulses may be neglected. This applies to the analysis of photon echo collisional relaxation.

IV. Elastic Scattering Relaxation

IV.1. Decay of the echo signal

In section III the sample is an atomic vapor where the pressure is so low that no collision can occur on the experiment time scale. Foreign gas is now admixed with the initial vapor and elastic collisions can occur between active atoms and foreign perturbers. The binary collision approximation is assumed to be valid. Elastic collisions act during the intervals between light pulses. Their effect is twofold. They destroy the quantum mechanical coherence between states a and b (b and c) and this results in the decay of ρ_{ab} (ρ_{bc}). They also produce velocity changes of atoms whether they are in a superposition state or in a pure state. The consequence of these effects on the photon echo signal formation is threefold. Collisions which destroy coherence between states add their phase-interruption rate γ_{ij}^{ph} to the natural decay rate γ_{ij} of coherences ρ_{ij}. Collisional velocity changes perturb the building of the velocity phase shift $kv_z t_{12}$ and they hamper the perfect rephasing of dipoles at the echo time. Finally collisional velocity changes wash out the periodic structure which is stored in the velocity distribution of level populations and which preserves the excitation memory in the stimulated photon echo scheme.

When collisions occur in a superposition state they affect both the internal coherence and the atomic velocity [14,15]. The superposition states which occur in the photon echo scheme combine states which are separated by optical distances. These states are different enough so that close encounters destroy the coherence between them. Thus short distance, large scattering angle collisions are completely accounted for by their contribution to γ_{ij}^{ph} while long distance, small scattering angle collisions which may preserve optical coherences are described by velocity changes of atoms in a superposition state [14,15]. In the following we assume that the velocity change δu which is accumulated by coherences during the time interval t_{12}, due to large distance encounters, is small enough so that :

$$k\delta u\ t_{12} << 1$$

Therefore the dipole rephasing is supposed not to be disturbed by collisional velocity changes.

Collisions which occur in a pure state obliterate the periodic structure stored in the corresponding velocity distribution. An atom no longer contributes to this structure as soon as its accumulated velocity change between t_2 and t_3 is larger than the modulation period $2\pi/kt_{12}$ (cf. fig.1). Thus a stimulated photon echo signal collects the contributions from the only atoms for which the collisional velocity change during the t_{23} interval does not exceed $2\pi/kt_{12}$ [11]. In present experiments where t_{12} is varied typically between 10 ns and 1 μs, this upper limit is contained between 50 m/s and .5 m/s. It should be compared with the thermal velocity which amounts several hundred meters/s. Thus only atoms which have undergone small velocity changes can contribute to the echo signal. We show in the next section that after its building at time t_2 the periodic structure in velocity space decays at a rate $\gamma_b(t_{12})$ which depends on the period $2\pi/kt_{12}$ of the structure.

The stimulated photon echo signal evolution reflects all these effects. The ratio of its intensity I in the presence of a perturber gas to its intensity I_o in the absence of perturbers is [11] :

$$\frac{I}{I_o} = \exp -2\left[\left(\gamma_{ab}^{ph}+\frac{k}{k'}\gamma_{bc}^{ph}\right)t_{12} + \gamma_b(t_{12})t_{23}\right] \quad (15)$$

Experimentally, $\gamma_{ab}^{ph} + \frac{k}{k'}\gamma_{bc}^{ph}$ is determined first by setting $t_{23} = 0$. Then $\gamma_b(t_{12})$ may be derived from the data obtained with $t_{23} \neq 0$.

IV.2. Collisional decay rate of the modulation in velocity space

The active atoms undergo velocity changes when they are elastically scattered by the perturbers. The density probability per unit time for scattering from velocity $\vec{v}'$ to velocity $\vec{v}$ is

$$W(\vec{v}',\vec{v}) = N\int d^3v_p d^3v_p' W_p(v_p')\frac{1}{u}\,\frac{d\sigma}{d\Omega}\delta(u-u')\delta(\vec{V}-\vec{V}') \quad (16)$$

This expression involves the integration of the differential scattering cross section $\frac{d\sigma}{d\Omega}$ over the undetected particle variables. Averaging over the perturber initial velocity $\vec{v}_p'$ is weighted by the perturber velocity distribution $W_p(\vec{v}_p')$. Summation over the perturber final velocity $\vec{v}_p$ is conditioned by the conservation laws of energy and momentum which are expressed in terms of the initial and final relative velocities u' and u and center-of-mass velocities $\vec{V}'$ and $\vec{V}$. The perturber density is N. The collisional evolution of the density in level b at the point $(\vec{r},\vec{v})$, $n_b(\vec{r},\vec{v},t)$, is expressed by the Boltzmann equation [16] which accounts for the diffusion at position $\vec{r}$ and which strikes a balance between departure from and arrival to velocity $\vec{v}$:

$$\frac{\partial}{\partial t} n_b(\vec{r},\vec{v},t)\Big|_{coll.} = -\vec{v}\vec{\nabla} n_b(\vec{r},\vec{v},t) - \int d^3v' \Big[W(\vec{v},\vec{v}')n_b(\vec{r},\vec{v},t) - W(\vec{v}',\vec{v})n_b(\vec{r},\vec{v}',t) \Big] \quad (17)$$

Since, according to eq.13, no spatial structure is built in the level population we may drop the diffusion term. Besides the active atom velocity distribution is perturbed by light pulses along the beam direction Oz only. We assume that the transverse velocity distribution is little disturbed by the axial perturbation and that it remains close to the equilibrium distribution $W(\vec{v}_\perp)$. The following form is thus assumed :

$$n_b(\vec{v},t) = n_b(v_z,t)W(\vec{v}_\perp) \quad (18)$$

Finally the transport equation 17 turns into :

$$\frac{\partial}{\partial t} n_b(v_z,t)\Big|_{coll.} = -\int dv_z' \Big[W(v_z,v_z')n_b(v_z,t) - W(v_z',v_z)n_b(v_z',t) \Big] \quad (19)$$

where $W(v_z,v_z') = \int d\vec{v}_\perp d\vec{v}_\perp' W(\vec{v}_\perp)W(\vec{v},\vec{v}') \quad (20)$

Immediately after the two pulse sequence which builds it, the perturbation $\delta n_b(v_z,t)$ in level b population exhibits the following shape :

$$\delta n_b(v_z,t_2) = \delta n_b(0,t_2)\frac{W(v_z)}{W(0)}\cos kv_z t_{12} \quad (21)$$

With this boundary value condition, eq.20 is solved under the assumption that the modulation period λ/t_{12} is much smaller than the active atom thermal velocity. One obtains :

$$\delta n_b(v_z,t) = \delta n_b(v_z,t_2)\exp-\Big[\gamma_b + \gamma_b(t_{12},v_z)\Big](t-t_2) \quad (22)$$

where $\gamma_b(t_{12},v_z) = \gamma_b^t(v_z) - \int dv_z' W(v_z',v_z)\cos k(v_z'-v_z)t_{12} \quad (23)$

and where $\gamma_b^t(v_z) = \int dv_z' W(v_z, v_z')$ (24)

These expressions show that :

i) collisions with velocity changes $|v_z - v_z'| << (kt_{12})^{-1}$ do not contribute to the *decay rate* $\gamma_b(t_{12})$ since, as $\cos k(v_z - v_z')t_{12} \simeq 1$, departure and restitution terms balance each other ;

ii) collisions with velocity changes $|v_z - v_z'| >> (kt_{12})^{-1}$ fully contribute to the decay rate since then their contribution to the *restitution term* averages to zero.

The v_z dependence of $\gamma_b^t(v_z)$ reflects the anisotropy of the relative velocity distribution. This anisotropy occurs because a definite value is assigned to the active atom velocity axial component. The collision kernel may be regarded as a function of $v_z' - v_z$ and of v_z. As far as small velocity changes are considered, the origin of the v_z dependence at fixed $(v_z' - v_z)$ is essentially the same as in $\gamma_b^t(v_z)$. In the light perturber limit, the relative velocity is close to the perturber velocity and its distribution function is nearly isotropic. Thus the v_z dependence in $\gamma_b^t(v_z)$ and the v_z dependence at fixed $(v_z - v_z')$ in $W(v_z, v_z')$ may be neglected. As a consequence $\gamma_b(t_{12}, v_z)$ no longer depends on v_z :

$$\gamma_b(t_{12}) = \gamma_b(t_{12}, 0) = \gamma_b(t_{12}, v_z) \qquad (25)$$

In the forthcoming discussion this approximation is assumed to be valid.

According to eq.22, the decay rate of the periodic structure in level b velocity distribution is added to the natural decay rate γ_b in eq.14. According to eq.25 the decay factor is taken out of the summation over velocity classes which occurs in the signal calculation (cf. eq.9) and eq.15 is eventually obtained.

The quantity $\gamma_b(t_{12}) - \gamma_b^t$ is the Fourier transform of the collision kernel which could be determined by inverse Fourier transform of experimental data. Existing experiments have not yet achieved this program. Available data are compared with calculated expressions starting with simple interaction potential models. Collision dynamics are analyzed in the frame of a semiclassical picture.

V. Collisional Relaxation in SPE Experiments

V.1. Population elastic scattering

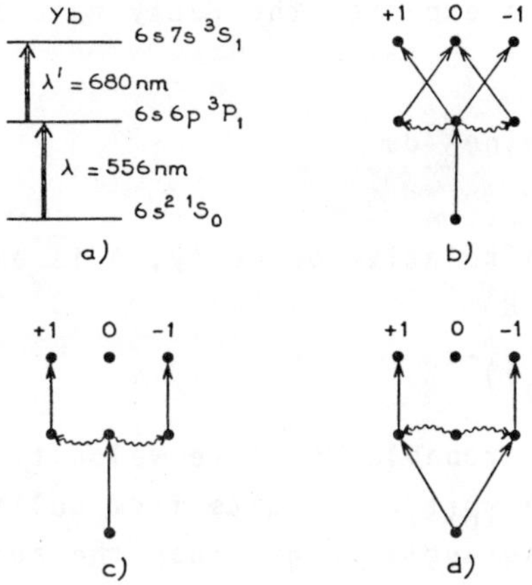

Figure 2 : a) three level scheme in Yb ; b),d) crossed polarization excitation ; c) parallel polarization excitation.

An example is provided by an experiment performed in Yb vapor using levels $6s^{2\,1}S_0$, $6s6p\,^3P_1$ and $6s7s\,^3S_1$ as levels a, b, c respectively (fig.2a) [17]. Echo is detected in the same linear polarization as the third laser pulse. The first two beams have the same polarization which can be set parallel or perpendicular to that of the third pulse. In the crossed polarization configuration (fig.2b) the SPE signal is built up from the modulation stored in the total population of the $5s6p\,^3P_1$ level (γ_b^{-1} = 875 ns). Experimental values of $d\gamma_b(t_{12})/dp$ are obtained for different perturbers (He, Ne, Xe) with

t_{12} ranging from 13.6 ns to 77.6 ns. Comparison with calculated expressions of $d\gamma_b(t_{12})/dp$ is achieved under the assumption of Van der Waals interaction. The corresponding scattering cross section is inserted in eq.16 in order to calculate the collision kernel. Then the decay rate $d\gamma_b(t_{12})/dp$ is derived from eq.23. This final expression depends on the single parameter σ_b which is the total scattering cross section in level b. Thus this quantity can be deduced from a fit of the calculated decay rate to the experimental data.

In the region that we explore, the decay rate takes on the simple form [17]

$$\gamma_b(t_{12}) = AN\bar{v}_r \int_{\theta_{12}}^{\pi} 2\pi \sin\theta \frac{d\sigma}{d\Omega} d\theta \qquad (26)$$

where $\bar{v}_r$ is the mean relative velocity, A is a constant which is close to unity and :

$$\theta_{12} = (k\bar{u}t_{12}m_p/(m+m_p))^{-1} \qquad (27)$$

where $\bar{u}$ is the most probable relative velocity. According to eq.26 the decay rate $\gamma_b(t_{12})$ results from collisions which occur at a scattering angle larger than the adjustable parameter θ_{12}. This equation demonstrates the information which may be gained on the elastic scattering angular dependence. As t_{12} increased, $\gamma_b(t_{12})$ tends asymptotically to the total collision rate γ_b^t. This is illustrated on fig.3 in the light perturber limit for a Van der Waals interaction. Our experimental data for Yb*-He collisions are plotted on the curve.

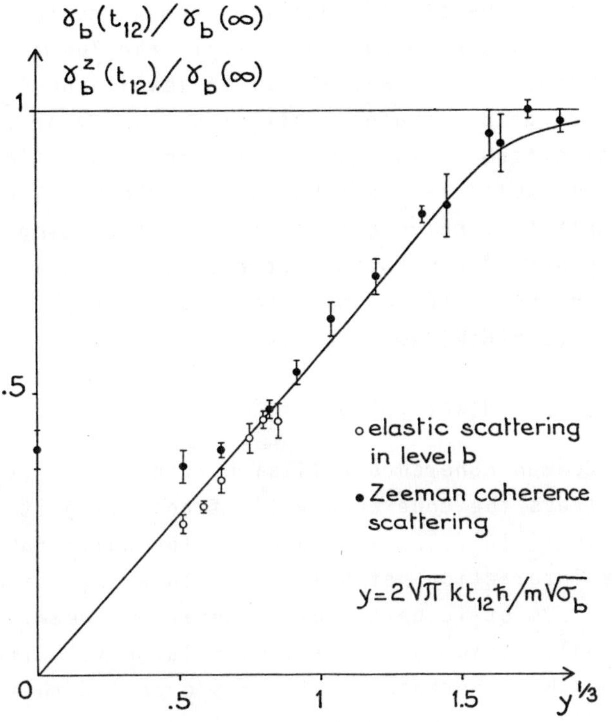

Figure 3 : decay rate of the periodic structure stored in the population ($\gamma_b(t_{12})$) or in the Zeeman coherence ($\gamma_b^z(t_{12})$) of Yb($6s6p\,^3P_1$) due to collisions with He. The full line corresponds to elastic scattering in a Van der Waals interaction potential. No fitting parameter is used.

V.2. Scattering of Zeeman coherences

In the previous example, the first two pulses prepare the atoms in a definite sublevel of level b. Instead atoms may be prepared in a Zeeman superposition state. This is accomplished using two crossed-polarized pulses. The experiment is performed on the $6s^2\,^1S_0$-$6s6p\,^3P_1$ transition of Yb [18]. The probe pulse is tuned to the same transition. It is polarized parallel to pulse 1 and the echo is detected along the pulse

2 polarization. The atomic quantity from which the third pulse builds a radiating dipole density is the Zeeman coherence between sublevels m = 0 and m = ±1 in level $6s6p\,^3P_1$. This quantity undergoes a twofold collisional decay process during its evolution from t_2 to t_3. Firstly, the modulation which is stored in its velocity distribution is deleted. Secondly, internal substrate coherence is destroyed. Accordingly the decay rate of the periodic structure stored in the level b velocity distribution should be changed from (eqs.23,25) :

$$\gamma_b(t_{12}) = \gamma_b^t(0) - \int d\zeta W(\zeta,0)\cos k\zeta t_{12} \qquad (28)$$

to :

$$\gamma_b^Z(t_{12}) = \gamma_b^t(0) - \int d\zeta W^Z(\zeta,0)\cos k\zeta t_{12} \qquad (29)$$

where the Zeeman coherence collision kernel $W^Z(v_z', v_z)$ partially destroys the coherence while transfering it from velocity class v_z' to velocity class v_z. The anisotropy of the long range interaction potential between a $^3P_1$ state and a spherically symmetric perturber is generally weak. As a consequence, only close encounters, with large velocity changes, can destroy the coherence. It is thus expected that :

$$\begin{aligned} W^Z(\zeta,0) &\simeq W(\zeta,0) \quad \text{when } \zeta < \zeta^Z \\ W^Z(\zeta,0) &< W(\zeta,0) \quad \text{when } \zeta > \zeta^Z \end{aligned} \qquad (30)$$

where ζ^Z is the velocity change at the internuclear distance where electronic orbitals begin to overlap. Zeeman coherences are elastically (inelastically) scattered when velocity changes are smaller (larger) than ζ^Z. Substitution of eq.30 into eq.29 leads to :

$$\begin{aligned} \gamma_b^Z(t_{12}) &\simeq \gamma_b(t_{12}) \text{ when } k\zeta^Z t_{12} > 1 \\ \gamma_b^Z(t_{12}) &> \gamma_b(t_{12}) \text{ when } k\zeta^Z t_{12} < 1 \end{aligned} \qquad (31)$$

Experimental values of $\gamma_b^Z(t_{12})$ and $\gamma_b(t_{12})$ behave as expected. This is illustrated on fig.3 where we have plotted the data for both elastic scattering and Zeeman coherence scattering in Yb*-He collisions.

In the experiment on Zeeman coherences, the collisional depolarizing process is analyzed through the decay of an atomic quantity. Inelastic processes are studied with better sensitivity when one detects a collision-induced signal instead of observing the gradual collisional obliteration of the signal.

V.3. Inelastic processes

Now the intermediate state b is assumed to be made up of two substates b_1 and b_2 with energy separation ε. The first two laser pulses are tuned to resonance with the transition a-b_1 and level b_2 is unpopulated at time t_2. Thus a periodic structure is stored in the velocity distribution of sublevel b_1 and some collisional transfer of this structure from b_1 to b_2 is expected. A quasi resonance condition is laid down which requires that the speed change $\varepsilon/m\bar{u}$ caused by a transfer between substates is much smaller than the modulation period λ/t_{12}. The transfer from substate b_1 and velocity class v_z' to substate b_2 and velocity class v_z is described by its density probability per unit time $W^{in}(v_z',v_z)$. A straightforward extension of eq.22 leads to :

$$\begin{cases} \frac{\partial}{\partial t} n_{b_1}(v_z,t) = -(\gamma_b+\gamma_b(t_{12}))\ \delta n_{b_1}(v_z,t)+\hat{W}^{in}(kt_{12})\delta n_{b_2}(v_z,t) \\ \frac{\partial}{\partial t} n_{b_2}(v_z,t) = -(\gamma_b+\gamma_b(t_{12}))\ \delta n_{b_2}(v_z,t)+\hat{W}^{in}(kt_{12})\delta n_{b_1}(v_z,t) \end{cases} \tag{32}$$

where $\hat{W}^{in}(kt_{12}) = \int d\zeta W^{in}(\zeta,0)\cos k\zeta t_{12}$ (33)

Solving eq.32 with initial condition $n_{b_2}(v_z,t_2) = 0$ one obtains :

$$\frac{\delta n_{b_2}(v_z,t)}{\delta n_{b_1}(v_z,t)} = \mathrm{th}(\hat{W}^{in}(kt_{12})(t-t_2)) \tag{34}$$

Since the signal intensity is proportional to the squared atomic density, the ratio of the signal intensity obtained by probing on resonance with transition b_2-c to that obtained by probing on resonance with transition b_1-c is :

$$I_2/I_1 = th^2(\hat{W}^{in}(kt_{12})t_{23}) \qquad (35)$$

The quantity $\hat{W}^{in}(kt_{12})$ which is derived from measurement of I_2/I_1 may be regarded as the rate for collisional transfer between substates at a scattering angle smaller than $\theta_{12} = (k\bar{u}t_{12}m_p/(m+m_p))^{-1}$.

Depolarizing collisions in Yb vapor offer an example for the determination of $\hat{W}^{in}(kt_{12})$ [17]. Now $\hat{W}^{in}(kt_{12})$ stands for the collisional transfer rate between m = 0 and m = ±1 substates in $6s6p\,^3P_1$ at scattering angle smaller than θ_{12}.

In section VI the signal intensity $I_\perp$ is recorded in the case when the third pulse and the echo are crossed-polarized with the first two pulses. Then the third pulse probes the substates initially excited by the first two pulses (fig. 2b,d). In the parallel-polarization case (fig.2c), the signal intensity is denoted $I_{/\!/}$ and the third pulse probes the only atoms which have undergone a collisional transition between Zeeman substates . Therefore, the signal intensity ratio $I_{/\!/}/I_\perp$ is connected with the transfer rate $\hat{W}^{in}(kt_{12})$. The corresponding expression is [17]

$$\frac{I_{/\!/}}{I_\perp} = \left[\frac{2-2\exp-[3\hat{W}^{in}(kt_{12})t_{23}]}{2+\exp-[3\hat{W}^{in}(kt_{12})t_{23}]}\right]^2 \qquad (36)$$

which is the three-substate equivalent to eq.35. The transfer rate $\hat{W}^{in}(kt_{12})$ is derived from the measurement of $I_{/\!/}/I_\perp$ as a function of the foreign gas pressure for different t_{12} values. These data are complemented by the determination of the total collisional transfer rate $\gamma_b^{in} = \int W^{in}(\zeta,0)\,d\zeta = \hat{W}^{in}(0)$. For that purpose we produce a conventional photon echo signal on the upper transition (fig.4). An initial pulse, tuned on resonance with a-b transition irradiates the sample at time t_1. Its polarization is selected in order to populate either m = 0 or m = ±1 substates in b. At a delay T after t_1 a sequence of two pulses, tuned on resonance with b-c transition is delivered in order to build a regular pho-

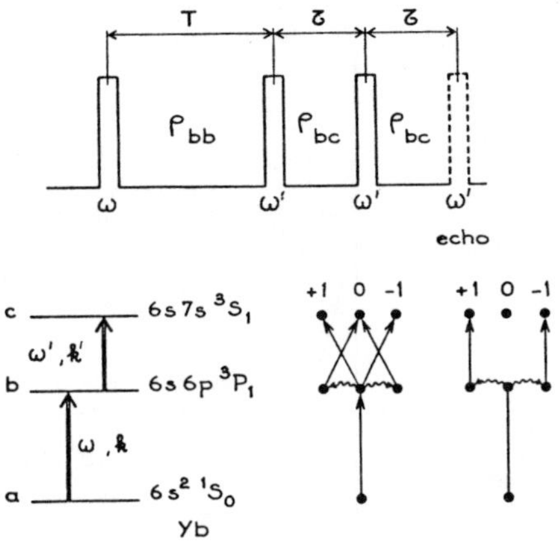

Figure 4 : conventional photon echo on the upper transition of Yb three level system.

ton echo on this transition. The polarization of these pulses is selected in order to only probe the m = ±1 substates in level b. The rate $W^{in}(0)$ is simply deduced from the signal intensity ratio measurement

$$\frac{I_{//}}{I_{\perp}} = \left[\frac{2-2\exp-3\hat{W}^{in}(0)t_{12}}{2+\exp-3\hat{W}^{in}(0)t_{12}}\right]^2 \tag{37}$$

For a potential with long range form $V(r) = -C/r^s$, the quantity $R(x) = \hat{W}^{in}(kt_{12})/\hat{W}^{in}(0)$ is defined as a function of

$$x = (\gamma_b^t/\gamma_b^{in})^{s/2}\hbar k t_{12}/m\sqrt{\sigma_b} \tag{38}$$

A standard model calculations of R(x) [17] leads to an expression which is independent of the potential anisotropy and, to a large extend, of the particle masses. The corresponding curve is drawn on fig.5 for s = 6. Experimental data for Yb*-Ne are plotted on the same diagram (fig.5). Horizontal error bars result from uncertainties in the experimental determination of the total rates γ_b^t and γ_b^{in}. Satisfactory agreement is obtained between experiment and calculation.

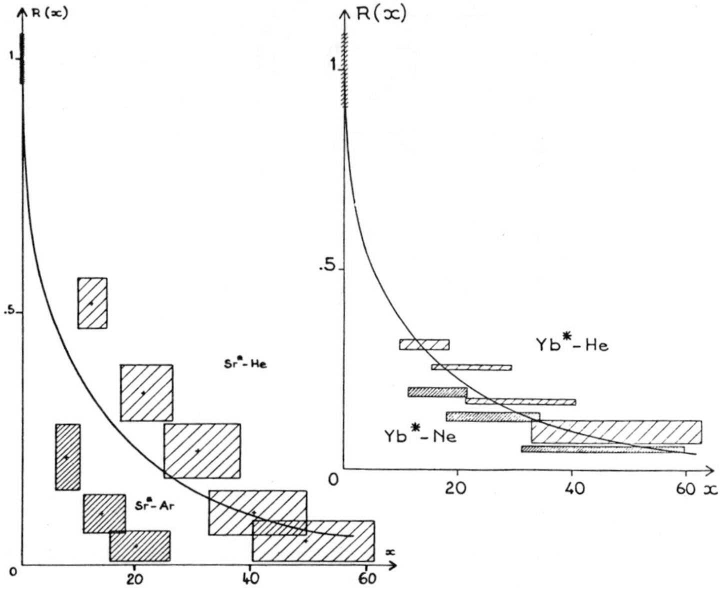

Figure 5 : collisional transition between Zeeman sublevels in $Yb(6s6p\,^3P_1)$ and $Sr(5s5p\,^3P_1)$.

Transfer between Zeeman sublevels in $Sr(5s5p\,^3P_1)$ is studied in the same way [19]. Experimental data are compatible with calculation for Sr*-He collisions ; this is no longer the case for Sr*-Ar collisions. This discrepancy suggests that the potential curve may depart from the simple $1/r^6$ form.

All the experiments which are discussed in section V are performed with laser pulses in the nanosecond range. The pulse duration τ_L is always larger than Ω_D^{-1}. In opposition to our initial assumption in section III, these pulses are unable to coherently excite the velocity distribution over its full width. Carving a modulation inside a coherently excited velocity domain then requires that the modulation period λ/t_{12} is much smaller than coherently excited domain width λ/τ_L. The modulation visibility conditions is $\tau_L \ll t_{12}$. Should it be concluded that nanosecond pulses can only carve a short-period modulation inside the velocity distribution ? This point is questioned in the next section.

VI. Coherent Transients after Incoherent Excitation

VI.1. The temporal coherence requirement

In the previous sections the excitation pulses fullfil both conditions of spatial and of temporal coherences. The requirement of temporal coherence is now questioned.

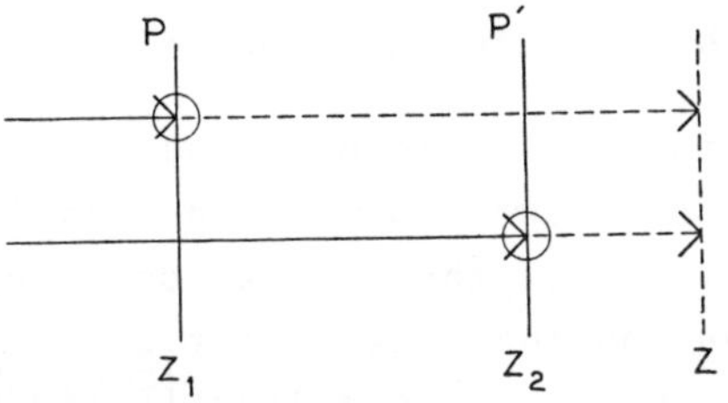

Figure 6 : echo signal build up.

Let us examine the echo electric field which arrives at position z at time t (fig.6). This field is a plane wave which includes the contributions emitted by the different points of the sample throughout its depth. For instance, the field at (z,t) includes contributions

- emitted at time $t_1 = t-(z-z_1)/c$ by the sample slice located at z_1 ;
- emitted at time $t_2 = t-(z-z_2)/c$ by the sample slice located at z_2.

Then the external field wave front which excited atoms at (z_1,t_1) reaches position z_2 at time t_2. Thus the two contributions to the echo field originate from excitations by the *same* wavefront at *different* positions z_1 and z_2. Thus the additive interference of contributions from all the sample slices does not require the coherence of external field wavefronts located at different positions at a given time. In other words the signal build up does not require the temporal coherence of the excitation pulses. This contrasts with other phenomena such as holography where making different wavefronts interfere demands that the wavetrain length is large. It should be noted that the expression "coherent transient" does not refer to the coherence of the field. It rather means that the signal is emitted by atomic quantities - the nondiagonal density matrix elements or "atomic coherences" - which are excited in phase with the external light field.

In the previous sections limitations arise from the duration τ_L of the temporally coherent light pulses. As indicated in section V, the visibility of the modulation stored in the level velocity distribution requires that $t_{12} >> \tau_L$. More generally the accuracy on the pulse delays is limited by the pulse duration τ_L. This prevents from studying relaxation processes with decay rates larger than τ_L^{-1}. The degree of temporal coherence of a field is characterized by the coherence time τ_c. A single characteristic time is relevant for a coherent pulse since then $\tau_c \sim \tau_L$. On the contrary two time constants occur for incoherent light excitation since then $\tau_c << \tau_L$ and the question arises : which one of these two characteristic times determines the time resolution in a coherent transient experiment ? This point is examined on the specific example of stimulated photon echo.

VI.2. Excitation by two pulses of incoherent light

We examine the perturbation which is caused in the level population difference of a two level system by a sequence of two pulses of incoherent light ($\tau_c << \tau_L$). The pulses are mutually coherent. This means that the second pulse is the delayed replica of the first one. Calculation is performed in the small pulse area limit and all relaxation processes are ignored.

We start with eq.4 where both χ and φ are now time varying quantities. Since a small angle is allowed between the pulse wave vectors $\vec{k}_1$ and $\vec{k}_2$, χ is given by : $\chi = \chi_1 \exp i\vec{K}\vec{r}/2 + \chi_2 \exp -i\vec{K}\vec{r}/2$ where $\vec{K} = \vec{k}_2 - \vec{k}_1$. The complex Rabi frequency $\underline{\chi}_i$ is defined by : $\underline{\chi}_i = \chi_i \exp i\varphi_i$. To second order in the field strength the perturbation caused by combined action of the first two pulses is :

$$n_{ab}^{(2)}(\vec{r},\vec{v}) = -\frac{1}{2}\int_{-\infty}^{+\infty} dt' \int_{-\infty}^{+\infty} d\tau \exp[i\vec{k}\vec{v}\tau - i\vec{K}\vec{r}] \times \times \underline{\chi}_1^*(t')\underline{\chi}_2(t'-\tau) n_o(\vec{v}) + c.c. \tag{40}$$

where $\vec{k} = (\vec{k}_1 + \vec{k}_2)/2$. According to their mutual coherence the two fields are connected by :

$$\underline{\chi}_2(t) = \underline{\chi}_1(t - t_{12}) \tag{41}$$

Taking advantage of this property we obtain :

$$n_{ab}^{(2)}(\vec{r},\vec{v}) = -n_o(\vec{v}) \cos[\vec{k}\vec{v}t_{12} + \vec{K}\vec{r}] S(\vec{k}\vec{v}) \tag{42}$$

where

$$S(\vec{k}\vec{v}) = \int d\tau \exp -i\vec{k}\vec{v}\tau \int dt' \underline{\chi}_1(t')\underline{\chi}_1^*(t'-\tau) \tag{43}$$

Since $\tau_c \ll \tau_L$ the integral over t' identifies with the autocorrelation function of the quasistationnary function $\underline{\chi}_1(t)$. According to the Wiener-Khinchine theorem, its Fourier transform $S(\vec{k}\vec{v})$ equals the power spectrum of $\underline{\chi}_1(t)$.

The perturbation in the population difference, as expressed by eq.42, exhibits a grating modulation in both position and velocity spaces. The diffusion decay of the spatial grating is ignored. Three characteristic times appear in this equation. These are the inverse width of $S(\Delta)$, τ_c, the pulse delay t_{12} and the inverse Doppler width Ω_D^{-1}. As long as $\tau_c \ll \Omega_D^{-1}$ the modulation in velocity space may spread over the entire velocity distribution and the modulation is visible as long as its period λ/t_{12} does not exceed the width Ω_D/k of the velocity distribution. The shape of the perturbation depicted by eq.42 is identical to that of the structure which should be carved in the velocity distribution by coherent pulses with duration τ_c. Thus the pulse duration no longer sets any limitation on t_{12} values and the time resolution for the periodic structure build up is τ_c.

The crucial condition of mutual coherence between the light pulses underlies eqs.42,43. When this condition is fulfilled the pulses may be regarded as trains of contiguous elementary pulses with duration τ_c. Each elementary component inside pulse 1 coherently excites the sample over a spectral range τ_c^{-1}. Its replica inside pulse 2 irradiates the sample with the delay t_{12}. The perturbation in level population difference results from the addition of contributions from all the couples of elementary pulses. The atomic coherences which are implied in the perturbation build up, evolve freely during the time separation t_{12} between elementary pulses and this interval is known with an accuracy τ_c. When relaxation destroys the coherences during the pulse interval at a rate γ_{ab}, the decay factor which multiplies $n_{ab}^{(2)}(\vec{r},\vec{v})$ is $\exp-\gamma_{ab}t_{12}$ as long as $t_{12} >> \tau_c$, independently of the pulse duration. Some experiments, with long duration pulses, make use of that property to study fast coherence decays under conditions where $t_{12} << \tau_L$ [20,21]. The collisional relaxation of the modulation stored in the velocity distribution could be probed when $\tau_c << t_{12} < \tau_L$ in a way similar to that outlined in section V. In the next paragraph we propose another application of SPE method to the study of gas phase dynamics.

VI.3. Stimulated photon echoes with incoherent light

In the following we assume that $\tau_c << \Omega_D^{-1} << \tau_L$. This condition is easily fulfilled by light pulses in the nanosecond range. The sample is illuminated by a third pulse with wavevector $\vec{k}_3$ after the extinction of the first two pulses. The following coherence term is originated :

$$\rho_{ab}^{(3)}(\vec{r},\vec{v},t) = \frac{i}{2}\int_{-\infty}^{t} dt'\chi_3(t')\exp(-i\vec{k}_3\vec{r}+i\vec{k}_3\vec{v}(t-t'))n_{ab}^{(2)}(\vec{r},\vec{v}) \quad (44)$$

A stimulated photon echo is radiated by this coherence. Its field amplitude $E(\vec{r},t)$ is proportional to $\int d\vec{v}\rho_{ab}^{(3)}(\vec{r},\vec{v},t)$. Thus, substituting eq.43 into eq.44 one obtains :

$$E(\vec{r},t) \propto \int_{0}^{\infty} d\tau\chi_3(t-\tau)\Big[\hat{n}_o(k(\tau-t_{12}))\exp-i\vec{k}_e\vec{r}+\hat{n}_o(k(\tau+t_{12}))\times \times\exp-i\vec{k}'_e\vec{r}\Big]S(0) \quad (45)$$

where $\hat{n}_o(k(\tau\pm t_{12})) = \int d^3vn_o(\vec{v})\exp[ik(\tau\pm t_{12})v_z]$ and $\vec{k}_e = \vec{k}_3+\vec{k}_2-\vec{k}_1$, $\vec{k}'_e = \vec{k}_3+\vec{k}_1-\vec{k}_2$. The z axis is taken along $\vec{k}_3$ and we assume that $K\Omega_D t_{12} << k$. This guarantees that no transverse Doppler dephasing can occur. The quantity $S(\Delta)$ is replaced by $S(0)$ since $\tau_c << \Omega_D^{-1}$. Two echoes are emitted in directions $\vec{k}_e$ and $\vec{k}'_e$. The energy radiated in both direction is given by :

$$W_{\vec{k}_e} \propto \int_{-\infty}^{\infty}dt\int_{0}^{\infty}d\tau\int_{0}^{\infty}d\tau'\chi_3(t-\tau)\chi_3^*(t-\tau')\hat{n}_o(k(\tau-t_{12}))\hat{n}_o^*(k(\tau'-t_{12})) \quad (46)$$

$$W_{\vec{k}'_e} \propto \int_{-\infty}^{\infty}dt\int_{0}^{\infty}d\tau\int_{0}^{\infty}d\tau'\chi_3(t-\tau)\chi_3^*(t-\tau')\hat{n}_o(k(\tau+t_{12}))\hat{n}_o^*(k(\tau'+t_{12}))$$

These expressions involve the autocorrelation function of $\chi_3(t)$. Owing to the smallness of the width τ_c of this function with respect to that of $\hat{n}_o(k\tau)$ one finally obtains :

$$W_{\vec{k}_e} \propto \int_{-t_{12}}^{\infty} d\tau|\hat{n}_o(k\tau)|^2$$
$$W_{\vec{k}'_e} \propto \int_{t_{12}}^{\infty} d\tau|\hat{n}_o(k\tau)|^2 \quad (47)$$

Derivation of eq.47 with respect to t_{12} leads to

$$\frac{d}{dt_{12}}W_{\vec{k}_e} \propto |\hat{n}_{ab}^{(o)}(-kt_{12})|^2$$
$$\frac{d}{dt_{12}}W_{k'_e} \propto -|\hat{n}_{ab}^{(o)}(kt_{12})|^2 \quad (48)$$

The rate of variation of the echo signal as a function of t_{12} identifies with the square of the Fourier transform of the velocity distribution. Thus, measurement of the echo signal intensity as a function of t_{12} offers a new way for the fast sampling of atomic and molecular velocity distribution. We have obtained preliminary results [22] in Cs vapor at normal equilibrium (fig.7). The SPE signal is generated on the

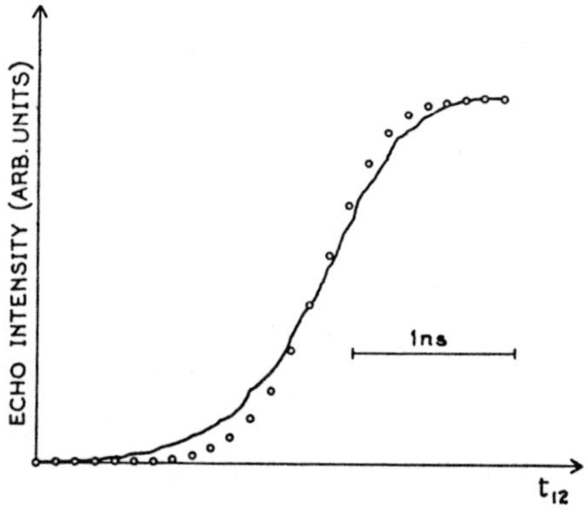

Figure 7 : stimulated photon echo in direction k_e in Cs vapor. Circles correspond to the calculated signal intensity assuming normal equilibrium.

$6s\,^2S_{1/2}$-$7p\,^2P_{1/2}$ transition and the pulse time constant are τ_C = 200 ps, τ_L = 7 ns. The experiment feasibility is demonstrated but the technique will acquire its optimum efficiency when the signal quality enables us to directly derive the square root of the Fourier transform of $[dW_{k_e}(t_{12})/dt_{12}]$. The proposed technique should be an efficient tool to analyze the translational motion of the nascent products issuing from photodissociation and photofragmentation processes. This technique is the time domain analog of the Doppler spectroscopy method which has already been used in the frequency domain to determine fragment velocity distribution [23,24].

VII. Conclusion

The atomic quantities involved in the photon echo build up have been specified. Their collisional relaxation have been described. Attention is focussed on the laser carving of a periodic structure inside the atomic velocity distribution. The destruction of this structure by elastic scattering is monitored. Its transfer from level to level by inelastic collisions is tracked. In this way some information is derived on differential scattering cross sections. However, this way of collisional process exploration is still in its infancy, especially as far as inelastic processes are concerned. Only depolarizing collisions have given rise to some detailed investigations. In present experiments, a stringent limitation is imposed by the largeness of the pulse separation t_{12}. Due to the resulting smallness of its period λ/t_{12}, the structure stored in the atomic velocity distribution is too fragile and does not survive to large scattering angle collisions. Thus its transfer by short range inelastic processes can not be tracked. Use of incoherent light pulses should permit to reduce t_{12} dramatically and to store less fragile structures in the velocity distribution. Besides, stimulated photon echo with incoherent light is shown to offer a new way for fast sampling of atomic velocity distribution.

* Permanent address : Département de Physique, Faculté des Sciences, Tunis, Tunisie.

References

1. Nienhuis, G. : this book.
2. Gel'mukhanov, F.Kh. : this book.
3. McCall, S.L., Hahn, E.L. : Phys. Rev. 183, 457 (1969).
4. Miles, R.B., Harris, S.E. : IEEE J. QE-9, 470 (1973).
5. MacGillivray, J.C., Feld, M.S. : Phys. Rev. A 14, 1169 (1976).

6. Berman, P.R. : Phys. Rev. A 5, 927 (1972).
7. Brewer, R.G. : Frontiers in laser spectroscopy, Les Houches, Balian et al. ed., p. 342, North Holland Publishing Company, 1977.
8. Jackson, J.D. : Classical Electrodynamics, Second Edition, John Wiley 1962.
9. Kurnit, N.A., Abella, I.D., Hartmann, S.R. : Phys. Rev. Lett. 13, 567 (1964) ; Phys. Rev. 141, 391 (1966).
10. Gordon, J.P., Wang, C.H., Patel, C.K.N., Slusher, R.E., Tomlinson, W.J. : Phys. Rev. 179, 294 (1969).
11. Mossberg, T., Flusberg, A., Kachru, R., Hartmann, S.R. : Phys. Rev. Lett. 42, 1665 (1979).
12. Mossberg, T.W., Kachru, R., Hartmann, S.R., Flusberg, A. M. : Phys. Rev. A 20, 1976 (1979).
13. Fujita, M., Nakatsuka, H., Nakanishi, H., Matsuoka, M. : Phys. Rev. Lett. 42, 974 (1979).
14. Kachru, R., Chen, T.J., Hartmann, S.R., Berman, P.R. : Phys. Rev. Lett. 47, 902 (1981).
15. Berman, P.R., Mossberg, T.W., Hartmann, S.R. : Phys. Rev. A 25, 2550 (1982).
16. Résibois, P., de Leener, M. : Classical Kinetic Theory of Fluids (Wiley - New York 1977).
17. Keller, J.-C., Le Gouët, J.-L. : Phys. Rev. Lett. 52, 2034 (1984) ; Phys. Rev. A 32, 1624 (1985).
18. Yodh, A.G., Golub, J., Mossberg, T.W. : Phys. Rev. A 32, 844 (1985).
19. Saïdi, M., Keller, J.-C., Le Gouët, J.-L., to be published.
20. Asaka, S., Nakatsuka, H., Fujiwara, M., Matsuoka, M. : Phys. Rev. A 29, 2286 (1984).
21. Nakatsuka, H., Tomita, M., Fujiwara, M., Asaka, S. : Opt. Commun. 52, 150 (1984).
22. Defour, M., Keller, J.-C., Le Gouët, J.-L., to be published in J.O.S.A. (B) march 1986.
23. Schmiedl, R., Dugan, H., Meier, W., Welge, K.H. : Z. Phys. A 304, 137 (1982).
24. Gerber, G., Möller, R. : Phys. Rev. Lett. 55, 814 (1985).

NEW TOPICS IN FIELD ASSISTED COLLISIONS. PHOTON CORRELATION EFFECTS

F. Trombetta[+], R. Daniele[+], F. Morales[°] and G. Ferrante[+]

+ Istituto di Fisica dell'Università, via Archirafi 36, 90123 Palermo, Italy.

° Dipartimento di Energetica e di Applicazioni della Fisica, viale delle Scienze, 90128 Palermo, Italy.

1. INTRODUCTION

The last few years have witnessed the observation of new atomic multiphoton phenomena, made possible by the availability in different research laboratories of very intense laser fields[1-5]. As an instance, Rodhes[2] has reported on studies of atomic multiphoton absorption at ultraviolet wavelengths, exhibiting anomalous behaviour in terms of energy transfer. These studies were done with the 193 nm ArF^{*} laser focussed to generate intensities in the range of 10^{15} — 10^{17} W/cm^2 in the experimental volume. At these high intensities, the theoretical description of the new observed phenomena may be hardly based on standard perturbation theory for the laser-atom interaction or on the model of a laser as a purely coherent field (homogeneous, single-mode, parameter-stabilized field). So it is generally accepted that a proper theoretical description of the observed new phenomena must go beyond the (appropriate lowest order of) perturbation theory and include, through different models, as much as possible of the information of intense real lasers (multimode operation, fluctuations of the parameters, pulsed regime, focussing, and so on).

The above considerations apply to the multiphoton free-free transitions (MFFT) as well[6,7], with, however, two bonuses. First, highly non linear behaviour is observed in MFFT at relatively modest intensities[8] (about 10^9 W/cm^2 with a CO_2 laser). This because in a continuum-continuum transition the scattered electrons move under the action of two fields, with the static scattering field playing the role of a weak perturbation. Second, for MFFT a treatment exact in the particle-laser interaction is available, making the theoretical study of multiphoton phenomena in this elementary process immediately possible. In our opinion, are just these features which make interesting and useful a detailed study of the several aspects of MFFT.

This Lecture deals with the study of photon correlation effects in multiphoton non-resonant free-free transitions. Considering that in this elementary process the particle-laser interaction is treated exactly (within the classical approximation for the field), the present material is intended to serve three scopes. First, it is expected that the results reported below be representative of the photon correlation effects in non-resonant particle-atom scattering in the presence of strong fluctuating laser fields, when field dressed scattering states are used. Second, it is expected that the information reported below be useful in general also for non-resonant multiphoton laser-atom processes. As a matter of fact, as it will be shown below, in the limit of weak particle-field coupling, contact between MFFT and other multiphoton processes is readily established. Third, it is hoped that the consideration of models of laser fluctuations in the framework of an exact treatment of particle-field interaction may prove useful to assess the validity of these models and of the underlying approximations in high-intensity domains.

Investigations on photon correlation effects in field-assisted collisions are at their beginning, and the available literature is represented by ref.s 8, 9, 10, 19, 21.

Below we will be specifically concerned with the theory of a charged particle scattered by a finite-range, static potential $V(\underset{\sim}{r})$ in the presence of a strong radiation field. The radiation field will be supposed to undergo stochastic fluctuations of some of their parameters.

In Sec. 2 we describe two statistical models of the radiation field and in Sec. 3 the theory of the potential scattering in the presence of a laser is briefly reviewed; in Sec. 4 we apply it, in the Born approximation, to the case of a fluctuating field, within the models discussed in the Sec. 2, and give some limiting expressions together with corrections to the results of the lowest order perturbation theory. Sec. 5 is devoted to selected numerical calculations of the exact formulas. Sec. 6 contains a number of final remarks and indication of the lines of possible future developments.

2. THE FIELD MODELS

In this Section we review two laser statistical models which, under different assumptions, account for field fluctuations. Here, we shall be mainly concerned with the Chaotic Field Model and the Phase Diffusion Model. The first describes a multimode laser as the limit of an infinite number of uncorrelated modes; the second describes a single-mode laser, with stabilized electric field amplitude and the phase as a random quantity.

Of course, other models may be profitably developed, each of them accounting for some particular aspects of fluctuations of the field parameters. For instance, the Random Telegraph Signal Model[11] is likely to exhibit flexibility to simulate the field fluctuations in field-assisted collision problems. Similarly, the Gaussian Amplitude Model too is expected to be useful. In the latter, only the field amplitude is fluctuating, the phase being stabilized. Compared with other models, this model is useful to understand the separate effects of phase and amplitude fluctuations[12].

Besides, for comparison purposes, we shall refer frequently to the results obtained within the Purely Coherent Field Model (the idealized model of a field with well-stabilized parameters).

In what follows we assume a classical electric field, taken in dipole approximation and written as

$$\underset{\sim}{E}(t) = \hat{e}_L\, 2^{-1} \left[\bar{\mathcal{E}}(t)\, e^{-i\omega t} + c.c. \right] \tag{2.1}$$

where $\hat{e}_L$ is the linear polarization unit vector and ω the spectrum center frequency. The complex amplitude $\bar{\mathcal{E}}(t)$ can further be written as

$$\bar{\mathcal{E}}(t) = \mathcal{E}(t)\, \exp\{-i\,\varphi(t)\} \tag{2.2}$$

$\mathcal{E}(t)$ and $\varphi(t)$ being respectively the field real amplitude and phase.

(a) The Chaotic Field Model

In the Chaotic Field Model (CFM), $\bar{\mathcal{E}}(t)$ has a gaussian statistics, arising as a multimode laser statistics in the limit of an infinite number of uncorrelated modes. Usually, $\bar{\mathcal{E}}(t)$ is further assumed to be a unidimensional projection of an n-dimensional complex Markov process, in order to be able to use some powerful technics which permit the computation of a large class of averages of stochastic quantities[13,14].

This means to assume that $\bar{\mathcal{E}}(t)$ obeys the Langevin equation

$$\frac{d^m \bar{\mathcal{E}}}{dt^m} = -\sum_{K=1}^{m} b_K \frac{d^{K-1}\bar{\mathcal{E}}}{dt^{K-1}} + F(t) \tag{2.3}$$

where the random force $F(t)$ is gaussian, zero-mean and white spectrum, i.e.

$$\langle F(t)\rangle = 0 \quad ; \quad \langle F(t)\, F^*(t)\rangle = 2\,\Gamma\,\delta(t-t') \tag{2.4}$$

Eq.s (2.3) and (2.4) guarantee that the n-dimensional process

$$\left\{ \bar{\mathcal{E}}(t), \frac{d\bar{\mathcal{E}}}{dt}, \ldots\ldots, \frac{d^{n-1}\bar{\mathcal{E}}}{dt^{n-1}} \right\}$$

is markovian, so that one can write for its conditional probability a Kramers-like equation. Besides, as $\bar{\mathcal{E}}(t)$ it is gaussian, all its higher order correlation functions can be obtained by the first-order one through

$$\langle \bar{\mathcal{E}}^*(t_1) \ldots \bar{\mathcal{E}}^*(t_m) \bar{\mathcal{E}}(t'_1) \ldots \bar{\mathcal{E}}(t'_k) \rangle = \delta_{mk} \sum_P \prod_{s=1}^{m} \langle \bar{\mathcal{E}}^*(t_s) \bar{\mathcal{E}}(t_{P(s)}) \rangle \quad (2.5)$$

P being the index permutation operator.

In a perturbative treatment of the field, the probability of the n-photon process depends, at the first non vanishing order, by the Fourier transform of the two-times n-th order correlation function, which for the CFM is given by

$$\langle \bar{\mathcal{E}}^n(t) \bar{\mathcal{E}}^{*n}(t') \rangle = n! \left(\langle \bar{\mathcal{E}}(t) \bar{\mathcal{E}}^*(t') \rangle \right)^n \quad (2.6)$$

As an instance, for a two-dimensional Markov process (which describes $\bar{\mathcal{E}}(t)$ as a non markovian process), the first order correlation function can be written as

$$\langle \bar{\mathcal{E}}(t) \bar{\mathcal{E}}^*(t') \rangle = \langle |\mathcal{E}^2| \rangle \left[\beta e^{-b|t-t'|} - b e^{-\beta|t-t'|} \right] / (\beta - b) \quad (2.7)$$

where, with reference to the notations of (2.3) and (2.4), we have

$$b\beta = b_1 \quad ; \quad b + \beta = b_2 \quad ; \quad \langle |\mathcal{E}^2| \rangle = \Gamma / b_1 b_2 \quad (2.7a)$$

For $\beta \gg b$, eq. (2.7) becomes an exponential, giving a lorentzian field spectrum, so that the correlation time β^{-1} of the electric field derivative describes the departure from the lorentzian spectrum. In the case of a lorentzian spectrum for a chaotic field, the transition probability of an n-photon process, calculated with the first nonvanishing order of perturbetion theory, will have a lorentzian lineshape broadened, as compared to the field bandwidth, by a factor n (the photon multiplicity).

(b) The Phase Diffusion Model

The Phase Diffusion Model (PDM) describes a single-mode, amplitude-stabilized laser ($\mathcal{E}(t)=\mathcal{E}_0$ in (2.2)) whose phase derivative is an Ornstein-Uhlenbeck (OU) stochastic process, obeying

$$\ddot{\varphi} = -\beta\dot{\varphi} + \mathcal{F}(t) \tag{2.8}$$

where β^{-1} represents the correlation time of $\dot{\varphi}$ (depending on the cavity width and on decay constants of the lasering atoms) and $\mathcal{F}(t)$ is a gaussian force, with

$$\langle \mathcal{F}(t)\rangle = 0 \quad ; \quad \langle \mathcal{F}(t)\mathcal{F}(t')\rangle = 2b\beta^2\delta(t-t') \tag{2.9}$$

For $\beta^{-1}\rightarrow 0$, the OU process becomes the Wiener-levy process, in which the phase is a markovian process obeying

$$\dot{\varphi} = F(t) \tag{2.10}$$

with

$$\langle F(t)F(t')\rangle = 2\Delta\omega\,\delta(t-t') \quad ; \quad \Delta\omega = \lim_{\beta\to\infty} b \tag{2.11}$$

describing the brownian diffusion of the phase around its initial value, with diffusion time $(\Delta\omega)^{-1}$. If the phase does not diffuse (b=0), one recovers the purely coherent model of the field.

By solving (2.8) and using the gaussian property of $\varphi(t)$ we obtain the first order $\bar{\mathcal{E}}(t)$ correlation function as

$$\langle \bar{\mathcal{E}}(t)\bar{\mathcal{E}}^*(t')\rangle = \mathcal{E}_0^2 \exp\left\{-b\left[|t-t'| - \beta^{-1}\left(1-e^{-\beta|t-t'|}\right)\right]\right\} \tag{2.12}$$

showing that only neglecting the $\dot{\varphi}$ correlation time β^{-1} one recovers the lorentzian field spectrum. Of interest here is the two-times n-th order correlation function given by

$$\langle \bar{\mathcal{E}}^{n}(t)\bar{\mathcal{E}}^{*n}(t')\rangle = \mathcal{E}_0^{2n} \exp\left\{-n^2 b\left[|t-t'| - \beta^{-1}\left(1-e^{-\beta|t-t'|}\right)\right]\right\}, \quad (2.13)$$

so that, at the lowest non vanishing order of the perturbation theory in the field, the n-photon transition probability has a lineshape similar to that of the spectrum, but broadened by a factor n^2.

By defining a coherence factor as the ratio between the n-photon transition probabilities in the presence of the stochastic field and of the purely coherent (PC) one,

$$\chi_n^{(x)} = W_n^{(x)} / W_n^{PC} \quad (2.14)$$

one finds, at the lowest order of the perturbation theory, using (2.6) and (2.13)

$$\chi_n^{CFM} = n! \quad (2.15)$$

and

$$\chi_n^{PDM} = 1 \quad (2.16)$$

We observe that these well-known perturbative results are not changed by the present non markovian treatments of the field fluctuations. Thus, resuming a number of points of this Section, we conclude that, in the weak field domain, the shape of the transition line results to be sensitive to the non markovian features of the fluctuating quantities while the coherence factors do not.

In what follows, by using a non perturbative theory for treating the field-particle interaction (multiphoton free-free transitions) and a non markovian treatment of the field fluctuations, we shall show how these results change in the domain of strong fields; in particular, we shall show that not only the lineshapes but also the coherence factors become sensitive to the non markovian features of the fluctuations and that the validity of the markovian limit becomes suspect for very strong fields.

3. CHARGED PARTICLE SCATTERING IN THE PRESENCE OF A STRONG FLUCTUATING RADIATION FIELD

Let us consider a particle of mass m and charge e moving in the presence of a static finite-range potential $V(\underset{\sim}{r})$ and of a radiation field, taken in dipole approximation and described by its vector potential $\underset{\sim}{A}(t)$.

Its Schrodinger equation is

$$\frac{1}{2m}\left[\frac{\hbar}{i}\underset{\sim}{\nabla}-\frac{e}{c}\underset{\sim}{A}(t)\right]^2\psi(\underset{\sim}{r},t)+V(\underset{\sim}{r})\psi(\underset{\sim}{r},t)=i\hbar\dot{\psi}(\underset{\sim}{r},t) \tag{3.1}$$

For strong assisting fields, it is appropriate to assume the potential V as a perturbation causing the transition, as in the conventional scattering theory. The unperturbed initial and final states, then, are solutions of

$$\frac{1}{2m}\left[\frac{\hbar}{i}\underset{\sim}{\nabla}-\frac{e}{c}\underset{\sim}{A}(t)\right]^2\chi(\underset{\sim}{r},t)=i\hbar\dot{\chi}(\underset{\sim}{r},t) \tag{3.2}$$

which are, whatever be $\underset{\sim}{A}(t)$, the non-relativistic Volkov waves

$$\chi_{\underset{\sim}{K}}(\underset{\sim}{r},t)=\exp\left\{i\underset{\sim}{K}\cdot\underset{\sim}{r}-\frac{i\hbar}{2m}\int^t\left[K^2-\frac{2e}{\hbar c}\underset{\sim}{K}\cdot\underset{\sim}{A}(\tau)\right]d\tau\right. \times$$

$$\times\exp\left\{-\frac{i}{\hbar}\int^t\frac{e^2}{2mc^2}A^2(\tau)d\tau\right\} \tag{3.3}$$

labelled by $\hbar\underset{\sim}{K}$, the particle momentum averaged over the field period or statistics, as the case requires.

The S-matrix for the transition from initial momentum $\hbar\underset{\sim}{K}_i$ to $\hbar\underset{\sim}{K}_f$ is

$$S=-\frac{i}{\hbar}\left(\chi_{\underset{\sim}{K}_f},V\psi_i^+\right) \tag{3.4}$$

where $\chi_{\underset{\sim}{K}_f}$ is a Volkov wave and Ψ_i^+ is the exact solution of (3.1) for the incident channel, with standard causal boundary conditions. Round brackets for the scalar product indicate both space and time integrations.

As

$$\Psi_i^+ = \chi_{\underset{\sim}{K}_i} + G^+ V \chi_{\underset{\sim}{K}_i} \tag{3.5}$$

G^+ being the retarded Green function in the presence of both V and $\underset{\sim}{A}$, (3.4) yields

$$S = (-i/\hbar)\left(\chi_{\underset{\sim}{K}_f}, V\chi_{\underset{\sim}{K}_i}\right) + (-i/\hbar)\left(\chi_{\underset{\sim}{K}_f}, VG^+V\chi_{\underset{\sim}{K}_i}\right) \tag{3.6}$$

The first term in (3.6) gives the Born approximation to the exact S-matrix for the field-assisted potential scattering; the second term accounts for higher order terms in the scattering potential. In fact, expanding G^+ in powers of V as

$$G^+ = G_o^+ + G_o^+ V G_o^+ + \ldots.. \tag{3.7}$$

G_o^+ being the retarded Green function in the absence of V one obtains the expansion

$$S = \sum_{\nu=1}^{\infty} S^{(\nu)} \tag{3.8}$$

where the ν-th term contains ν times V and (ν-1) times the field-assisted propagator G_o^+. In the coordinate representation, this latter is given by

$$G_o^+ = (-i/\hbar)\,\Theta(t-t')\int d\underset{\sim}{K}_m\, \chi_{\underset{\sim}{K}_m}(\underset{\sim}{r},t)\,\chi^*_{\underset{\sim}{K}_m}(\underset{\sim}{r}',t') \tag{3.9}$$

where Θ is the step function and $\chi_{\underset{\sim}{K}_m}$ are Volkov waves.

The first two orders of the expansion (3.8) are given explicitly by

$$S^{(1)} = (-i/\hbar) \int_{-T}^{T} dt \int d\underset{\sim}{r}\, \chi^*_{\underset{\sim}{K}_f}(\underset{\sim}{r},t)\, V(\underset{\sim}{r})\, \chi_{\underset{\sim}{K}_i}(\underset{\sim}{r},t) \quad T\to\infty \tag{3.10}$$

$$S^{(2)} = \lim_{T \to \infty} (-i/\hbar)^2 \int_{-T}^{T} dt \int_{-T}^{t} dt' \int d\underset{\sim}{K}_m \int d\underset{\sim}{r}\, \chi^*_{\underset{\sim}{K}_f}(\underset{\sim}{r},t)\, V(\underset{\sim}{r})\, \chi_{\underset{\sim}{K}_m}(\underset{\sim}{r},t) \times$$

$$\times \int d\underset{\sim}{r}'\, \chi^*_{\underset{\sim}{K}_m}(\underset{\sim}{r}',t')\, V(\underset{\sim}{r}')\, \chi_{\underset{\sim}{K}_i}(\underset{\sim}{r}',t') \tag{3.11}$$

By (3.8)-(3.9), for any ν (i.e., at any order in the scattering potential) the field enters to all orders, through the initial and final states and the intermediate states. Accounting for the field fluctuations amounts to take the average

$$\langle |S|^2 \rangle = \langle | \sum_{\nu=1}^{\infty} S^{(\nu)} |^2 \rangle \tag{3.12}$$

depending on the specific statistics assumed. Here, we shall use the Born approximation (ν =1 only).

4. FIRST ORDER THEORY

Using (3.12), (3.10) and (3.3), the first order probability per unit time of transition from the initial momentum $\hbar \underset{\sim}{K}_i$ to $\hbar \underset{\sim}{K}_f$, averaged over the stationary field fluctuations, is found as

$$W = 2\hbar^{-2}\, \mathrm{Re} \int_0^{\infty} dt\, \exp\{i\omega_{fi} t\} \langle X(t) \rangle\, |\tilde{V}(\underset{\sim}{\Delta})|^2 \tag{4.1}$$

where

$$\omega_{fi} = \frac{\varepsilon_f - \varepsilon_i}{\hbar} \quad ; \quad \varepsilon_J = \frac{\hbar^2 K_J^2}{2m} \quad (J = i, f) \tag{4.2}$$

$$\underset{\sim}{\Delta} = \underset{\sim}{K}_i - \underset{\sim}{K}_f \tag{4.2a}$$

$$X(t) = \exp\left\{ i\alpha \int_0^t d\tau\, A(\tau) \right\} \tag{4.3}$$

$$\alpha = (e/mc)\,\hat{e}_L \cdot \underset{\sim}{\Delta} \quad ; \qquad \hat{e}_L = \underset{\sim}{A}(t)/A(t) \tag{4.4}$$

$$\tilde{V}(\underset{\sim}{\Delta}) = \int d\underset{\sim}{r}\; e^{i\underset{\sim}{\Delta}\cdot\underset{\sim}{r}}\, V(\underset{\sim}{r}) \tag{4.5}$$

As it follows from the previous Section, also in this first order treatment of the scattering potential, the radiation field is included exactly, whatever be the time dependence (deterministic or stochastic) of the potential $\underset{\sim}{A}(t)$. In the particular approximation considered here, the field enters only the exponential (4.3). As the chaotic field is a gaussian process, the vector potential

$$A(t) = -c\int^{t} E(\tau)\,d\tau \tag{4.6}$$

too is a gaussian process, its statistical properties following by (4.6). It must be pointed out that, generally speaking, the stationarity of the electric field (i.e. the independence of its statistical properties on the choice of the origin of time) does not extend automatically to the vector potential $A(t)$. Stationarity is shared by $A(t)$ only in the limit of vanishing field bandwidth. However, it can be shown[15] that the non stationary features of $A(t)$ can be safely neglected when its amplitude is a slowly varying function in the time scale of the field period (quasi-coherent field); in this case, eq. (4.1) is valid and one has to compute the average of (4.3).

The evaluation of this average and of its half-sided Fourier transform in (4.1) requires different treatments, depending on the field statistical model. Details can be found elsewhere[9,16]; here, we report only the final results, in order to show some peculiar features and limiting behaviours.

(a) The Chaotic Field Case

For the Chaotic Field Model, assuming for simplicity the complex amplitude to be a projection of a two-dimension complex Markov process, we write eq. (2.3) for n=2 in the decoupled form[17]

$$\dot{\bar{\mathcal{E}}} = -b\bar{\mathcal{E}} + \mathcal{J}(t) \quad ; \quad \dot{\mathcal{J}}(t) = -\beta\mathcal{J} + F(t) \tag{4.7}$$

with

$$\langle F(t)F^*(t')\rangle = 2\langle|\bar{\mathcal{E}}|^2\rangle b\beta(b+\beta)\,\delta(t-t') \tag{4.8}$$

Then, for $b \ll \omega$ one obtains, by eq.s(4.1)-(4.5)[15]

$$W = 2\hbar^{-1}\sum_{n=-\infty}^{+\infty}|\tilde{V}(\underset{\sim}{\Delta})|^2 e^{-\rho}\sum_{K=0}^{\infty} f_{\nu K}(\lambda^2/2)\,\frac{\Gamma_{\nu K}\cos n\varphi_0 - \varepsilon_n \sin n\varphi_0}{\Gamma_{\nu K}^2+\varepsilon_n^2} \tag{4.9}$$

where $\nu = |n|$ and

$$\lambda = \lambda_{CFM} = (e/m\omega)\left[\langle|\mathcal{E}^2|\rangle/(\gamma^2+\Omega^2)\right]^{1/2}\hat{e}_L\cdot\underset{\sim}{\Delta} \tag{4.10}$$

$$\rho = \lambda^2/2\,\cos\varphi_0$$

$$\gamma = b\,\frac{\beta(b+\beta)\,\omega^2}{(b\beta)^2+\omega^2(b+\beta)^2} \tag{4.11}$$

$$\Omega = \omega\,\frac{(b+\beta)\,\omega^2}{(b+\beta)^2+\omega^2(b+\beta)^2}$$

$$\Gamma_{\nu K} = \hbar\gamma\,(\nu+2K+\lambda^2/2) \quad ; \quad \varepsilon_n = \varepsilon_f-\varepsilon_i-n\hbar\Omega \tag{4.12}$$

$$\varphi_0 = tg^{-1}\left[2\gamma\Omega/(\Omega^2-\gamma^2)\right]$$

$$f_{\nu\kappa}(x) = \frac{1}{\kappa!\,(\nu+\kappa)!} \left(\frac{x}{2}\right)^{\nu+2\kappa} \tag{4.13}$$

In the present process λ is the particle-field coupling parameter and plays a key role in identifying the coupling regimes. So $\lambda \ll 1$, loosely speaking, corresponds to the weak coupling and is expected to cover the regime of the lowest first orders of perturbation theory. Non perturbative behaviour appears at $\lambda > 1$. $\lambda \gg 1$ covers the regime of very strong coupling. Writing λ as

$$\lambda = \frac{\Delta E}{\hbar\omega}$$

we can view ΔE as the energy variation associated to an interacting particle suffering a momentum change $\underset{\sim}{\Delta}$, and λ as a measure of ΔE in units of the field photon $\hbar\omega$. Thus, it is intuitive that if we are asking for the probability that, say, n photons are exchanged in a scattering event in which $\lambda \simeq n$, the answer should be that that probability is the highest as compared to the probabilities of photon exchanges of other multiplicities. It is just what happens with the purely coherent field, for which the electric field is constant, and accordingly λ is a parameter not fluctuating during the scattering event. The above considerations suggest that the interplay between λ and n too is important in defining the probabilities of scattering with multiphoton exchanges. That it is the case will be shown below.

From (4.9) we see that the lineshape, centered at $\varepsilon_n \simeq 0$, or when $\varepsilon_n \ll \Gamma_{\nu\kappa}$ and $n\varphi_0 \simeq 0$, is similar to a lorentzian for weak fields (k=0). These conditions amount to be close to the line peaks, to have narrow field spectra and few photons exchanged. At $\varepsilon_n \simeq \Gamma_{\nu\kappa}$ (in the wings) the line deforms. In particular, the dependence of (4.9) on the sign of ε_n implies an asymmetry of the line and one can show that towards no photon exchanged the line is higher.

The dependence of the linewidth $\Gamma_{\nu\kappa}$ on the field-particle coupling λ

(eq. (4.12)) and on the order k of the λ-power series shows that for sufficiently strong fields the various multiphoton processes may be not resolved, as a consequence of the higher order contributions yielding the same final number of exchanged photons. In this case the picture of a given number of photons exchanged during a collision looses its meaning.

For $b \to 0$, the k-summation in (4.9) can be performed giving[18,19]

$$W = (2\pi/\hbar) \sum_{n=-\infty}^{+\infty} e^{-\lambda^2/2} I_n(\lambda^2/2) \, |V(\underset{\sim}{\Delta})|^2 \, \delta(\varepsilon_f - \varepsilon_i - n\hbar\omega) \tag{4.14}$$

I_n being the Bessel function of imaginary argument.

The purely coherent result is instead given by[20]

$$W = (2\pi/\hbar) \sum_{n=-\infty}^{+\infty} J_n^2(\lambda) \, |V(\underset{\sim}{\Delta})|^2 \, \delta(\varepsilon_f - \varepsilon_i - n\hbar\omega) \tag{4.15}$$

where now

$$\lambda = \lambda_{PC} = \left(\frac{e E_0}{m\omega^2}\right) \hat{e}_L \cdot \underset{\sim}{\Delta} \tag{4.16}$$

with the constant value of the electric field E_0 replacing the square root of the electric field variance $(\langle |\mathcal{E}|^2 \rangle)^{1/2}$. Below, when comparing results corresponding to the same quantity and within different models (like, for instance, in the coherence factors), as a rule a value of E_0 common to the compared models will be taken:

$$\lambda = \lambda_{CFM} = \lambda_{PC} = \lambda_{PDM}$$

Now, comparing (4.14) and (4.15), we see that already at vanishing field bandwidths (at which the markovian or non markovian character of the fluctuations is unimportant), the coherence factor (2.12) for the chaotic field becomes a quantity much more involved than the relatively simple $n!$, depending on all the parameters of the scattering process:

$$\chi_n^{(CFM)} = \exp\{-\lambda^2/2\} \, I_n(\lambda^2/2) / J_n^2(\lambda) \tag{4.17}$$

This expression has already been analized elsewhere[21]; here we only recall that the n! behaviour is recovered from (4.17) for $\lambda \ll 1$, and for $\lambda \gg 1$ as well, provided the additional constraint $\lambda \ll n$ fulfilled. Reminding the interpretation of λ , given above, it is easy to see that this apparently double result has actually the same physical content. Rewriting $\lambda \ll 1$ and $\lambda \ll n$, respectively as $\Delta E \ll \hbar\omega$ and $\Delta E \ll n\hbar\omega$, we see that both correspond to a physical situation in which the energy variation associated to the interacting scattered particle ΔE is much smaller than the number of photons we ask to be exchanged. For $\lambda \gg 1$ but $n \lesssim \lambda$, we have instead[16]

$$\eta_n^{CFM} \approx \sqrt{\pi} \;\; \exp\{-n^2/\lambda^2\} \left[1 - n^2/\lambda^2\right]^{1/2} \tag{4.18}$$

showing a strong reduction of the probability of the process with n photons exchanged in the case of a chaotic field as compared to a purely coherent field. It is particularly true for $n \approx \lambda$.

Finally, for $\lambda \gg n$, η_n^{CFM} is an oscillating function (the oscillations being controlled by $J_n^2(\lambda)$), with $\sqrt{\pi}$ as the predicted average value. Numerical evaluations often confirm the basic information obtained with the help of rough estimates, based on the asymptotic forms of the Bessel functions entering eq. (4.17). However, it should be pointed out that situations may be encountered when the values of physical parameters reach the limit of validity on which the theoretical model is based, but it does not correspond yet to the asymptotic behaviour of the Bessel functions. In that case it is simply wrong to use asymptotic forms of these functions, and only precise numerical evaluation must be trusted. As shown below, this caution applies to the predicted average value of $\sqrt{\pi}$ (see Fig. 2). For weak coupling ($\lambda \ll 1$), retaining the k=0 term only in (4.9) gives

$$W = 2\hbar^{-1} \sum_{n=-\infty}^{+\infty} |V(\underset{\sim}{\Delta})|^2 \left(\frac{\lambda^2}{2}\right)^{\nu} \frac{1}{n!} \; \frac{\hbar\gamma(\nu+\lambda^2/2)\cos n\varphi_0 - \varepsilon_n \sin n\varphi_0}{[\hbar\gamma(\nu+\lambda^2/2)]^2 + \varepsilon_n^2} \tag{4.19}$$

$(\nu = |n|)$

Considering that λ^2 is proportional to the electric field variance, it is also proportional to I, the mean field intensity. Thus, eq. (4.19) reproduces the well-known I^n-dependence of an n-photon process, typical of the lowest order of perturbation theory. Similarly, eq. (4.19) contains the n-dependence of the lineshape. Additionally, eq. (4.19) contains new features as the line asymmetry and the dependence of the lineshape on the particle-field coupling λ .

(b) The Phase Diffusion Case

For the Phase Diffusion Model, let us turn now to the average of $X(t)$, eq. (4.3). In this case, the transition probability per unit time is found as[25]

$$W = 2\hbar^{-2} \operatorname{Re} \left[U^+ M U \right]_{00} |\tilde{V}(\underset{\sim}{\Delta})|^2 =$$

$$= 2\hbar^{-2} \operatorname{Re} \sum_{\ell,\kappa} J_\ell(\lambda)\, M_{\ell\kappa}(b,\beta,\lambda)\, J_\kappa(\lambda)\, |\tilde{V}(\underset{\sim}{\Delta})|^2 \qquad (4.20)$$

where J_n is the Bessel function of the first kind of argument $\lambda = \lambda_{PDM} = \lambda_{PC}$, and the infinite matrices M and U are given by

$$M = \left[iD + A \frac{b\beta}{\beta - iD - A \dfrac{2b\beta}{2\beta - iD - A \ldots\ldots} A} A \right]^{-1} \qquad (4.21)$$

$$U_{m\kappa} = J_{m+\kappa}(\lambda)$$

and

$$A_{mk} = i\omega\left[\lambda/2\left(\delta_{m,k-1} + \delta_{m,k+1}\right) - m\,\delta_{mk}\right]\;; \qquad D = \omega_{fi}\, I + \omega B \qquad (4.22)$$

I is the identity matrix and $B_{mk} = -m\,\delta_{mk}$. As A is nondiagonal, also the matrix-continued fraction M is not diagonal and the double sum in (4.20) cannot be reduced.

If the field bandwidth is zero (monochromatic field) one recovers the purely coherent field limit. In fact, by (4.20)-(4.22) one easily obtains the eq. (4.15) remarking that

$$\mathrm{Re}\, M_{\ell k} = \pi\, \delta_{\ell k}\, \delta(\omega_{fi} - \ell\omega) \qquad (4.23)$$

En passant, one has the useful representation of the delta functions comb

$$\pi\hbar \sum_{m=-\infty}^{+\infty} J_m^2(\lambda)\, \delta(\varepsilon_f - \varepsilon_i - m\hbar\omega) = \lim_{b\to 0} \mathrm{Re}\left[U^{+} M U\right]_{00} \qquad (4.24)$$

which can avoid the calculation of the Bessel functions summation, particularly for $\beta \to 0$, as no Bessel functions appears in the right hand term[9].

For weak particle-field coupling ($\lambda \ll 1$), the matrix A^2 is approximately diagonal, giving

$$W = 2\hbar^{-2}\, \mathrm{Re} \sum_{\ell\ (\lambda\ll 1)} J_\ell^2(\lambda)\, M_\ell^{(0)}\, |\tilde{V}(\underset{\sim}{\Delta})|^2 \qquad (4.25)$$

where

$$M^{(0)} = \left[\, iD + A^{(0)} \cfrac{b\beta}{\beta - iD - A^{(0)} \cfrac{2b\beta}{2\beta - iD - A^{(0)} \cdots}} \,\right]^{-1} \qquad (4.26)$$

and

$$A_{mk}^{(0)} = -\delta_{mk}\left(n^2 + \lambda^2/2\right) \qquad (4.27)$$

Thus, for weak coupling and arbitrary bandwidth the J_1^2-dependence typical of the purely coherent assisting field is found to hold also for the PDM field, but now the various multiphoton channels are not resolved: each of them exhibits a lineshape $M_\ell^{(0)}$ depending on b , β and λ .

(c) The Quasi-markovian Limit

An interesting limit of the eq. (4.20) is the so-called quasi-markovian limit, in which the matrix M can be truncated at the first order including a finite β . As discussed in a different context by Dixit et al.[26], this amounts to include at the first-order the memory effects due to the non markovian character of the phase fluctuations. By (4.21)-(4.22) this limit is valid if

$$\frac{b\beta}{(\beta+\omega)^2}\left(n^2+\lambda^2/2\right) \ll 1 \qquad (4.28)$$

For vanishing β^{-1} (i.e., neglecting the correlation time of the phase derivative), the eq. (4.28) is trivially fulfilled; but for finite β , the short memory condition (4.28) is satisfied only for sufficiently small λ . Wanting to use this quasi-markovian limit, one is faced with the restriction to relatively small field intensities. Thus it would make of little use the advantage of having an exact treatment of the particle-field interaction. This limitation on the intensity for fluctuating fields is a new result, which may have far-reaching consequences on models and approximations used to describe fluctuations in high intensity domains.

If the (4.28) is fulfilled, the matrix M becomes

$$M = -\left[iD + A\,\frac{b\beta}{\beta - iD}\,A\right]^{-1} \qquad (4.29)$$

In this limit, and if $\lambda \ll 1$, M becomes diagonal, yielding for the transition probability (4.20)

$$W^{(I)} = 2\hbar^{-2} \sum_{\ell \atop (\lambda \ll 1)} J_\ell^2(\lambda) \frac{(\ell^2 + \lambda^2/2)b}{[\omega_{fi} - \ell\omega]^2 + [(\ell^2 + \frac{\lambda^2}{2})b + \beta^{-1}(\omega_{fi} - \ell\omega)]^2} |\tilde{V}(\underset{\sim}{\Delta})|^2 \tag{4.30}$$

This expression describes a superposition of lineshapes, each centered on the exchange of ℓ photons and weighted by a Bessel function $J_\ell^2(\lambda)$ (proportional to I^ℓ for $\lambda \ll 1$, I being the field intensity). In the markovian limit, each lineshape is a lorentzian, whose bandwidth $(\ell^2 + \lambda^2/2)b$ reproduces for $\lambda \to 0$ known perturbative results[24,27]; for non markovian fluctuations, each line becomes asymmetric, as already found for the chaotic field.

By defining the "perturbative" bandwidth

$$\Gamma^{(1)}(\lambda) = (\ell^2 + \lambda^2/2)b \tag{4.31}$$

the condition (4.28) for the applicability of the quasi-markovian limit reads

$$\frac{\beta}{(\beta + \omega)^2} \Gamma^{(1)}(\lambda) \ll 1 \quad , \tag{4.32}$$

showing the connection, already discussed for the chaotic field, between the line broadening and the markovian features of the field fluctuations.

5. SELECTED NUMERICAL RESULTS AND COMMENTS

In the previous Section we have derived the exact (in the field) first-order transition probabilities for two laser models. Besides, using approximate or asymptotic analytical expressions, we have shown that several results known in the literature are contained in our general expressions as limiting cases. In particular, it applies to the small coupling limit $\lambda \ll 1$, which yields known "perturbation theory" results, or new results belonging properly to weak field regimes. At this point, it must be stressed that numerical evaluation of the exact expressions, performed at arbitrary values of the relevant parameters (especially of λ) yields results strongly departing from those based on perturbation theory treatments (small λ). More explicitly, as anticipated by some rough estimates in the previous Section, the results of the exact calculations often contradict the expectation based on (lowest order of) perturbation theory treatments. A useful insight to some of the numerical results reported below may be gained by the following consideration. As compared to the small coupling case ($\lambda \ll 1$), the case of arbitrary λ may be thought as corresponding to a physical situation, in which a net final energy exchange between scattered particles and fields takes place through several virtual processes with repeated exchanges of photons. It is obvious that in such cases the bandwidth $\Delta\omega$ or the intensity fluctuations interfere strongly to disturb the "energy-matching" conditions, which may give larger transition probabilities or scattering cross sections.

Unless stated otherwise, the calculations are performed for: i) an incident particle energy of 100 eV; ii) a field linearly polarized, with the polarization $\hat{e}_L$ parallel to the incoming particle beam ($\hat{e}_L \parallel \underset{\sim}{k}_i$); iii) a finite-range, Yukawa type scattering potential (screening radius $r_o = 50\ a_o$, with a_o the Bohr radius); iv) the scattering angle $\vartheta = 45^\circ$; v) the field bandwidth $\hbar\Delta\omega = 1.0^{-4}$ eV.

As the results reported in Fig.s 1-8 are sufficiently self-explanatory, we may confine ourselves to only few concise comments.

Fig. 1 shows the ratio of the overall width of the scattering lines to the field bandwidth plotted against the electric field amplitude for various numbers of exchanged photons. The full curves refer to the PDM case, while the broken ones to the CFM. A number of points deserve to be noted: i) for (relatively) weak fields, the curves exhibit the behaviour predicted by the (lowest order of) perturbation theory; namely, the n^2- behaviour for

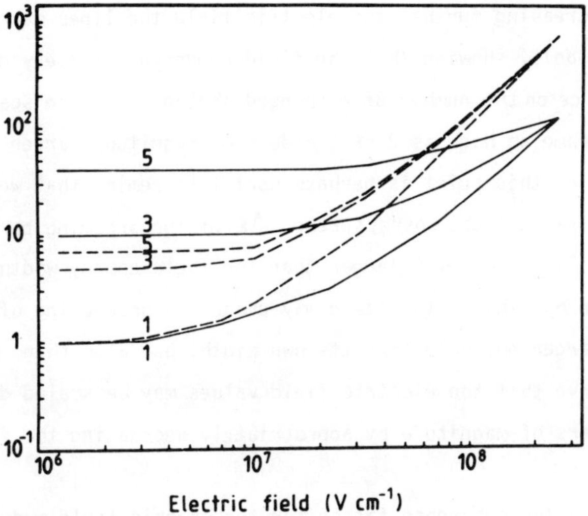

Fig. 1 The ratio of the overall width of the scattering line Γ to the field bandwidth $\hbar\Delta\omega$ vs the electric field (in V/cm) for various numbers of absorbed photons (n=1, 3 and 5, numbers on the curves) and two laser models: i) Phase Diffusion Model (continuous lines) and ii) Chaotic Field Model (dashed lines). Values of the used parameters: laser photon energy $\hbar\omega$ =1.17 eV; field energy bandwidth $\hbar\Delta\omega = 10^{-4}$ eV; screened coulomb potential with unit charge and screening radius r_o=50 a_o, with a_o the Bohr radius; incident particle beam energy $\mathcal{E}_i$ =100 eV; field polarization $\hat{e}_L$ parallel to the incoming particle momentum $\hat{e}_L \parallel \underline{k}_i$; scattering angle $\vartheta = 45^\circ$. (After ref. 9).

the PDM case and the n-behaviour for the CFM case[23,24,27]. ii) Increasing the electric field gives rise to a field-dependent broadening larger than predicted by perturbation theory, with the CFM broadening larger than the PDM one. iii) Increasing further the electric field the lines merge in one line (for each model), showing that the field-dependence largely dominate over the dependence on the number of exchanged photons. iv) The scattering linewidth may become as high as 2 or 3 orders of magnitude larger than the field bandwidth. At this point is perhaps useful to remind that we have neglected in our treatment the energy spread $\Delta\varepsilon$ of the incoming particle beam which in many realistic cases is larger than the field energy width $\hbar\Delta\omega$. However, as shown by Fig. 1, the field may produce a broadening of the scattering line much larger not only than its own width, but also than $\Delta\varepsilon$. Finally, we observe that the electric field values may be scaled down by 2 or 3 or more orders of magnitude by appropriately decreasing the field frequency.

Fig. 2 shows the coherence factor for the chaotic field model vs the intensity for n=1 and 5 and zero bandwidth. The first portion of the curves exhibits the n! - behaviour; then a decreasing behaviour is observed up to a first minimum (which is the deepest one) corresponding roughly to $n \approx \lambda$. It means that in this region a purely coherent field gives cross sections much larger than the chaotic field. Increasing further the intensity, an oscillating behaviour is found. Calculations are for $\hbar\omega$ =1.17 eV, with intensities up to 10^{15} Watt/cm^2. No indication is still observed towards the end of the curves of $\sqrt{\pi}$ as the average of the oscillating behaviour.

The Figures 3-8 report differential and total cross sections. The differential cross section (DCS) of a particle being scattered at a given angle and a final energy corresponding to n photons exchanged is defined as

$$\left(\frac{d\sigma}{d\Omega}\right)_n = \int_{\delta_n^-}^{\delta_n^+} d\varepsilon_f \, \frac{d^2\sigma}{d\Omega\, d\varepsilon_f} = \frac{m^2}{(2\pi\hbar)^3} \int_{\delta_n^-}^{\delta_n^+} d\varepsilon_f \left(\frac{K_f}{K_i}\right) W \tag{5.1}$$

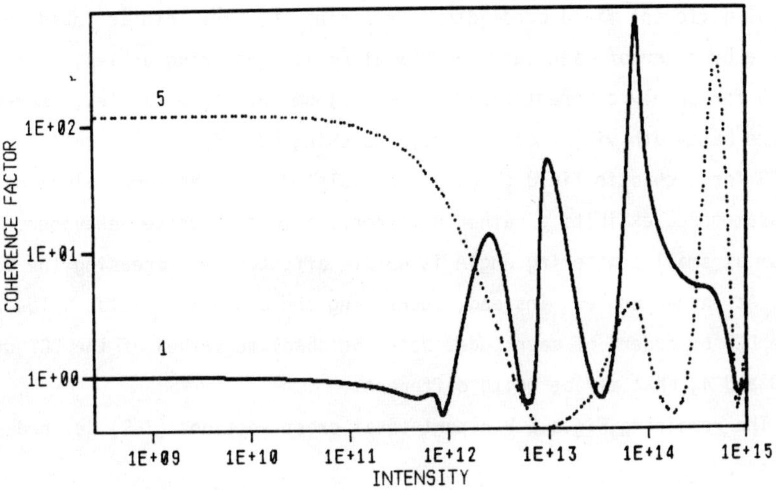

Fig. 2 The coherence factor χ_m^{cFM} for the chaotic field model (formula (4.17) of the main text) vs the intensity (in Watt/cm^2); the numbers on the curves are the absorbed photons. The chaotic field is zero bandwidth, while the other parameters are as in Fig. 1.

where

$$\delta_m^{\pm} = \varepsilon_i + m\hbar\omega \pm \hbar\omega/2$$

and n positive or negative, as the case requires, depending on whether emission or absorption is considered.

Fig. 3 reports DCS as a function of the scattering angle for the PDM field at different field bandwidths. Increasing the bandwidth is found to increase by order of magnitude the DCS at small scattering angles, where the DCS for purely coherent field has a minimum. At larger angles, increasing the bandwidth yields essentially a washing out of the oscillations. The DCS for a chaotic field (Fig. 4), calculated with the same values of the parameters, exhibits a rather different, almost opposite behaviour. The minimum at small scattering angle is hardly affected by increasing the bandwidth. At larger angles, instead, increasing the bandwidth yields a lowering of the DCS by orders of magnitude. Note the absolute values of the DCS of Fig. 3 and 4, that may be quite different.

The remaining Fig.s 5-8 report total cross sections (TCS) defined as usually

$$\sigma_m = \int d\Omega \left(\frac{d\sigma}{d\Omega} \right)_m \tag{5.2}$$

The Figures show TCS for the purely coherent field (Fig. 5), the PDM field (Fig. 6), the chaotic field (Fig. 7), and comparison among TCS in different models (Fig.s 6 and 8). Apart from the absolute values, the reported TCS exhibit remarkable features vs the intensity. It is especially true for the purely coherent field. A detailed explanation of the physical aspects corresponding to the various portions of the curves shown in the Figures is reported elsewhere[29]. Here we confine ourselves to remark only that the maximum in the TCS of purely coherent field and the following oscillatory structure correspond to a range of the physical parameters, when the oscillatory velocity of the particles in the field is equal or larger than the incident velocity.

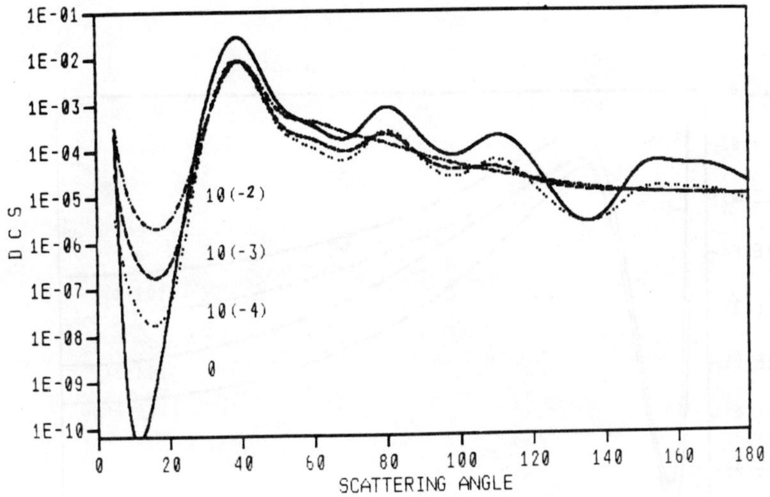

Fig. 3 Differential Cross Sections, DCS, (in a_o^2/ster. units) for a phase diffusion field vs the scattering angle (in degrees). The field intensity is $I=10^{13}$ Watt/cm^2; other parameters as in Fig. 1.

—·—·—·—·—: $\hbar\Delta\omega = 10^{-2}$ eV; - - - - - - -: $\hbar\Delta\omega = 10^{-3}$ eV;

··········: $\hbar\Delta\omega = 10^{-4}$ eV; ———— : $\hbar\Delta\omega = 0$.

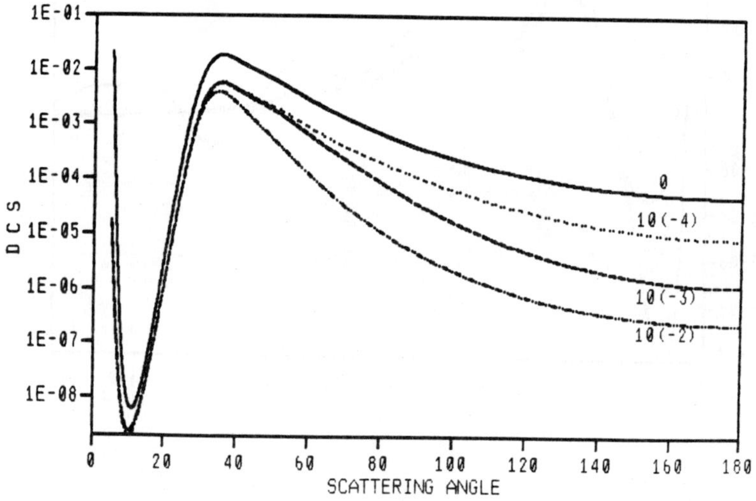

Fig. 4 As in Fig. 3, but for a chaotic field. The intensity has now to be meant as the mean intensity.

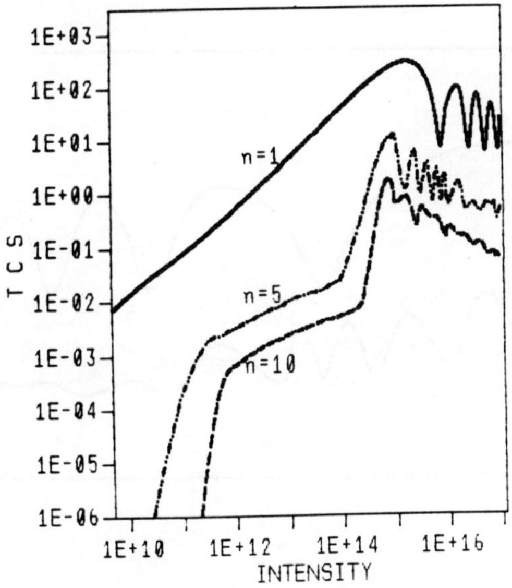

Fig. 5 Total Cross Sections, TCS, (in πa_0^2 units), for a purely coherent field <u>vs</u> the field intensity (in Watt/cm^2) for various numbers of absorbed photons (numbers on the curves). Other parameters as in Fig. 1.

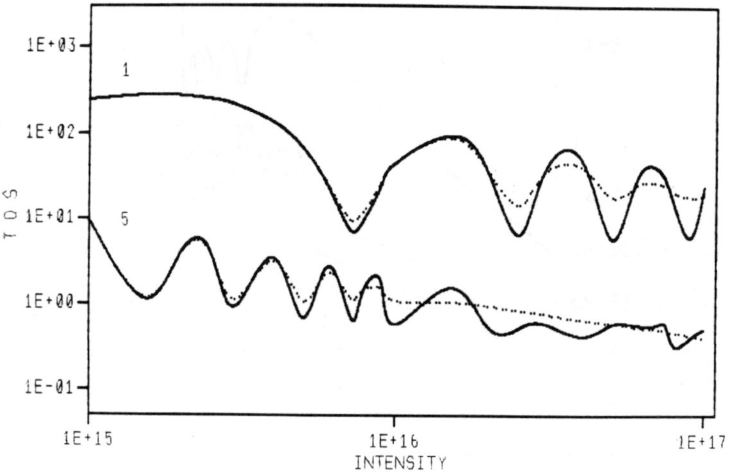

Fig. 6 TCS (in πa_o^2 units) vs the field intensity (in Watt/cm^2) for a phase diffusion field (............., $\hbar\Delta\omega=10^{-3}$ eV) and a purely coherent field (————). The numbers on the curves indicate the absorbed photons. Other parameters as in Fig. 1.

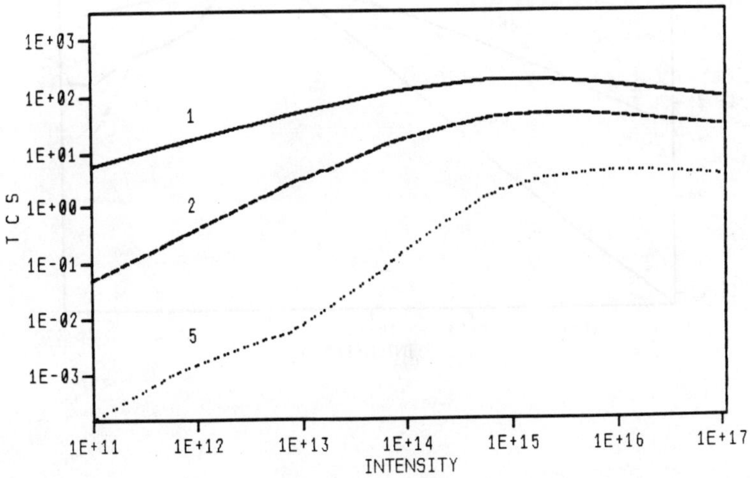

Fig. 7 As in Fig. 5, but for a chaotic field.

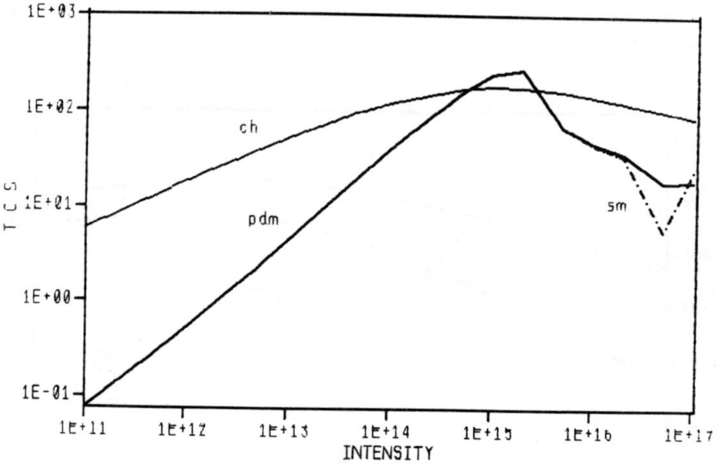

Fig. 8 TCS (in πa_0^2 units) vs the field intensity (in Watt/cm^2) for various field models. The curves refer to 5 photons absorbed. Other parameters as in Fig. 1.

ch : chaotic field model;

pdm : phase diffusion model;

sm : purely coherent field.

When such a condition is reached (in the parallel geometry $\hat{\varepsilon}_L \,//\, \underset{\sim}{K}_i$ considered here) for times of the order of the field period the particles have a small resulting velocity and the time of interaction with the scattering potential is considerably increased. Moreover, we may think in terms of the field forcing the particles to go back and forth inside the range of action of the scattering potential.

6. CONCLUDING REMARKS

In this Lecture we have discussed photon correlation effects as they manifest in collision processes assisted by strong laser fields.

In spite of being restricted only to one collision process (free-free transitions) and two field models (chaotic field and phase diffusion field) a lot of new information has been obtained, specific to high intensity regimes. It has been made possible by the fortunate circumstance that in the considered collision process, the particle-field interaction is treated exactly. It is our opinion that this kind of analysis is well worth to be continued, as it may give unique information on radiation-matter interaction in elementary processes in very high intensity domains; on the modifications of the field-assisted scattering parameters; on the role of the statistical fluctuations of the radiation fields in extremal conditions. Finally, this analysis may stimulate advances also in the assessement of the range of validity of some of the time-honoured approximations of the statistical theory in the case of very strong fluctuating fields.The work on the topic is at its very beginning, up to now having been confined to a couple of laser models, free-free transitions, first-order treatments in the scattering potential, and selected numerical calculations. It to say that in practice all the conventional directions of a mature theoretical understanding are still unexplored.

ACKNOWLEDGEMENTS

The authors express their thanks to the University of Palermo Computational Centre for the computer time generously provided to them.

This work was supported in part by the Italian Ministry of Education, the National Group of Structure of Matter and the Sicilian Committee for Nuclear and Structure of Matter Researches.

REFERENCES

1. See, for a comprehensive review and updated references: Multiphoton Processes, ed.s P. Lambropoulos and S.J. Smith, Springer, Berlin, 1984; Fundamentals of Laser Interactions, ed. F. Ehlotzky, Springer, Berlin, 1985.
2. F.K. Rodhes in: Multiphoton Processes, ed.s P. Lambropoulos and S.J. Smith, Springer, Berlin, 1984, pag. 31.
3. L.A. Lompré, G. Mainfray, C. Manus and J.P. Marinier, J. Phys. B: At. Mol. Phys. 14, 4307, (1981).
4. L.A. Lompré, A. L'Huillier, G. Mainfray and C. Manus, Phys. Lett.s 112A, 319, (1985).
5. A. L'Huillier and M Trahin, Phys. Lett.s 112A, 377, (1985).
6. G. Ferrante in: Fundamentals of Atomic Collisions, ed.s H. Kleinpoppen, H.O. Lutz and J.S. Briggs, Plenum Press, New York, 1985.
7. C. Leone, S. Bivona, R. Burlon and G. Ferrante, "On the Exact Treatment of Radiation Matter Interaction in Free-Free Transitions", Europhysics Letters (submitted).
8. See, for instance, A. Weingartshofer, J.K. Holmes, J. Sabbagh and S.L. Chin, J. Phys. B: At. Mol. Phys. 16, 1805, (1983).
9. F. Trombetta, G. Ferrante, K. Wodkiewicz and P. Zoller, J. Phys B: At. Mol. Phys. 18, 2915, (1985).

10. F. Trombetta, R. Daniele, F. Morales and G. Ferrante, "Field Coherence in Laser Assisted Charged Particle Scattering", Il Nuovo Cimento D (submitted).
11. J.H. Eberly, K. Wodkiewicz and B.W. Shore, Phys. Rev. A30, 2381, (1984); K. Wodkiewicz, B.W. Shore and J.H. Eberly, Phys. Rev. A30, 2390, (1984); K.Wodkiewicz and J.H. Eberly, Phys. Rev. A32, 992, (1985).
12. F. Trombetta, G. Ferrante and K. Wodkiewicz, "Electron Scattering in the Presence of Strong Stochastic Fields. The Gaussian Amplitude Model", (in preparation).
13. N.G. van Kampen, Phys. Rep. 24, 171, (1976).
14. P. Zoller in: Multiphoton Processes, ed.s P. Lambropoulos and S.J. Smith, Springer, Berlin, 1984, pag. 68.
15. F. Morales, R. Daniele, F. Trombetta and G. Ferrante, "Field Fluctuations in Multiphoton Free-Free Transitions. The Chaotic Model and the Non-Lorentzian Spectra", (in preparation).
16. R. Daniele, G. Ferrante, F. Morales and F. Trombetta in: Fundamentals of Laser Interactions, ed. F. Ehlotzky, Springer, Berlin, 1985, pag. 51.
17. P. Zoller, G. Alber and R. Salvador, Phys. Rev. A24, 398, (1981).
18. P. Zoller, J. Phys. B: At. Mol. PHys. 13, L249, (1980).
19. R. Daniele and G. Ferrante, J. Phys. B: At. Mol. Phys. 14, L635, (1981).
20. F.V. Bunkin and M.V. Fedorov, Sov. Phys. JETP 22, 844, (1966).
21. F. Trombetta, C.J. Joachain and G. Ferrante in: Collisions and Half-Collisions with Lasers, ed.s N.H.K. Rahman and C. Guidotti, Harwood, London, 1984.
22. A.T. Georges and P. Lambropoulos, Phys. Rev. A20, 991, (1979).
23. B.R. Mollow, Phys. Rev. 175, 1555, (1968).
24. G.S. Agarwal, Phys. Rev. A1, 1445, (1970).
25. F. Trombetta, G. Ferrante and P. Zoller, "High Intensity Effects in Multiphoton Processes Induced by Non-Markovian Fluctuating Fields", (in preparation).

26. S.N. Dixit, P. Zoller and P. Lambropoulos, Phys. Rev. A21, 1289, (1980).

27. D.S. Elliott, M.W. Hamilton, K. Arnett and S.J. Smith, Phys. Rev. A32, 887, (1985).

28. R. Daniele, G. Ferrante, F. Morales and F. Trombetta, "On the Validity of the Sum Rule for Multiphoton Free-Free Transitions", J. Phys. B: At. Mol. Phys. (in press).

29. R. Daniele, G. Ferrante, F. Morales and F. Trombetta, "High Intensity Effects in Field-Assisted Potential Scattering", (in preparation).

NEW TOPICS IN FIELD ASSISTED COLLISIONS. GAUGE ASPECTS.

C. Leone[+] and G. Ferrante[x]

+ Dipartimento di Energetica e
di Applicazioni della Fisica
Viale delle Scienze 90128 Palermo, Italy

x Istituto di Fisica dell'Università
Via Archirafi 36, 90123 Palermo, Italy

&1 INTRODUCTION

This Lecture will be primarily concerned with a number of questions regarding the gauge in which are described the strong electromagnetic fields assisting an atomic collision. This kind of questions is not new in the context of optical transitions and other related elementary processes[1-11]. So an effort will be made in what follows to generalize previous results to the extent they are pertinent to our purposes, and to present a unified discussion of the gauge aspects in both optical transitions and field-assisted collisions. Gauge questions in optical transitions have been extensively investigated[1-11], while in the context of field-assisted collisions they are a relatively new topic. In a recent paper by Leone <u>et al.</u>[12] the interested reader may found details concerning how gauge aspects appear in field-assisted collisions, the motivations to investigate them and a number of preliminary results. Here we confine only to emphasize an important difference compared with optical transitions. While generally in the latter, the field enters the transition operators, which couple bare atomic (or molecular) states, in atomic collisions assisted by a strong field, the latter frequently enters instead the unperturbed dressed scattering states, the transition operators being those of the conventional field-free theory. Thus gauge aspects appear in a qualitatively new manner. Besides, frequently, in the dressed states the field enters in a non-perturbative way.

Our discussion will be based on a number of rather elementary notions of quantum mechanics: on the requirement that the physical quantities be unaffected by unitary transformations of the corresponding operators; on the correct identification of which operators correspond to physical quantities and which do not, and so on. Next, starting from the general requirement that the physical quantities must be independent of the choice of the gauge, we outline a gauge independent treatment of the field-assisted collisions. In particular we show that the transition probabilities, calculated exactly or within approximate methods, are gauge independent when the initial and final states correspond, in any gauge, to the same physical observable and the operators are expanded in powers of the same quantity.

Concerning collisions, below we restrict our discussion to scattering by a structureless target. It is not the most general case to discuss gauge aspects, but it is a study case where the differences with the optical transitions are the most neat. Thus this discussion is expected to complement what is presently known on the gauge questions in elementary atomic processes.

The level of presentation is rather simple. This is felt appropriate because, due to their relevance, gauge questions should be treated as clearly as possible, and made immediately accessible.

&2 UNITARY TRANSFORMATIONS

We start by rewiewing a number of basic notions concerned with unitary transformations in quantum mechanics, which will be needed in what follows.

As it is well known, wavefunctions and operators of physical quantities, when subjected to any unitary transformation T (which may be time-dependent as well),

$$T\,T^{+} = T^{+}T = 1 \tag{2.1}$$

transform according to

$$\Psi \rightarrow \tilde{\Psi} = T\Psi \tag{2.2}$$

$$O \rightarrow \tilde{O} = T O T^{+} \tag{2.3}$$

Let us consider now how changes the Schroedinger equation

$$i\hbar \frac{\partial}{\partial t}\Psi = H\Psi \tag{2.4}$$

when subjected to an explicitly time-dependent unitary transformation T(t). From (2.4) we have

$$i\hbar T\frac{\partial}{\partial t}\Psi \equiv i\hbar \frac{\partial}{\partial t}(T\Psi) - i\hbar\left(\frac{\partial}{\partial t}T\right)\Psi = TH\Psi \tag{2.5}$$

which using (2.1) and (2.2), is rewritten as

$$i\hbar \frac{\partial}{\partial t}\tilde{\Psi} = H''\tilde{\Psi} \tag{2.6}$$

with

$$H'' = \tilde{H} + i\hbar\left(\frac{\partial T}{\partial t}\right)T^{+} \tag{2.7}$$

and

$$\tilde{H} = T H T^{+} \tag{2.8}$$

Eq.(2.6), with (2.7) and (2.8), gives the unitarily transformed Schroedinger equation. The first point to observe is that the Hamiltonian H of Eq.(2.4) has transformed not according to the general rule (2.3), which must be obeyed by any quantum mechanical operator of a physical quantity. Next, we observe that in the transformed equation (2.6) we may deal with two operators ($\tilde{H}$ and H"), replacing, as we will see, the initial Hamiltonian H.

To elucidate this point and the role of the new operators $\tilde{H}$ and H" we remind that the Hamiltonian H plays, in general, two roles. First, it gives the time evolution of the wavefunction of the system, according to the Schroedinger equation. Second, if it does not dependent explicitly on time, H is additionally an integral of motion, and its eigenvalues have immediate physical meaning, being the eigenenergies of the system. On the contrary, if H(t) depends explicitly on time, it looses the second role. In fact, in this case the energy of the system is not conserved and the eigenvalues of the istantaneous Hamiltonian H(t) do not have, generally speaking, a precise physical meaning.

If the initial Hamiltonian H does not dependent explicitly on time and, accordingly, plays simultaneously the two above mentioned roles, after a time-dependent unitary transformation, these roles distribute between the operators $\tilde{H}$ and H". The role of time-evolution operator is taken up by H", while $\tilde{H}$ is the transformed operator of the conserving physical quantity. Of course, the spectra of the eigenvalues of H and $\tilde{H}$ coincide. That the operator $\tilde{H}$ represents an integral of motion is proved by an immediate calculation:

$$\frac{d\tilde{H}}{dt} \equiv \frac{\partial \tilde{H}}{\partial t} + i\hbar \left[H'', \tilde{H} \right] = 0 \tag{2.9}$$

Finally, if the initial Hamiltonian H(t) depends explicitly on time, it does not possess the property of integral of motion. Accordingly, the operator $\tilde{H}$ too has no immediate physical meaning, while H" as before is the evolution operator, giving the time evolution of the wavefunction of the system.

In conclusion, the Hamiltonian, generally speaking, is not the operator of a physical quantity, and accordingly, is not bound to obey the rule (2.3). It answers the first observation made after Eq.s(2.7) and (2.8).

In many problems the Hamiltonian H may split into two parts

$$H = H_0 + H_1 \tag{2.10}$$

with H_1 an interaction with an external agent.

The Schroedinger equation of the unperturbed system is

$$i\hbar \frac{\partial}{\partial t} \Psi_0 = H_0 \Psi_0 \tag{2.11}$$

A formal solution to Eq.(2.4) with H given by (2.10) and casual boundary conditions is

$$\Psi^{+} = \Psi_0 + G^{+}_{H_1}\Psi_0 \tag{2.12}$$

with

$$G^{+} = \left[i\hbar \frac{\partial}{\partial t} - H + i\eta \right]^{-1} \tag{2.13}$$

the retarded full Green function.The spectral representation of G^{+} is

$$G^{+}(t,t_0) = (-i/\hbar)\sum_j \Psi_j(t)\,\Psi_j^{*}(t_0) \tag{2.14}$$

Performing a unitary transformation on the Schroedinger equation with the complete Hamiltonian (2.10), for the transformed full Green function we have

$$\tilde{G}^{+} = \left[i\hbar \frac{\partial}{\partial t} - H'' + i\eta \right]^{-1} \tag{2.15}$$

which is related to G^{+} by the general relation (2.3). In fact,

$$\begin{aligned} \tilde{G}^{+} &= -(i/\hbar)\sum_j \tilde{\Psi}_j(t)\,\tilde{\Psi}_j^{*}(t_0) \\ &= -(i/\hbar)\sum_j T\Psi_j(t)\,\Psi_j^{*}(t_0)\,T^{+} \\ &= T\,G^{+}\,T^{+} \end{aligned} \tag{2.16}$$

We have also

$$\tilde{\Psi}^{+} = \tilde{\Psi}_0 + \tilde{G}^{+}_{H_1}\tilde{\Psi}_0 \tag{2.17}$$

Finally, it is well known that the physical properties of the system are not affected by a unitary transformation. From this general requirement follows that the propability of the system to undergo a transition from the unperturbed state Ψ_{oi} to Ψ_{of} must remain unaffected by any unitary transformation T.

&3 ELECTROMAGNETIC GAUGES AND OPTICAL TRANSITIONS

We now use some of the notions outlined in the previous section to discuss the optical transitions, and to show how to have transition probabilities independent of the gauge used to describe the electromagnetic fields.

Let us consider the Schroedinger equation of an electron moving in a static potential $V(\underset{\sim}{r})$ and in e.m. field, described by the vector potential $\underset{\sim}{A}^G(\underset{\sim}{r},t)$ and the scalar potential $U^G(\underset{\sim}{r},t)$:

$$i\hbar \frac{\partial}{\partial t} \Psi^G(\underset{\sim}{r},t) = \left[K + V_1^G(\underset{\sim}{r},t) + V(\underset{\sim}{r}) \right] \tag{3.1}$$

where

$$K = p^2/2m \tag{3.2}$$

$$V_1^G(\underset{\sim}{r},t) = (1/2m)\Big[-(e/c)\big(\underset{\sim}{A}^G(\underset{\sim}{r},t)\cdot\underset{\sim}{p} + \underset{\sim}{p}\cdot\underset{\sim}{A}^G(\underset{\sim}{r},t)\big) + (e^2/c^2)\big(\underset{\sim}{A}^G(\underset{\sim}{r},t)\big)^2\Big] + e\,U^G(\underset{\sim}{r},t) \tag{3.3}$$

In Eq.s (3.1)-(3.3) $\underset{\sim}{p}$ is the operator of the canonical momentum, which is related to that of the kinetic momentum by

$$\underset{\sim}{\pi} = \underset{\sim}{p} - (e/c)\,\underset{\sim}{A}^G(\underset{\sim}{r},t) \tag{3.4}$$

The superscript "G" is to indicate an arbitrary e.m. gauge. Over a gauge transformation $G \rightarrow G'$ the vector and scalar potentials are transformed according to

$$\left.\begin{aligned} \underset{\sim}{A}^{G'}(\underset{\sim}{r},t) &= \underset{\sim}{A}^G(\underset{\sim}{r},t) + \underset{\sim}{\nabla} f_{GG'}(\underset{\sim}{r},t) \\ U^{G'}(\underset{\sim}{r},t) &= U^G(\underset{\sim}{r},t) - \frac{1}{c}\frac{\partial}{\partial t} f_{GG'}(\underset{\sim}{r},t) \end{aligned}\right\} \tag{3.5}$$

$f_{GG'}(\underset{\sim}{r},t)$ being an arbitrary function of $\underset{\sim}{r}$ and t. From the standpoint of quantum mechanics a gauge transformation amounts to perform a unitary transformation with the help of the time-dependent operator

$$T(t) = \exp\left\{\frac{ie}{\hbar c} f_{GG'}(\underset{\sim}{r},t)\right\} \tag{3.6}$$

Inserting (3.6) into (2.2), (2.6) and (2.7), omitting the subscripts in $f(\underset{\sim}{r},t)$, one obtains

$$\Psi^{G'}(\underset{\sim}{r},t) = \exp\left\{\frac{ie}{\hbar c} f(\underset{\sim}{r},t)\right\} \Psi^{G}(\underset{\sim}{r},t) \tag{3.7}$$

and the transformed Scroediger equation

$$i\hbar\frac{\partial}{\partial t}\Psi^{G'}(\underset{\sim}{r},t) = \Big\{ (1/2m)\Big[\underset{\sim}{p} - (e/c)\big(\underset{\sim}{A}^{G}(\underset{\sim}{r},t) + \underset{\sim}{\nabla} f(\underset{\sim}{r},t)\big)\Big]^2 + e\Big[U^{G}(\underset{\sim}{r},t) - \frac{1}{c}\frac{\partial}{\partial t} f(\underset{\sim}{r},t)\Big] + V(\underset{\sim}{r})\Big\} \Psi^{G'}(\underset{\sim}{r},t) \tag{3.8}$$

Using the expressions (3.5), Eq.(3.8) becomes

$$i\hbar\frac{\partial}{\partial t}\Psi^{G'}(\underset{\sim}{r},t) = \Big[K + V_1^{G'}(\underset{\sim}{r},t) + V(\underset{\sim}{r})\Big]\Psi^{G'}(\underset{\sim}{r},t) \tag{3.9}$$

with

$$V_1^{G'}(\underset{\sim}{r},t) = (1/2m)\Big[(-e/c)\big(\underset{\sim}{A}^{G'}(\underset{\sim}{r},t)\cdot\underset{\sim}{p} + \underset{\sim}{p}\cdot\underset{\sim}{A}^{G'}(\underset{\sim}{r},t)\big) + (e^2/c^2)\big(\underset{\sim}{A}^{G'}(\underset{\sim}{r},t)\big)^2\Big] + e\,U^{G'}(\underset{\sim}{r},t) \tag{3.10}$$

Eq.s (3.9) and (3.10) show that the Schroedinger equation is forminvariant.

When the atomic dimensions are much smaller than the radiation wavelength the long wavelength approximation (LWA) may be used, where the spatial dependence of $\underset{\sim}{A}^{G}(\underset{\sim}{r},t)$ is neglected

$$\underset{\sim}{A}^{G}(\underset{\sim}{r},t) \rightarrow \underset{\sim}{A}^{G}(t) \tag{3.11}$$

In the optical transitions these conditions are fulfilled and the LWA will be adopted in the following.

A common gauge is the so called radiation gauge (R-gauge), where one has in LWA

$$U^{G}(\underset{\sim}{r},t) = 0 \tag{3.12}$$

In R-gauge Eq.(3.3) becomes

$$V_1^R(\underset{\sim}{r},t) = (-e/mc)\,\underset{\sim}{A}^R(t)\cdot\underset{\sim}{p} + (e^2/2mc^2)\,(A^R(t))^2 \tag{3.13}$$

A special gauge may be obtained from the R-gauge in the LWA performing the gauge transformation (3.5) with the function

$$f_{RE}(\underset{\sim}{r},t) = -\,\underset{\sim}{A}^R(t)\cdot\underset{\sim}{r} \tag{3.14}$$

In the new gauge, called electric gauge (E-gauge), the vector potential vanish and the scalar potential is

$$U^E(\underset{\sim}{r},t) = -\,\underset{\sim}{E}(t)\cdot\underset{\sim}{r} \tag{3.15}$$

with $\underset{\sim}{E}(t)$ the electric field, where the spatial variation is neglected (dipole approximation). In E-gauge Eq.(3.3) becomes

$$V_1^E(\underset{\sim}{r},t) = -e\,\underset{\sim}{E}(t)\cdot\underset{\sim}{r} \tag{3.16}$$

which is the electron-radiation interaction; from Eq.(3.4) one obtains

$$\underset{\sim}{\pi} = \underset{\sim}{p} \tag{3.17}$$

and so K, Eq.(3.2), in E-gauge is the kinetic energy. Thus, the three terms of Hamiltonian in Eq.(3.1), in E-gauge, are operators of physical quantities and any combination of these terms corresponds to a physical observable. In particular

$$H_0 = K + V(\underset{\sim}{r}) \tag{3.18}$$

is the energy operator of an electron moving in a static potential (but non interacting with the radiation field). Besides H_0 coincides with the field-free Hamiltonian. The eigenvalues and eigenstates of operator H_0 are thus the energy values E_n and the states u_n of the system in the absence of the radiation field:

$$H_0\, u_n(\underset{\sim}{r},t) = E_n\, u_n(\underset{\sim}{r},t) \tag{3.19}$$

In an optical transition, before the perturbation is switched on, the system is in an eigenstate with energy E_i. As in the presence of a radiation field the energy is not a constant of motion, one has a non vanishing probability to find at time t the system in a state u_n with energy $E_n \neq E_i$. In E-gauge the transition probability is

$$P_{if} = \left| < u_f(\underset{\sim}{r},t) \,|\, \Psi_i^E(\underset{\sim}{r},t) > \right|^2, \tag{3.20}$$

Ψ_i^E being the exact solution of (3.1) in E-gauge. In Eq.(3.20) the brackets indicate space integration.

Over an arbitrary gauge transformation the operator H_0 transforms to

$$\tilde{H}_0^G = \left[\underset{\sim}{p} - (e/c)\underset{\sim}{A}^G(t)\right]^2 + V(\underset{\sim}{r}) \tag{3.21}$$

Its eigenvalues are again E_n , while its eigenfunctions $\tilde{u}_n^G(\underset{\sim}{r},t)$ are related to $u_n(\underset{\sim}{r},t)$ by

$$\tilde{u}_n^G(\underset{\sim}{r},t) = u_n(\underset{\sim}{r},t)\exp\left\{i\frac{e}{\hbar c}\underset{\sim}{A}^G(\underset{\sim}{r},t)\cdot\underset{\sim}{r}\right\} \tag{3.22}$$

The transition probability in the new gauge is

$$P_{if} = \left| < \tilde{u}_f^G (\underset{\sim}{r},t) \,|\, \Psi_i^G (\underset{\sim}{r},t) \right|^2 \tag{3.23}$$

Eq.(3.23) is coincident with (3.20), as $\tilde{u}_f^G$ and Ψ_i^G are obtained from u_f and Ψ_i^E applying the same unitary transformation, Eq.(3.6).

Frequently the transition probability in a G-gauge is written transforming in Eq.(3.20) only Ψ_i^E, so one has

$$\tilde{P}_{if}^G = \left| < u_f (\underset{\sim}{r},t) \,|\, \Psi_i^G (\underset{\sim}{r},t) > \right|^2 \tag{3.24}$$

which is evidently gauge dependent. In other words it is taken that the unperturbed states in any gauge are eigenstates of the energy operator H_o , Eq.(3.19). It must be emphasized that it is not true. In an arbitrary gauge, $\underset{\sim}{p}$ is the canonical momentum; as a consequence, K is not the kinetic energy operator; the operator H_o looses the meaning it has in the E-gauge; $\tilde{P}_{if}^G$ too ceases to be a probability with a well defined meaning, due to the ambiguity characterizing the observables of the initial and final states. Only in same special cases, in the R-gauge, Eq.(3.24) yields the same results as those given by the correct expression, Eq.(3.20). This occurs for pulses, when the system is prepared at $t \to -\infty$ and the measurement is performed at $t \to +\infty$,as the vector potential vanishes at $t = \pm\infty$ for a wavepacket with finite frequency bandwith. This is also true for monochromatic fields, no matter whether or not they are adiabatically switched on and off, provided the field photon $\hbar\omega$ is equal to $(E_f - E_i)/n$ with n = 1,2,3... Proves of the last statements may found in ref.s 3,10 and 11.

&4 CHARGED PARTICLE SCATTERING IN THE PRESENCE OF AN INTENSE LASER FIELD

Let us now discuss how the problem of the gauge independence stands in the context of the theory of a scattering process assisted by a strong laser field. As anticipated in the Introduction, in the most popular formulation of the theory of charged particle scattering by a structureless target in the presence of a laser, the transition operator is that of the conventional theory (the static potential), while the scattering states are free-states embedded in the radiation field. It sets the basic difference and the interesting peculiarity of field assisted collisions as compared to the optical transitions, as far as gauge aspects are concerned[12].

The Schroedinger equation for electron scattering by a static potential $V(\underset{\sim}{r})$ in the presence of an intense laser field, treated classically and in dipole approximation, in E-gauge is written as

$$i\hbar \frac{\partial}{\partial t} \Psi^E(\underset{\sim}{r},t) = \bar{H}^E(\underset{\sim}{r},t)\Psi^E(\underset{\sim}{r},t) \tag{4.1}$$

with

$$\bar{H}^E(\underset{\sim}{r},t) = p^2/2m + V(\underset{\sim}{r}) - e\underset{\sim}{E}(t)\cdot\underset{\sim}{r} \tag{4.2}$$

Neglecting in Eq.(4.1) the static potential $V(\underset{\sim}{r})$ one obtains the equation of a free particle embedded in a radiation field

$$i\hbar \frac{\partial}{\partial t} \Phi^E_k = H^E \Phi^E_k \tag{4.3}$$

with

$$H^E = p^2/2m - e\underset{\sim}{E}(t)\cdot\underset{\sim}{r} \tag{4.4}$$

Omitting, instead, in Eq.(4.1) the laser-electron interaction one has

$$i\hbar \frac{\partial}{\partial t} u(\underset{\sim}{r},t) = H_0 u(\underset{\sim}{r},t) \tag{4.5}$$

with

$$H_0 = p^2/2m + V(\underset{\sim}{r}) \tag{4.6}$$

As in E-gauge the canonical momentum $\underset{\sim}{p}$ coincides with the kinetic momentum $\underset{\sim}{\pi}$, Eq.(4.6) is again the energy operator of the field-free electron and additionally is the operator of time evolution of the wavefunction of the field free system according the Schroedinger equation (4.5).

From Eq.(4.1), (4.3) and (4.5) we may define respectively three retarded Green functions:

$$\bar{G}^{+E} = \left[i\hbar \frac{\partial}{\partial t} - \bar{H}^E + i\eta \right]^{-1} \tag{4.7}$$

$$G^{+E} = \left[i\hbar \frac{\partial}{\partial t} - H^E + i\eta \right]^{-1} \tag{4.8}$$

$$G_0^+ = \left[i\hbar \frac{\partial}{\partial t} - H_0 + i\eta \right]^{-1} \tag{4.9}$$

For a monochromatic radiation field, the Hamiltonian (4.2) and (4.4) are periodic in time and solutions to Eq.s (4.1) and (4.3) can be written as[13] :

$$\Upsilon_\varepsilon^E(\underset{\sim}{r},t) = \exp\left(\frac{i}{\hbar}\varepsilon t\right) \bar{\chi}_\varepsilon^E(\underset{\sim}{r},t) \tag{4.10}$$

$$\Phi_k^E(\underset{\sim}{r},t) = \exp\left(\frac{i}{\hbar}\varepsilon_k t\right) \chi_k^E(\underset{\sim}{r},t) \tag{4.11}$$

with the ε's the so-called quasi-energies of the system, and $\bar{\chi}_\varepsilon^E$ and χ_k^E periodic functions, satisfying the equations

$$\left[\varepsilon - \bar{H}^E + i\hbar \frac{\partial}{\partial t} \right] \bar{\chi}_\varepsilon^E(\underset{\sim}{r},t) = 0 \tag{4.12}$$

$$\left[\varepsilon_k - H^E + i\hbar \frac{\partial}{\partial t} \right] \chi_k^E(\underset{\sim}{r},t) = 0 \tag{4.13}$$

The functions $\bar{\chi}_\varepsilon^E$ and χ_k^E are defined in the composite Hilbert space $\mathcal{R} + \mathcal{P}$, where $\mathcal{R}$ is the usual space of stationary funtions $f(\underset{\sim}{r})$ and $\mathcal{P}$ is the space, which consists of all the possible periodic functions of time t with period $(2\pi/\omega)$.

When the laser-electron interaction is neglected Eq(4.12) becomes

$$\left[\varepsilon_n^0 - H_0 + i\hbar \frac{\partial}{\partial t} \right] \varphi_n^E(\underset{\sim}{r},t) = 0 \tag{4.14}$$

where $\varphi_m^E(r,t)$ are the field-free states in the $\mathcal{R}+\mathcal{T}$ Hilbert space and a complete set of solutions to (4.14) is

$$\varphi_m^E(\underset{\sim}{r},t) = u_m(\underset{\sim}{r}) \exp\left\{\frac{i}{\hbar}\left(\varepsilon_m^\circ + k\hbar\omega\right)t\right\} \tag{4.15}$$

$(k = 0, \pm 1, \pm 2, \dots)$

ε_m° and $u_n(\underset{\sim}{r})$ are, respectively, the eigenvalues and the eigenfunctions for the energy operator (4.6).

From Eq.s (4.12)-(4.14) in the $\mathcal{R}+\mathcal{T}$ Hilbert space we may define the following retarded Green functions

$$\bar{g}^{+E} = \left[\varepsilon - \bar{H}^E + i\hbar\frac{\partial}{\partial t} + i\eta\right]^{-1} \tag{4.16}$$

$$g^{+E} = \left[\varepsilon_k - H^E + i\hbar\frac{\partial}{\partial t} + i\eta\right]^{-1} \tag{4.17}$$

$$g_0^+ = \left[\varepsilon_0 - H_0 + i\hbar\frac{\partial}{\partial t} + i\eta\right]^{-1} \tag{4.18}$$

Over an arbitrary gauge transformation the wavefunctions (4.10)-(4.11) transform according to Eq.(2.2) and similarly do the periodic functions $\bar{\chi}_\ell^E$, χ_k^E and φ_m^E. According to Eq.(2.7) the transformed Hamiltonians are

$$\bar{H}^G = (1/2m)\left[\underset{\sim}{p} - (e/c)\underset{\sim}{A}^G(t)\right]^2 + eU^G(\underset{\sim}{r},t) + V(\underset{\sim}{r}) \tag{4.19}$$

$$H^G = (1/2m)\left[\underset{\sim}{p} - (e/c)\underset{\sim}{A}^G(t)\right]^2 + eU^G(\underset{\sim}{r},t) \tag{4.20}$$

$$\left.\begin{aligned} H_0^G &= (1/2m)\left[\underset{\sim}{p} - (e/c)\underset{\sim}{A}^G(t)\right]^2 + V(\underset{\sim}{r}) - e\frac{\partial}{\partial t} f_{EG}(\underset{\sim}{r},t) \\ &= (1/2m)\left[\underset{\sim}{p} - (e/c)\underset{\sim}{A}^G(t)\right]^2 + V(\underset{\sim}{r}) + eU^G(\underset{\sim}{r},t) + e\underset{\sim}{E}(t)\cdot\underset{\sim}{r} \end{aligned}\right\} \tag{4.21}$$

being the function transforming from E-gauge to the arbitrary G-gauge. Eq.(4.21) shows that in an arbitrary G-gauge the non-interacting part of total Hamiltonian of a particle in the presence of a radiation field is generally not the same as the Hamiltonian of the same particle in the absence of any radiation field, Eq.(4.6). Identity occurs only in

E-gauge. Further H_o is not the energy operator as the gauge transformation operator is generally a time dependent unitary operator.

By means of Eq.s(2.15) and (2.16) we may transform the Green functions defined in Eq.s(4.7)-(4.9) and (4.16)-(4.18). Again, we remark that the retarded Green function for a particle non interacting with the radiation field is not the same as that in absence of field. According to Eq.(3.21) one has

$$G_o^{+G} = \left[i\hbar\frac{\partial}{\partial t} - \frac{1}{2m}\left(\underset{\sim}{p} - \frac{e}{c}\underset{\sim}{A}^G(t)\right)^2 - V(\underset{\sim}{r}) - eU^G(\underset{\sim}{r},t) - e\underset{\sim}{E}(t)\cdot\underset{\sim}{r} + i\eta \right]^{-1} \qquad (4.22)$$

$$g_o^{+G} = \left[\varepsilon_o - \frac{1}{2m}\left(\underset{\sim}{p} - \frac{e}{c}\underset{\sim}{A}^G(t)\right)^2 - V(\underset{\sim}{r}) - eU^G(\underset{\sim}{r},t) - e\underset{\sim}{E}(t)\cdot\underset{\sim}{r} + i\hbar\frac{\partial}{\partial t} + i\eta \right]^{-1} \qquad (4.23)$$

Assuming the static potential $V(\underset{\sim}{r})$ as the perturbation responsible for the transition, the initial and final states are given in a G-gauge by Eq.(4.3) transformed to that gauge. So the exact S-matrix is

$$S = (-i/\hbar)\,(\Phi_{k_f} | V | \Psi_i^+) \qquad (4.24)$$

with

$$\Psi_i^+ = \Phi_{ki} + G^+ V \Phi_{ki}. \qquad (4.25)$$

In Eq.s(4.24) and (4.25) the index G is omitted and the round brackets indicate both space and time integrations.

According to Eq.(2.2), the exact S-matrix, Eq.(4.24), is gauge invariant as the static potential $V(\underset{\sim}{r})$ and the unitary operator T(t) commute:

$$\left.\begin{aligned} S &= (-i/\hbar)(\Phi_{k_f} T^+ | V | T\Psi_i^+) \\ &= (-i/\hbar)(\Phi_{k_f} | V | T^+ T \Psi_i^+) \\ &= (-i/\hbar)(\Phi_{k_f} | V | \Psi_i^+) \end{aligned}\right\} \qquad (4.26)$$

As the exact wavefunction Ψ_i^+ is not known, approximate expressions of S may be obtained expanding G^+ in powers either of the static potential $V(\underset{\sim}{r})$ or of the laser interaction. Expanding G^+ in powers of $V(\underset{\sim}{r})$ gives the Born series for scattering of field-dressed plane waves

$$\bar{G}^+ = G^+ + G^+ V G^+ + G^+ V G^+ V G^+ + \cdots \tag{4.27}$$

Inserting (4.27) into (4.24), and assuming a monochromatic radiation field, after the time integration one obtains

$$S = (-2\pi i)\sum_n \delta(\varepsilon_{k_f} - \varepsilon_{k_i} - n\hbar\omega)$$
$$\times \left\{ \langle\langle \chi_{k_{f},n} | V | (1 + g^+ V + g^+ V g^+ V + \dots) \chi_{k_i} \rangle\rangle \right\}$$
$$= (-2\pi i)\sum_n \delta(\varepsilon_{k_f} - \varepsilon_{k_i} - n\hbar\omega) \left\{ \langle\langle \chi_{k_{f},n} | V | (1 + \bar{g}^+ V) \chi_{k_i} \rangle\rangle \right\} \tag{4.28}$$

with

$$\chi_{k_{f},n} = \exp(in\hbar\omega t)\, \chi_{k_f} \tag{4.29}$$

$$\langle\langle \cdots \rangle\rangle = (2\pi/\omega)^{-1} \int_0^{2\pi/\omega} dt \, \langle \cdots \rangle \tag{4.30}$$

and $\langle\ldots\rangle$ indicating as before only space integration. ε_k is the quasi-energy of the free-particle in a laser field, Eq.(4.11), ω is the radiation frequency and n characterizes the scattering channel in which the energy $n\hbar\omega$ is exchanged between the scattered particle and the field ($n > 0$ for absorption, $n < 0$ for emission); χ_k, $\bar{g}^+$ and g^+ are given by Eq.s (4.13), (4.16) and (4.17) transformed to the G-gauge. According to Eq.(2.2) and (2.16) the terms of expansion (4.28) are gauge invariant. Let us to consider a generic term of (4.28)

$$\langle\langle \chi_{k_{f},n} T^+ | V | (T g^+ T^+ V \ldots T g^+ T^+ V) T \chi_{k_i} \rangle\rangle$$

$$= \langle\langle \chi_{k_{f},n} | V | (T^+ T g^+ V \ldots T^+ T g^+ V) T^+ T \chi_{k_i} \rangle\rangle$$

$$= \langle\langle \chi_{k_f,n} | V | (g^+ V \cdots g^+ V) \chi_{k_i} \rangle\rangle \qquad (4.31)$$

For a linearly polarized radiation field in the First Born Approximation (FBA) for the scattering potential the S-matrix is

$$S^{(1)} = (-2\pi i) \sum_n \delta(\varepsilon_{k_f} - \varepsilon_{k_i} - n\hbar\omega)\, J_n(\underset{\sim}{\alpha}_0 \cdot \underset{\sim}{k}_{fi}) \langle \underset{\sim}{k}_f | V | \underset{\sim}{k}_i \rangle \qquad (4.32)$$

with J_n the Bessel function of first kind,

$$\left.\begin{aligned} |\underset{\sim}{k}\rangle &= \exp(i\underset{\sim}{k}\cdot\underset{\sim}{r}) \\ \underset{\sim}{\alpha}_0 &= (e/m\omega^2)\underset{\sim}{E}_0\,; \quad \underset{\sim}{k}_{fi} = \underset{\sim}{k}_f - \underset{\sim}{k}_i \end{aligned}\right\} \qquad (4.33)$$

The S-matrix series, Eq.(4.28), which is gauge independent at all orders in V, is in practice too involved to be used in actual calculations beyond the FBA, so that is of use to consider some approximation of $\underset{\sim}{g}^+$ appearing in Eq.(4.28).

&5 LOW FREQUENCY APPROXIMATION

G^+ is the exact propagator of a particle moving in the potential $V(\underset{\sim}{r})$ and in the radiation field. The expansion of G^+ in powers of $V(\underset{\sim}{r})$ has been just considered above. Usually, other approximations to the exact propagator G^+ imply manipulations of the particle-field interaction. At this point it is important to realize that approximating the particle-field interaction makes gauge-dependent a theory which is gauge independent in its usual formulation. Among the different approximations to the particle-field interaction, particular attention is generally paid to the low-frequency approximation (LFA). This approximation is important and useful, because it corresponds to frequently encountered physical conditions at which real scattering processes occur, and because it allows to obtain formal results of general interest (see ref. 14 for details).

The basic assumptions of the LFA are:
(i) the initial particle energy is much greater than the photon energy;

(ii) the field couples strongly to the particle only in the initial and final states;
(iii) in the intermediate states the field couples weakly to the particle.

In what follows we shall operate in the composite Hilbert space $\mathcal{R} + \mathcal{S}$.

In R-gauge $\bar{g}^{+}$ of Eq.(4.28) is

$$\bar{g}^{+R} = \left[\varepsilon_{K_i} - \frac{p^2}{2m} + \frac{e}{mc} \underset{\sim}{A} \cdot \underset{\sim}{p} - \frac{e^2}{2mc^2} A^2 - V + i\hbar \frac{\partial}{\partial t} + i\eta \right]^{-1} \tag{5.1}$$

Expanding g^{+R} in powers of

$$W^R = -\frac{e}{mc} \underset{\sim}{A} \cdot \underset{\sim}{p} + \frac{e^2}{2mc^2} A^2 \tag{5.2}$$

gives

$$\bar{g}^{+R} = g_0^{+E} + g_0^{+E} W^R g_0^{+E} + g_0^{+E} W^R g_0^{+E} W^R g_0^{+E} + \cdots \tag{5.3}$$

with g_0^{+E} given by Eq.(4.18). Inserting (5.3) into (4.28) one obtains

$$\begin{aligned} S = & (-2\pi i) \sum_n \delta(\varepsilon_{K_f} - \varepsilon_{K_i} - n\hbar\omega) \\ & \times \Big\{ \langle\langle \chi^R_{K_f,n} | V | \chi_{K_i} \rangle\rangle + \langle\langle \chi^R_{K_f,n} | V g_0^E V | \chi^R_{K_i} \rangle\rangle \\ & + \sum_{\nu=3}^{\infty} \langle\langle \chi^R_{K_f,n} | V | \left[g_0^{+E} W^R \right]^{\nu-2} g_0^{+E} V \chi^R_{K_i} \rangle\rangle \Big\} \end{aligned} \tag{5.4}$$

Combining the first three terms of (5.4), and taking only the terms proportional to ω^0 and ω^1, for the non resonant part the Kroll and Watson result[15] is recovered:

$$\begin{aligned} S = & (-2\pi i) \sum_n \delta(\varepsilon^0_{K_f} - \varepsilon^0_{K_i} - n\hbar\omega) J_n(\underset{\sim}{\alpha}_0 \cdot \underset{\sim}{K}_{fi}) \\ & \times \langle \underset{\sim}{K}_f - \underset{\sim}{\delta} k | T(\varepsilon^0_{K_i} - \delta(\varepsilon_i)) | \underset{\sim}{K}_i - \underset{\sim}{\delta} k \rangle \end{aligned} \tag{5.5}$$

with

$$\left.\begin{aligned} \varepsilon^{o}_{k} &= \hbar^2 k^2/2m \\ \delta(\varepsilon_i) &= (n\hbar\omega)\, \underset{\sim}{k_i}\cdot\underset{\sim}{d_o}/(\underset{\sim}{k_{fi}}\cdot\underset{\sim}{d_o}) \\ \underset{\sim}{\delta k} &= (n\hbar\omega)\, m\, \underset{\sim}{d_o}/(\hbar^2 \underset{\sim}{k_{fi}}\cdot\underset{\sim}{d_o}) \end{aligned}\right\} \tag{5.6}$$

In E-gauge, $\bar{g}^{+}$ of Eq.(4.28) is given by $\bar{g}^{+E}$,Eq.(4.16). Expanding in powers of electron-radiation interaction

$$W = -e\, \underset{\sim}{E}(t)\cdot\underset{\sim}{r} \tag{5.7}$$

after some algebra we derive an expansion of S-matrix similar to (5.5) where χ^{R}_{k} and W^{R} are substituded respectively with χ^{E}_{k} and W. To obtain the same result of the R-gauge (Eq.(5.5)) is now sufficient to combine only the first two terms of the new S-matrix series, confining to take only the parts proportioanl to ω^{0} and ω^{1} (see ref. 12 for details).

In conclusion the terms of the S-matrix, derived in R-gauge and E-gauge, are not gauge invariant, and considering the same number of terms in the two gauges different approximate expressions of S-matrix are obtained. The differences may be traced back to the expansions of the Green functions $\bar{g}^{+R}$ and $\bar{g}^{+E}$ in powers of operators corresponding to quantities having different meaning and physical content. W^{R}, which is a function of not observable quantities (canonical momentum and vector potential), is an operator representing no precise physical observable, while W, which is a function of observable quantities (such as the vector position and the electric field) is the operator of the energy interaction between the particle and the radiation field.

&6 GAUGE INVARIANT S-MATRIX SERIES

The comment concluding the previous Section contains the key for having a gauge-independent formulation of the S-matrix series, valid at any order. Namely to have a gauge independent series of S-matrix we need expand $\bar{g}^{+G}$, Eq.(4.28), in the same quantity. According to the point (iii) before formula (5.1), it is appropriate to expand $\bar{g}^{+G}$ in powers of W, which is an operator corresponding to a well-defined physical quantity:

$$\bar{g}^{+G} = g^{+G} + g^{+G} W g^{+G} + g^{+G} W g^{+G} W g^{+G} + \cdots \tag{6.1}$$

Inserting (6.1) into (4.28), yields the series:

$$S = (-2\pi i) \sum_n \delta\left(\varepsilon_{k_f} - \varepsilon_{k_i} - n\hbar\omega\right) \times \Big\{ \langle\langle \chi^G_{k_f,n} | V | \chi^G_{k_i} \rangle\rangle + \langle\langle \chi^G_{k_f,n} | V g_0^{+G} V | \chi_{k_i} \rangle\rangle + \sum_{\nu=3}^{\infty} \langle\langle \chi^G_{k_f,n} | V \left[g_0^{+G} W \right]^{\nu-2} g_0^{+G} V \chi^G_{k_i} \rangle\rangle \Big\} \tag{6.2}$$

By using Eq.s (2.2) and (2.16) the independence of any single term of (6.2) with respect of the choice of the gauge is easily proved. Choosing the E-gauge, (6.2) gives, of course, the same results previously obtained in this gauge, while (6.2) written in R-gauge is different from (5.4), though the summations of all terms of two series give identical results. In conclusion the results obtained with (6.2) coincide term by term only with those derived in E-gauge.

&7 CONCLUSIONS

We have reviewed a number of basic notions concerned with unitary transformations of wavefunctions and operators of quantum mechanics, and with correspondence of operators to physical quantities. It was required to discuss how gauge aspects appear both in optical transitions and in laser assisted collisions. The latter were of main concern in the present Lecture. It is found that the S-matrix formulation of a field assisted atomic collision in which states embedded in the field are used as scattering states exhibits the interesting peculiarity of being gauge independent at any order of the scattering potential, provided exact and correctly transformed wavefunctions are used. The same theoretical formulation becomes generally gauge-dependent as soon as the particle-radiation interaction is treated in some approximate way. For this latter case, it is shown how proceed to have a gauge-invariant S-matrix series. It basically requires in any gauge to expand the exact electron propagator in the same physical quantity. In this Lecture we have confined our discussion to potential scattering, as it appears the study case where the differences with respect to optical transitions are the most neat. Inclusion of the internal structure of the colliding systems will complicate the analysis. In that case, however, it may be anticipated that the problem will exhibit features which are typical of field assisted collisions, and features which are shared by the optical transitions as well.

ACKNOWLEDGMENTS

This work was supported in part by the Italian Ministry of Education, The National Group of Structure of Matter and The Sicilian Regional Committee for Nuclear and Structure of Matter Researches.

REFERENCES

1. W.L. Peticolas, R. Norris and K.F. Rieckoff, J. Chem. Phys. 42, 4164 (1968).
2. F. Bassani, J.J. Forney and A. Quattropani, Phys. Lett. 39, 1070 (1977).
3. J.J. Forney, A. Quattropani and F. Bassani, Nuovo Cimento B37, 78 (1977).
4. E.A. Power in: Multiphoton Processes, edited by J. Eberly and P. Lambropulos (1978).
5. Y. Aharonov and C.K. Au, Phys. Rev. A20, 1553 (1979) ———— Phys. Lett. A86, 269 (1981).
6. A. Quattropani, F. Bassani and S. Carillo, Phys. Rev. 25A, 3079 (1982.
7. M. Zukowski, Phys. Lett. 90A, 169 (1982).
8. A. Quattropani and F. Bassani, Phys. Rev. Lett. 50, 1258 (1983).
9. C.K. Au, J. Phys. B: At. Mol. Phys. 16, L563 (1983).
10. A. Quattropani and R. Girlanda, Rivista del Nuovo Cimento 6, 1 (1983).
11. R.R. Schlicher, W. Becker, J. Bergou and M.O. Scully in: Quantum Electrodynamics and Quantum Optics, Ed. A.O. Barut (Plenum Publishing Corporation) (1984).

12. C. Leone, P. Cavaliere, G. Ferrante and M. Zukowski, J. Phys. B: At. Mol. Phys. 18, 4225 (1985).
13. H. Sambe, Phys. Rev. A7, 2203 (1973).
14. G. Ferrante in: Fundamentals of Atomic Collisions, Ed.s H. Kleinpoppen, H.O. Lutz and J.S. Briggs, Plenum, New York (1985).
15. N.M. Kroll and K.M. Watson, Phys. Rev. A8, 804 (1973).

Drift and diffusion of laser-excited atoms

G. Nienhuis

Fysisch Laboratorium, Rijksunversiteit Utrecht, Postbus 80 000,
3508 TA, The Netherlands

1. Introduction

An intense light beam traversing an atomic gas can create gas-kinetic flows, which can be directly understood in a qualitative fashion. Let us assume that the light frequency lies within the Doppler width of an atomic resonance line, and that the position and velocity of an atom changes only slightly during the lifetime of the excited state. Then the density of excited atoms is spatially non-uniform, reflecting the spatial distribution of the light intensity. Moreover the velocity components of the excited atoms have a narrow distribution around the Doppler-selected value, and the excited atoms experience a drift. Of course, the spatial inhomogeneity and the velocity selection are largely compensated by a complementary distribution of ground-state atoms. Hence the total atomic density, irrespective of the internal state will remain uniform in general, and likewise the total velocity distribution is expected to be Maxwellian.

However, the gas-kinetic properties of excited and ground-state atoms are different in general. Therefore one could expect gas-kinetic flows of the two atomic states that do not exactly counterbalance each other, leaving an effective net flow resulting from all atoms combined.

Until now, the main emphasis has been placed on the simplest example of gas-kinetic flow induced by light, which is the light-induced drift (LID). This effect was predicted by Gel'mukhanov and Shalagin [1,2], and a first experimental observation was reported by Antsigin et al. [3]. The LID effect occurs when the vapor of active atoms is immersed in a buffer gas, and when the frequency of the incident light is slightly off-resonance, but still within the width of the Doppler-broadened line. Then due to the selection of a non-zero preferential velocity in the beam direction, the excited-atom density experiences a flow, whereas the ground-state atoms have a Bennett hole in their velocity distribution, corresponding to an opposite flow. Usuallly the excited atoms present a larger cross section for velocity-changing collisions (VCC) to the buffer-gas particles than the ground-state atoms, so that the flow of excited atoms suffers the greater diffusive friction. The net result is an effective flow of the total density of active atoms in a direction opposite to the direction of the selected velocity. This is illustrated in fig. 1. For excitation in the red Doppler wing, the atoms are pushed forward by the light in its propagation direction, whereas excitation in the blue wing has the effect of pulling the atoms in the upstream direction.

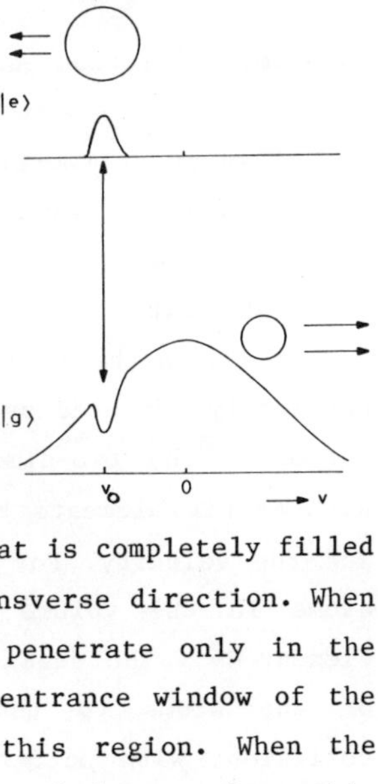

Fig. 1. Scheme of light-induced drift. Radiative excitation creates excited atoms with a preferential velocity. The net drift of these atoms is not exactly counterbalanced by the opposite drift of the ground-state atoms, since the excited atoms present a larger cross section for velocity-changing collisions to the buffer-gas particles.

In an elongated cell with a cross section that is completely filled by the light, the atoms cannot escape in the transverse direction. When the system is optically thick, the light can penetrate only in the region of a few optical dephts just after the entrance window of the cell, and the LID effect can occur only in this region. When the frequency is in the red wing, the atoms are pushed away from this region, so that the light can penetrate further, giving rise to LID further down the cell. For sufficiently high intensity the atoms can be effectively swept away towards the dark end of the cell. The light acts in this way as a semipermeable piston, which compresses the active atoms, while leaving the buffer-gas particles untouched. This effect of the optical piston was predicted by Gel'mukhanov and Shalagin [2], and it was demonstrated in an experiment by Werij et al. [4].

In a previous paper [5] we presented a theoretical treatment of LID and the optical piston within the weak-collision model for VCC, and by treating the collisions and the free flow as a perturbation. In the present paper we give a more systematic treatment, which allows direct generalization to higher-order flows (such as heat flow), with a less restrictive treatment of the collisions. We derive explicit expressions for the drift velocity and for the maximum density that can be sustained by the piston, in terms of the various parameters of the system. The results should be helpful to optimize the system in which LID can be studied experimentally, and they illustrate the details of the mechanism of the effect.

2. Atomic evolution equations.

A vapor of two-state atoms is immersed in a buffer gas. The atoms are supposed to have two internal states, the ground state $|g\rangle$ and the excited state $|e\rangle$, which are coupled by a radiation field, which we shall describe classically. The density of the buffer gas is high compared with the atomic density, so that the state of the buffer gas is negligibly affected by the atomic density and velocity distribution. We introduce the two-dimensional density matrix $\sigma(\vec{r},\vec{v},t)$ of the atoms, with matrix elements that are distribution functions over the position and the velocity. For instance, $\sigma_e(\vec{r},\vec{v},t)d\vec{r}d\vec{v}$ is the number of excited atoms in the volume element $d\vec{r}$, and with a velocity within the element $d\vec{v}$. The density matrix σ varies with time due to the coupling of the atoms to the driving field, to spontaneous emission, to collisions with buffer-gas particles that change the atomic velocity, and to the free flow of atoms with a velocity $\vec{v}$. The radiative effects give rise to terms as occurring in the optical Bloch equations, and the motional effects are expressed by terms of the same type as in the Boltzmann equation. The full evolution equations for the matrix elements are

$$\frac{\partial}{\partial t}\sigma_e = -A\sigma_e + \tfrac{1}{2}i\Omega(\sigma_{ge} - \sigma_{eg}) + L_e[\sigma_e] - \vec{v}.\mathrm{grad}\,\sigma_e$$

$$\frac{\partial}{\partial t}\sigma_g = A\sigma_e + \tfrac{1}{2}i\Omega(\sigma_{eg} - \sigma_{ge}) + L_g[\sigma_g] - \vec{v}.\mathrm{grad}\,\sigma_g \qquad (2.1)$$

$$\frac{\partial}{\partial t}\sigma_{eg} = -[\tfrac{1}{2}A+\gamma - i(\Delta - \vec{k}.\vec{v})]\sigma_{eg} + \tfrac{1}{2}i\Omega(\sigma_g - \sigma_e) - \vec{v}.\mathrm{grad}\,\sigma_{eg}$$

where $\sigma_{eg} = \sigma_{ge}{}^*$. Here A is the rate of spontaneous decay, Ω is the Rabi frequency, measuring the atom-field coupling strength, $\Delta = \omega - \omega_o$ is the detuning of the light frequency ω from the atomic resonance frequency ω_o and the term $\vec{k}.\vec{v}$ reflects the Doppler shift. The collisional effects are contained in the damping rate γ of the optical coherences, and in the Boltzmann-type terms L_e and L_g for excited atoms and ground-state atoms, which are of the form

$$L[\sigma] = -\kappa(v)\,\sigma(\vec{v}) + \int d\vec{v}'\,K(\vec{v} \leftarrow \vec{v}')\,\sigma(\vec{v}') \quad . \qquad (2.2)$$

The kernel $K(\vec{v} \leftarrow \vec{v}')$ gives the transition rate from an initial velocity $\vec{v}'$ to a final velocity $\vec{v}$, by binary VCC. The total collision rate $\kappa(v)$ is related to this kernel by the obvious sum rule

$$\int d\vec{v}\,K(\vec{v} \leftarrow \vec{v}') = \kappa(v') \quad . \qquad (2.3)$$

It is the difference in the collision kernels K_e and K_g for the two states that is responsible for the LID effect. We have supposed that the coherence kernel can be ignored, since velocity changing collisions will tend to damp the coherence whenever the potentials for the two states interacting with a buffer gas particle are markedly different [6]. Hence VCC only enhance the collisional damping rate γ of the coherences σ_{eg} and σ_{ge}, just as dephasing collisions. Note that all collisional parameters κ_e, κ_g, K_e, K_g and γ are proportional to the buffer-gas density, and that collisions between active atoms are neglected. This makes the Boltzmann terms linear in the distribution σ_e and σ_g.

Eqs. (1) are a combination of the optical Bloch equations and the Boltzmann equation. These equations will serve as the basis for our treatment of LID.

3. Separation of time scales

An exact solution of the eqs. (2.1) will be very difficult, in particular when the light intensity is inhomogeneous. Then the Rabi frequency Ω is a function of the position $\vec{r}$ in the system, which makes all density-matrix elements σ_{ij} function of $\vec{r}$. In the present paper we are mainly interested in the time evolution of the total density of the atoms, due to the effect of LID. This time evolution occurs on a time scale governed by the spatial inhomogeneities of the system, such as diffusive flows. This macroscopic time scale is much slower than the time scale determined by the radiative transitons, and it seems reasonable to assume that the radiative transitions have driven the system to a steady state at each position $\vec{r}$ in the system, before the LID effect has had time to modify the local density of atoms. The macroscopic flow of atoms then occurs slowly compared with the radiative transition rates, so that the deviations of the local distribution of atoms over the internal states from the local steady state are small at each instant. A similar statement could be made concerning the local velocity distribution of atoms, since the thermalizing effect of VCC is expected to occur rapidly compared with the slow macroscopic time scale.

In order to put these general considerations on a more rigorous footing, we first give a formal systematic treatment of a separation of time scales. We consider a general linear evolution equation of the form

$$\frac{d}{dt}\,\sigma(t) = L\,\sigma(t) \quad , \tag{3.1}$$

where we can separate the evolution operator according to

$$L = L_0 + L_1 \quad . \tag{3.2}$$

In accordance with the considerations mentioned above we assume that L_0 drives the system to a steady state, before L_1 has had time to cause any appreciable change. Hence we assume that the non-zero eigenvalues of L_0 have a negative real part of the order of some scaling parameter λ, which is large compared with the size of L_1 . We define a projection operator P, which projects any state $\sigma(t)$ on the eigenspace of L_0 with eigenvector zero, which is by definition the space of steady states with respect to L_0. The defining property of this projection operator P is expressed by

$$PL_0 = 0 \qquad L_0P = 0 \quad , \tag{3.3}$$

and the standard property of projectors

$$P^2 = P \quad . \tag{3.4}$$

Note that we did not assume that L_0 is Hermitian, so that we should distinguish right and left hand eigenvectors of L_0, which makes the two relations (3.3) independent of each other. Eq. (3.3) indicates that states in P-space are steady states with respect to L_0. Furthermore we require explicitly that all these steady states are contained in P-space, which means that an arbitrary initial state σ approaches in a time of order λ^{-1} its projection on P-space, when its evolution is governed by L_0. Hence we write

$$e^{L_0\tau}\sigma \to P\sigma \tag{3.5}$$

for values of τ of the order of λ^{-1} or larger, since in this time all non-zero eigenvalues of L_0 cause a decay of their eigenvectors. Finally we introduce the complementary projection operator

$$Q = 1 - P \quad , \tag{3.6}$$

which obeys the equality

$$Q^2 = Q \quad . \tag{3.7}$$

Multiplying eq. (3.1) successively with P and Q, using the identity $\sigma = (P+Q)\sigma$, and applying the equalities (3.3), we obtain the pair of equations

$$\frac{d}{dt} P\sigma(t) = PL_1P\sigma(t) + PL_1Q\sigma(t) \quad , \tag{3.8}$$

$$\frac{d}{dt} Q\sigma(t) = QL_1P\sigma(t) + Q(L_0 + L_1)Q\sigma(t) \quad . \tag{3.9}$$

Our aim is to obtain an approximate equation for $P\sigma(t)$, which evolves only on the slow time scale defined by L_1. Therefore we wish to substitute in the last term of (3.8) an approximate solution for $Q\sigma(t)$, which we expect to be small at least after a small transient time of the order of λ^{-1}. A formal integration of (3.9) gives the identity

$$Q\sigma(t) = \int_0^\infty d\tau \exp\,[Q(L_0 + L_1)Q\tau]QL_1P\sigma(t-\tau) \quad . \tag{3.10}$$

This result does not depend on $Q\sigma$ in the infinite past, since the operator $Q(L_0 + L_1)Q$ causes rapid decay of each operator $Q\sigma$ in Q-space.

If we substitute (3.10) in the last term in (3.8), we obtain an exact, but formal evolution equation for $P\sigma$, in terms of a memory kernel, that represents the coupling with Q-space. Evolution equations of this type are well-known, both in collision physics [7] and in statistical mechanics of transport processes [8]. According to the assumption made above, we expect $Q\sigma$ to be small, and the evolution of $P\sigma$ to be slow, so that an expansion in L_1/λ is justified. Therefore, we evaluate $Q\sigma$, as given by (3.10) to first order in L_1. In the exponential of (3.10), L_1 can then be omitted, and we can use the equality

$$QL_0Q = L_0 \tag{3.11}$$

which follows from (3.3) and (3.5). Furthermore, we can take $P\sigma(t-\tau)$ to zeroth order in L_1. Obviously, in zeroth order $P\sigma$ is constant, so that we may substitute $(P\sigma(t-\tau) \to (P\sigma(t)$. This gives the result

$$Q\sigma(t) = \int_0^\infty d\tau\, e^{L_0\tau} QL_1\, P\sigma(t) \quad , \tag{3.12}$$

where the projector Q does not show up in the potential anymore. The operator L_0 has eigenvalues zero, and is therefore not invertible, but since the exponentiated operator acts only on a Q-part, it decays to zero on the time scale λ^{-1}. Eq. (3.12) confirms that $Q\sigma$ is of order L_1/λ, which is small, according to our suppositions. Substitution of (3.12) into (3.8) gives the final result

$$\frac{d}{dt} P\sigma(t) = PL_1\, P\sigma(t) + PL_1 \int_0^\infty d\tau\, \exp(L_0\tau)\, QL_1\, P\sigma(t) \quad , \tag{3.13}$$

which is a closed evolution equation for $P\sigma(t)$, valid to second order in L_1. One notices that this equation contains no memory effect. Of course, this result has a general validity, and it resembles the Chapman-Enskog method of kinetic theory [9]. In this paper we apply eq. (3.13) to the light-irradiated system outlined in section 2.

4. Diffusion equation for the density of active atoms

In order to apply the formalism of the previous section, in particular eq. (3.13), to the atomic evolution equations (2.1), we have to define a separation of the total evolution in a rapid part L_0 and a slow part L_1, and we must give the corresponding projector P on the eigenspace of L_0 with eigenvalue zero.

As argued in the introduction, the effects of LID result from the difference between the Boltzmann terms L_e and L_g, giving rise to a difference in the thermalization rate of excited and ground-state atoms, and the free-flow term $-\vec{v}.\mathrm{grad}\,\sigma$, which is responsible for diffusive flows arising from density gradients. Hence it is natural to define the perturbation part L_1 as the sum of $L_e - L_g$, acting on the excited-state distribution, and the free-flow term, and we write

$$\begin{aligned} (L_1\sigma)_e &= (L_e - L_g)\sigma_e - \vec{v}.\mathrm{grad}\,\sigma_e \\ (L_1\sigma)_g &= -\vec{v}.\mathrm{grad}\,\sigma_g \\ (L_1\sigma)_{eg} &= -\vec{v}.\mathrm{grad}\,\sigma_{eg}, \quad (L_1\sigma)_{ge} = -\vec{v}.\mathrm{grad}\,\sigma_{ge} \quad . \end{aligned} \tag{4.1}$$

The evolution operator L_0 is defined as the complementary evolution operator, such that (2.1) is equivalent to (3.1) and (3.2). Since L_0 contains only radiative transitions and a single Boltzmann operator L_g acting both on excited and ground-state atoms, the evolution described by L_0 does not affect spatial inhomogeneities, and it drives the velocity-dependence of σ at each position to a steady state, indicated by $\bar{\sigma}(\vec{v})$, and obeying the defining relations

$$L_0\ \bar{\sigma}(\vec{v}) = 0 \quad ; \qquad \int d\vec{v}\ \mathrm{Tr}\bar{\sigma}(\vec{v}) = 1 \quad . \tag{4.2}$$

The radiative transitions do not modify the atomic velocities, and we find

$$\mathrm{Tr}\ L_0\ \sigma = L_g[\mathrm{Tr}\sigma] \quad . \tag{4.3}$$

This means that Tr $\bar{\sigma}(\vec{v})$ is eigenvector of L_g, with eigenvalue zero, and therefore it must be equal to the Maxwell distribution $W(\vec{v})$ at the temperature of the buffer gas. Hence we may write

$$\mathrm{Tr}\ \bar{\sigma}(\vec{v}) = W(\vec{v}) \tag{4.4}$$

indicating that L_0 drives the total atomic velocity distribution to a Maxwell distribution.

The projection operator P on the eigenspace of L_0 with eigenvalue zero is now defined by

$$P\sigma(\vec{r},\vec{v},t) = \bar{\sigma}(\vec{v})\ n(\vec{r},\vec{t}) \tag{4.5}$$

with

$$n(\vec{r},t) = \int d\vec{v}\ \mathrm{Tr}\ \sigma(\vec{r},\vec{v},t) \tag{4.6}$$

the density of atoms, irrespective of their velocity and their internal state. The validity of the equalities (3.3) follows then from (4.2), and the fact that

$$\int d\vec{v}\ \mathrm{Tr}\ L_0\ \sigma(\vec{v}) = \int d\vec{v}\ L_g[\mathrm{Tr}\ \sigma(\vec{v})] = 0 \qquad (4.7)$$

reflectng that radiative transitons leave the total density of atoms in each velocity class unchanged, and that the Boltzmann operator L_g does not modify the atomic density. Eq. (3.4) can be checked for the definition (4.5) by using the normalization conditions (4.2) of $\bar{\sigma}(\vec{v})$.

If we substitute these definitions in the approximate evolution equation (3.13), we obtain an evolution equation for the P-part of σ, which is proportional to the density n of atoms. This equation is valid, provided that the time scale of the drift effect is slow compared with the rate of radiative transitions, and the rate of velocity thermalization. We note that P gives zero when it is followed by the Boltzmann operator L_e or L_g, since these operators merely redistribute the velocities, without modifying the density of atoms. Likewise we find the equality

$$P(\vec{v}.\mathrm{grad})P = 0 \quad , \qquad (4.8)$$

since the left projection operator effectively takes a velocity average, with a distribution that has been made into the Maxwell distribution by the right projection operator. We conclude that the first term on the right-hand side of (3.13) disappears, and that the difference L_e-L_g contributes only as part of the second operator L_1 in the second term. The effect of the operator L_0 in the exponent in (3.13) is simplified by noting that

$$\mathrm{Tr}\ e^{L_0\tau}\sigma(t) = e^{L_g\tau}\ \mathrm{Tr}\sigma(t) \quad . \qquad (4.9)$$

which generalizes (4.3), and which can be proven in the same fashion after differentiating with respect to τ. The final result for the evolution of the atomic density reads

$$\frac{\partial}{\partial t} n(\vec{r},t) = \frac{\partial}{\partial \vec{r}} \cdot [D \frac{\partial}{\partial \vec{r}} n(\vec{r},t) - \vec{u}(\vec{r},t)\ n(\vec{r},t)] \quad . \qquad (4.10)$$

The constant D is defined by the relation

$$D = \frac{1}{3}\int_0^\infty d\tau \int d\vec{v}\ \vec{v}.[e^{L_g\tau} W(\vec{v})\vec{v}] = \frac{1}{3}\int_0^\infty d\tau \langle\vec{v}(\tau).\vec{v}(o)\rangle \quad , \tag{4.11}$$

and has the significance of the diffusion constant of ground-state atoms in the buffer gas. The effective velocity $\vec{u}$ is given by

$$\vec{u}(\vec{r},t) = \int d\vec{v}\ \vec{v}\int_0^\infty d\tau\ e^{L_g\tau}(L_e - L_g)\ [\bar{\sigma}(\vec{v})] \quad , \tag{4.12}$$

and it serves as an effective drift velocity of the atoms. The dependence of $\vec{u}$ on $\vec{r}$ and t is due to a possible variation of the intensity of the incident light (and thereby of the Rabi frequency Ω) on position and time, which arises from the time dependence of the density of absorbing atoms.

The result (4.10) - (4.12) is exctly the evolution equation (3.13) for the special choice of P, L_0 and L_1. The effect of LID is continued in the effective drift velocity $\vec{u}$, which depends on the local intensity of the incident light, which determined the normalized velocity distribution of excited atoms. This drift velocity is due to a difference in the Boltzmann operators L_e and L_g for excited and ground-state atoms, which causes different diffusive frictions of both atomic states.

For simple models of the diffusive friction, an initial velocity $\vec{v}$ of the atoms in the buffer gas damps out at a uniform rate ζ, which may be different for excited atoms and ground-state atoms. This is true both for a weak-collision model, where L_e and L_g are approximated by a Fokker-Planck equation [5], and for a strong-collision model, which we shall outline in section 6. In these cases, we can write for an arbitrary distribution function $f(\vec{v})$

$$\int d\vec{v}\ \ \vec{v}\ L[f(\vec{v})] = -\ \zeta\int d\vec{v}\ \vec{v}\ f(\vec{v}) \quad , \tag{4.13}$$

where we may substitute L_e or L_g for L, and accordingly ζ_e or ζ_g for ζ. Then we find from (4.12)

$$\vec{u}(\vec{r},t) = -\ \frac{\zeta_e - \zeta_g}{\zeta_g}\ \int d\vec{v}\ \vec{v}\ \sigma_e(\vec{v}) \quad . \tag{4.14}$$

The damping rate ζ is related to the diffusion constants by

$$D = k_B T/m\zeta_g \quad , \quad D' = k_B T/m\zeta_e \quad , \tag{4.15}$$

with D and D' the diffusion constants for ground-state and excited atoms with mass m in a buffer gas at temperature T, and with k_B Boltzmann's constant.

Eqs. (4.10) with (4.12) or (4.14) constitute the principal result of this section. The atomic density is found to obey a diffusion equation, with a drift term added, where the drift flow $\vec{u}n$ is the flow of excited atoms, multiplied by the relative difference of the two damping rates (or the two diffusion constants). This result (4.10) is justified when the diffusive flow - D grad n and the light-induced flow $\vec{u}n$ are sufficiently small not to cause appreciable variations of the density n during radiative lifetimes and velocity thermalization times. This puts a lower limit to the density of the buffer gas, which must be high enough to justify diffusion-type equations. Obviously, there will be no appreciable drift effect when the buffer gas density is so low that excited atoms have a negligible probability to suffer a velocity change during their lifetime. On the other hand, when full thermalization occurs during the lifetime of the excited state, the excited-state velocity distribution $\bar{\sigma}_e$ will become a Maxwell distribution, and the drift velocity $\vec{u}$ will be negligible. Furthermore, the frequency of the incident radiation must be off-resonance by an amount of the order of a Doppler width in order to maximize the drift velocity $\vec{u}$ .This is in line with the experimental observation [4]. A larger detuning will tend to decrease the fraction of excited atoms, and thereby the distribution $\bar{\sigma}_e$ in (4.14), and a smaller detuning will cause the excited atoms to have a smaller average velocity. When ζ_e is large than ζ_g (or when D' is smaller that D), as will usually be the case, the drift velocity $\vec{u}$ will be positive for excitation in the red Doppler wing, and negative for excitation in the blue Doppler wing.

We shall illustrate these general conclusions in the case of a special model. First we derive a general equation for the variation of the intensity, and thereby of the drift velocity, with position.

5. Absorption equation

We consider a beam of light, propagating through the system in the z-direction, for a given density $n(\vec{r})$ of the active atoms. The absorption can be derived from the work $\vec{E}.d\langle\vec{\mu}\rangle/dt$ which the field performs on the atomic dipoles. It is reasonable to assume that the atomic density matrix σ may be replaced by the steady-state solution $\bar{\sigma}(\vec{v})$, , defined as the eigenvector of L_0 with eigenvalue zero, for each local value of the Rabi frequency. The net power absorbed by a unit volume of the system is then equal to

$$\frac{\partial}{\partial z} I(\vec{r},t) = - \hbar\omega\, A\, n(\vec{r},t) \int d\vec{v}\; \bar{\sigma}_e(\vec{v}) \quad , \tag{5.1}$$

where $\bar{\sigma}_e(\vec{v})$ is itself a function of the intensity at position $\vec{r}$ and time t. This equality reflects that the number of photons removed from the incident beam is equal to the photons reemitted as fluorescence radiation.

The pair of differential equations (4.10) and (5.1) determine in a self-consistent way both the density as a function of $\vec{r}$ and t, and the intensity, for a given value of the incident intensity. The intensity distribution I is a functional of the density at the same instant, and the drift velocity $\vec{u}$ determining the evolution of the density depends upon the local intensity. Together these equations constitute an interesting set of differential equations coupled in a highly non-linear manner.

6. Strong-collision model

As a special model for the Boltzmann operators L_e and L_g, which describe VCC, we assume that collisions are represented by a thermalizing process occurring by random jumps at a rate ζ, which is independent of the atomic velocity, such that each jump restores the Maxwell distribution. Hence we write

$$L[f(\vec{v})] = -\zeta\, f(\vec{v}) + \zeta\, W(\vec{v}) \int d\vec{v}' f(\vec{v}') \quad , \tag{6.1}$$

both for L_e and L_g. This model follows from the general expression (2.2) of the Boltzmann operator by putting $\kappa(\vec{v})$ equal to ζ, and the kernel $K(\vec{v} \leftarrow \vec{v}')$ to $\zeta W(\vec{v})$, indicating that collisions occur at a rate ζ, independent of the velocity, and that a single collision is sufficient for complete thermalization. Alternatively, we may consider ζ as an effective thermalization rate.

It is easy to check that with this model the general relations (4.15) between the rates ζ_e and ζ_g and the diffusion constants D' and D are valid, and that eq. (4.14) for the drift velocity is exact. Hence the explicit specification of the diffusion equation (4.10) with the expression (4.12) for the drift velocity, and the absorption equation (5.1) requires the evaluation of the steady-state density matrix $\bar{\sigma}(\vec{v})$, which is the normalized eigenvector of L_0 with eigenvalue zero. The zeroth-order evolution L_0 is obtained by substituting in (2.1) eq. (6.1) with $\zeta = \zeta_g$, both for L_e and L_g, while ignoring the free-flow terms $-\vec{v}.\text{grad}\,\sigma$. The resulting steady-state distribution of excited atoms can be shown to be

$$\bar{\sigma}_e(\vec{v}) = [Aa(\vec{v}) + \zeta_g\langle a\rangle]\, W(\vec{v})/[A + 2\zeta_g\langle a\rangle] \tag{6.2}$$

where we defined

$$a(\vec{v}) = s(\vec{v})/[1 + 2s(\vec{v}) + \zeta_g/A] \tag{6.3}$$

with $s(\vec{v})$ the saturation parameter

$$s(\vec{v}) = (bI/\pi A)\, \text{Re}[\tfrac{1}{2}A + \gamma - i(\Delta - \vec{k}.\vec{v})]^{-1} \quad . \tag{6.4}$$

This saturation parameter is the ratio between the effective rate of stimulated radiative transitions for atoms with velocity $\vec{v}$, and the rate of spontaneous decay. The Einstein coefficient b for stimulated transitons is related to the Rabi frequency by

$$\tfrac{1}{2}\pi\Omega^2 = bI \quad . \tag{6.5}$$

Notice that I, and thereby also Ω, s and a are also functions of $\vec{r}$ and t, although we suppressed this dependence in eq. (6.2) - (6.5). The averages in (6.2) are taken with respect to the Maxwell distribution $W(\vec{v})$.

The drift velocity $\vec{u}$ can be directly found as a function of the intensity I, by substituting (6.2) into (4.14), and we obtain

$$\vec{u} = -\frac{D-D'}{D'} A \langle \vec{v}a \rangle / [A+2\zeta_g \langle a \rangle] \tag{6.6}$$

Likewise we find for the absorption equation (5.1) the explicit result

$$\frac{\partial}{\partial z} I(\vec{r},t) = -\hbar\omega A\, n(\vec{r},t)\, (A+\zeta_g)\, \langle a \rangle / [A+2\zeta_g \langle a \rangle] \quad . \tag{6.7}$$

The pair of equations (4.10) and (6.7), together with (6.6) and (6.3) provide us now with the explicit basic equations for the effect of light-induced drift. For each density distribution n, eq. (6.7) determines the intensity distribution over the system, as soon as the incident intensity at the entrance window is given. On the other hand, eq. (4.10) gives the evolution equation for the density, with $\vec{u}$ a known function of the local intensity, as specified by eq. (6.6), together with (6.3) and (6.4).

7. Strong velocity selection

We make two additional assumptions in order to simplify the equations for the density and the intensity. First we consider an elongated cell, oriented in the z-direction, with a cross section that is completely filled by the light that propagates in the same direction. Then both n and I are functions of z only, and the equations (4.10) and (6.7) can be replaced by their one-dimensional versions. Second, we consider the case that the homogeneous width $\tfrac{1}{2}A+\gamma$ of the absorption line is narrow compared with the Doppler width. Then the saturation parameter s(v) as well as the parameter a(v) differ from zero only in narrow region around the Doppler selected velocity

$$v_o = (\omega-\omega_o)/k \quad . \tag{7.1}$$

Then we may write

$$\langle va(v)\rangle = v_o\langle a(v)\rangle \quad . \tag{7.2}$$

A comparison of (6.6) and (6.7) then yields the equation for the light-induced flow

$$u(z,t)n(z,t) = -Dq \frac{\partial}{\partial z} I(z,t) \quad , \tag{7.3}$$

where we introduce the parameter

$$q = -\frac{D-D'}{DD'} \frac{v_o}{\hbar\omega(A+\zeta_g)} \quad . \tag{7.4}$$

Substituting (7.3) in the one-dimensional version of (4.10) gives

$$\frac{\partial}{\partial t} n(z,t) = D \frac{\partial^2}{\partial z^2} [n(z,t) + qI(z,t)] \quad . \tag{7.5}$$

A simplified explicit form of the absorption equation is found if we use

$$\langle a\rangle = W_o(v_o)\ bI(z,t)/[k(A+\zeta_g)(1+I(z,t)/I_1)^{\frac{1}{2}}] \tag{7.6}$$

with

$$I_1 = \pi(A+\zeta_g)(\tfrac{1}{2}A+\gamma)/2b \tag{7.7}$$

an effective saturation intensity. Substituting (7.6) into (6.7) gives the absorption equation

$$\frac{\partial}{\partial z} I(z,t) = -\ QI(z,t)n(z,t)/[(1+I(z,t)/I_1)^{\frac{1}{2}} + I(z,t)/I_2] \quad , \tag{7.8}$$

where

$$Q = \hbar cbW_o(v_o) \tag{7.9}$$

is the absorption cross section, and

$$I_2 = \hbar\omega A(A+\zeta_g)/2\zeta_g Q \tag{7.10}$$

determines a second effective saturation intensity.

The pair of equations (7.5) and (7.8) for the density and the intensity clearly reveals the strong non-linear coupling between these two quantities. The two types of saturaton arising in the absorption equation (7.8) are paramatrized by the effective intensities I_1 and I_2. The parameter I_1 determines the threshold intensity for inhomogeneous saturation of atoms with velocities within a homogeneous width around the selected velocity v_o. This type of saturation leads to a square root variation of the absorption with the intensity, and not to a complete flattening of the absorption. This is due to the fact that at higher intensities atoms with velocities further away from the selected velocity v_o start taking part in the absorption process, since the far wing of their Doppler-shifted Lorentzian absorption profile is no longer negligible. In fact, this effect is equivalent to saturation broadening of the Bennett hole in the ground-state distribution. A high rate ζ_g of thermalization makes I_1 higher, so that this saturation becomes less effective. Conversely, a high value of ζ_g makes the other saturation intensity I_2 smaller. This saturation is of a homogeneous type. Due to the collisional redistribution of the atomic velocities over the Maxwell distribution, all atoms have a probability to attain a velocity within the Bennet hole, and thereby take part in he absorption. Therefore a high value of ζ_g makes this saturation threshold I_2 easier to reach.

The parameter q, given in (7.4), transforms the intensity I into an effective density qI, which we shall show to be the maximum density to which the LID effect can drive up the atoms. It is this parameter that characterizes the strength of LID. It is proportional to the selected velocity, at least as long as v_o lies within the Doppler distribution. When ζ_g is small compared with the spontaneous decay rate A, q is proportional to the buffer-gas density. (Recall that D and D' are inversely proportional to this density.) In the reverse case, when thermalization of the velocity of the atoms is fast compared with the spontaneous lifetime, q becomes independent of the buffer-gas density.

The absorption equation (7.8) determines the intensity as a function of z, for a given time t, in terms of the the density distribution n(z,t) at the same instant. In fact, for a given incident intensity I_o, the dependence of I on z and t is only an implicit dependence. If we introduce the quantity M(z,t) as the integrated density n(z',t) up to the value z', according to the equality

$$M(z,t) = \int^{z} dz' n(z',t) \quad , \tag{7.11}$$

then eq. (7.8) can be rewritten as

$$\frac{dI}{dM} = - f(I) \tag{7.12}$$

with

$$f(I) = QI/[(1+I/I_1)^{\frac{1}{2}} + I/I_2] \quad . \tag{7.13}$$

Obviously, (7.12) determines I as a function of M, for a given value I_o of the incident intensity, corresponding to M = 0. The dependence of I on M is formally expressed by the intensity

$$M(I) = \int_{I}^{I_o} dI' [f(I')]^{-1} \quad . \tag{7.14}$$

Since f is a positive function of I, M is a monotonously decreasing function of I, and conversely, I is a monotonously decreasing function of M, with $I = I_o$ for M = 0. The integral (7.14) can be performed analytically for the function f determined by (7.13), by making the substitution $s = (1+I/I_1)^{\frac{1}{2}}$, but the result is a bit awkward, and is not easy to invert into an expression for I as a function of M. Rather than pushing the possibilities for analytic solutions to their limits, we prefer to consider simple solutions for n and I in special cases.

8. Stationary solution

We consider a cell with closed ends, and we look for stationary solutions of (7.5) and (7.8). Then the flux of atoms must vanish, and integrating (7.5) twice gives

$$n(z) + qI(z) = F \tag{8.1}$$

with F an integration constant. Substitution of (8.1) for n in (7.8) gives a differential equation for I alone, that can be integrated to yield z as a function of I. We shall not write down this rather complex result. Rather we draw some general conclusions.

In the case of excitation in the red wing, the parameter q is positive. Since I is monotonously decreasing as a function of z, n must be increasing to keep the sum (8.1) constant. The total decrease in density equals qI_o for an optically thick system, when I(z) drops to zero over the cell length. Obviously, this effect is relatively small when qI_o is small compared with the average density. A dramatic effect can be expected when the decrease in qI is larger than the average density. Then the LID effect pushes the atoms to the dark side, until a maximum density qI is reached.

On the other hand, when q is negative, which is the case for excitation in the blue wing, both n and I are decreasing functions of z. The atoms are pulled upstream, and produce a density maximum at the entrance window. When F is positive, and the cell is sufficiently long, full absorption takes place, and the integration constant equals the density at the dark side of the cell. For negative values of F, eq. (8.1) allows a solution with an incident intensity obeying $|qI_o| > |F|$. Then the intensity I(z) must remain finite for all values of z, which means that the optical thickness is finite, and that the density decays to zero for large z values. Then the finite amount of atoms is enclosed by the LID effect just after the entrance window, even for a half infinite cell.

It is a simple matter to illustrate these general conclusions in the simple case of negligible saturation, where the absorption equation

(7.8) may be replaced by

$$\frac{\partial}{\partial z} I(z) = - QI(z)\, n(z) \tag{8.2}$$

For positive values of q, the solution of (8.1) and (8.2) is

$$\begin{aligned} qI(z) &= F/[1 + \exp(QF(z - z_o))] \\ n(z) &= F/[1 + \exp(- QF(z - z_o))] \quad , \end{aligned} \tag{8.3}$$

with z_o an additional integration constant, indicating the position where qI and n are both equal to F/2. This solution is plotted in figure 2. When the cell is located in the region with $z < z_o$, the system is optically thin, and no appreciable absorption takes place. The density then shows an exponential increase. When the cell is located in the region with $z > z_o$, the intensity is too low to modify the density, which remains practically uniform. When the position $z = z_o$ is located within the cell, there is sufficient light absorption to sweep the atoms to the dark side of the cell. We can say that the optical piston pushes the atoms, until a density F is sustained at the dark end, which is about equal to qI_o.

For negative q-values, corresponding to blue-wing excitation, the solution of (8.1) and (8.2) is

$$\begin{aligned} qI(z) &= F/[1 - \exp(QF(z - z_o))] \\ n(z) &= F/[1 - \exp(-QF(z - z_o))] \quad . \end{aligned} \tag{8.4}$$

Fig. 2. Stationary solution for the density and the intensity, for a positive value of q.

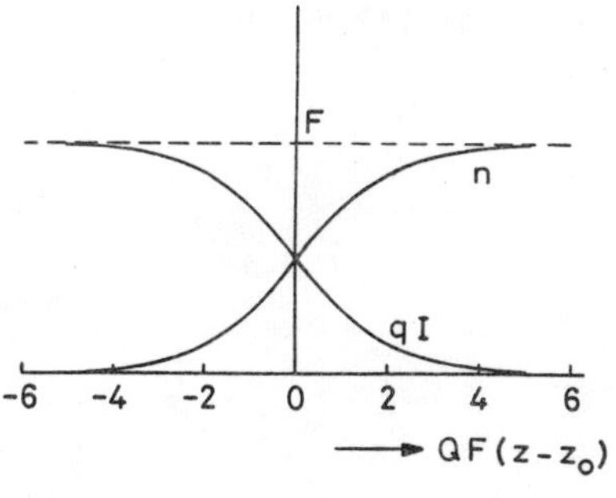

These solutions are shown in figure 3, both for a positive and a negative value of the constant F.

In this case without saturation, the drift velocity u is found by combining (7.3) with (8.2), and we obtain

$$u(z) = DqQI(z) \quad . \tag{8.5}$$

Hence the drift velocity has the same behavior as I(z). In the general case, also allowing saturation, the parameter q can be determined from the value of the maximum density behind the piston, which is equal to qI_o for an optically thick system, and sufficiently high intensity I_o.

The special symmetry of the solution (8.3), as displayed in figure 2, is typical for the case without saturation, and can be expressed by

$$n(z - z_o) = qI(z_o - z) \quad . \tag{8.6}$$

In the presence of saturation, the decrease of I on the light side of the piston (for $z < z_o$) is slower than predicted by the result (8.3). Accordingly, the low-density wing in this region is higher than in figure 3.

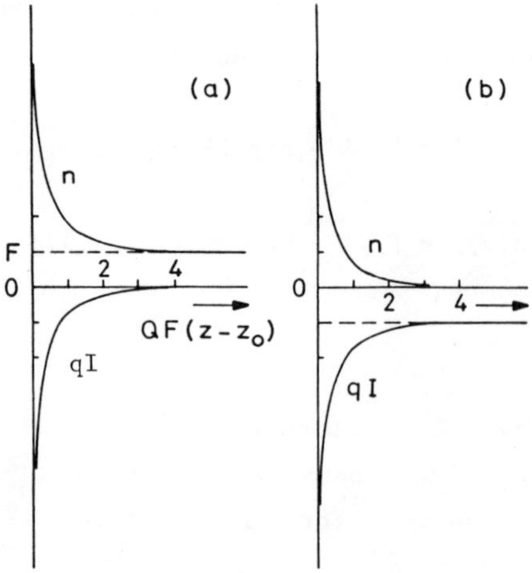

Fig. 3. Stationary solution for the density and the intensity, for a negative value of q. (a) F > 0. (b) F < 0.

9. Solitary packet

Another simple case of practical interest arises when a packet with a finite amount of atoms moves with uniform velocity w in an infinite cell. If we write

$$n(z,t) = n(z - wt)$$
$$I(z,t) = I(z - wt) \quad , \tag{9.1}$$

we can cast (7.5) in the form

$$- w \frac{d}{dz} n(z) = D \frac{d^2}{dz^2} [n(z) + qI(z)] \quad . \tag{9.2}$$

As in (7.11), we introduce the integrated density M(z) up to the value z in the moving frame, so that the absorption equation is replaced by (7.12) and (7.13).

If we integrate (9.2) we obtain

$$wn(z) = - D \frac{dn}{dz} + u(z)n(z) \quad , \tag{9.3}$$

where we used (7.3). This relation simply tells us that the total flow of atoms in the uniformly moving packet is the sum of the diffuse flow, determined by the diffusion constant, and the light-induced flow un. Hence, at the maximum of the packet, where the diffusion flow vanishes we must have

$$u_m = w = Dqf(I_m) \tag{9.4}$$

with f determined by (7.13), and I_m the intensity at the packet maximum.

If we integrate (9.1) twice, we obtain the differential equation

$$\frac{d}{dz} M(z) = - wM(z)/D + q(I_o - I(z)) \quad . \tag{9.5}$$

Notice that I(z) is a function of M(z) only, according to (7.14). The intensity which is left after passing through the entire packet is I_∞, which is determined by

$$M_\infty = \lim_{z\to\infty} M(z) \quad . \tag{9.6}$$

Since M_∞ must be finite in order to allow a uniform velocity, the density dM/dz must go to zero at infinity. Hence we find from (9.4) the expression for the packet velocity

$$w = qD(I_o - I_\infty)/M_\infty = \int dz\ u(z)\ n(z)/ \int dz\ n(z) \quad , \tag{9.7}$$

which has the significance of an average drift velocity for the entire packet. (This is also obvious from (9.3), since the average diffusive velocity is zero.)

For an optically thick packet, I_∞ is negligible compared with I_o, and we find

$$w = qDI_o/ M_\infty \quad , \tag{9.8}$$

which is inversely proportional to the optical thickness QM_∞. We may assume that at the dark side of the packet the light-induced drift is negligible, at least for sufficiently large QM_∞. Then (9.3) shows that the dark wing of the packet decays exponentially, with a characteristic length

$$l_d = D/w = M_\infty /qI_o \quad . \tag{9.9}$$

On the bright side of the packet, the drift velocity u is large compared with the packet velocity w, so that a characteristic length of the bright wing is

$$l_b = D/u_o = 1/qf(I_o) \quad . \tag{9.10}$$

In the absence of saturation, when f(I) = QI, tha ratio l_d/l_b amounts to the optical thickness QM_∞. With saturation, the bright wing will have a larger characteristic length.

At the maximum of the wavepacket, we find from (9.5)

$$n_m = - wM_m/D + q(I_o - I_m) \quad . \tag{9.11}$$

At large optical thickness QM_∞, and when I_m is not saturating, we find from (9.4) and (9.8)

$$I_m = I_o/QM_\infty \quad . \tag{9.12}$$

which is much smaller than I_o and can be neglected. If we substitute (9.8) again in the first term on the right-hand side of (9.11), we obtain

$$n_m = qI_o(1 - M_m/M_\infty) \quad . \tag{9.13}$$

This will be close to qI_o, since the dark wing is much longer than the bright wing of the packet.

In the absence of saturation we can give an explicit relation between I and M in the form

$$I(z) = I_o \exp(- QM(z)) \quad . \tag{9.14}$$

Then the relation (9.6) between w and u takes the form

$$w = DqI_o(1 - \exp(- QM_\infty))/M_\infty \quad , \tag{9.15}$$

and the drift velocity obeys the identity

$$u(z) = QDqI_o \exp(- QM(z)) \quad . \tag{9.16}$$

The maximum of the packet as determined by (9.4) has the M-value given by

$$QM_m = \log QM_\infty - \log(1 - \exp(- QM_\infty)) \quad . \tag{9.17}$$

which gives after substitution in (9.5)

$$n_m = qI_o \, [1 - \frac{w}{u_o}(1 + \log \frac{u_o}{w})] \quad , \tag{9.18}$$

with w and $u_o = u(o)$ given in (9.12) and (9.13). For large values of QM_∞ this value of n_m approaches qI_o, in accordance with the general conclusion that qI_o is the maximum density that can be sustained by the piston.

10. Stability of the solitary packet

An important problem in the theory of non-linear differential equations with soliton-like solutions is the question of their stability. In this section we shall prove that the solitary-packet solution of the previous section is stable. This is what one would intuitively expect, since a local increase of the density gives rise to increased absorption, leading to a decreased drift velocity further down the light beam.

We consider a solution of the equations of motion (7.5) and (7.8), transformed to the moving frame with uniform velocity w, as given by (9.7). Note that this velocity is uniquely defined for a given value of the total integrated density M_∞ and the incident intensity I_o. In this moving frame the integrated density M(z,t) obeys the equation

$$\frac{\partial}{\partial t} M(z,t) = w \frac{\partial}{\partial z} M(z,t) + D \frac{\partial}{\partial z} [n(z,t) + qI(z,t)] \tag{10.1}$$

with w given by (9.7), which follows from integrating (7.5). We introduce the quantity P(z,t) as the deviation of M(z,t) from the solitary-packet solution. Likewise

$$p(z,t) = \frac{\partial}{\partial z} P(z,t) \tag{10.2}$$

is the density deviation, and $J(z,t)$ is the intensity deviation. These quantities obey the equation

$$\frac{\partial}{\partial t} P(z,t) = w \frac{\partial}{\partial z} P(z,t) + D \frac{\partial}{\partial z} [p(z,t) + qJ(z,t)] \quad , \qquad (10.3)$$

which follows after subtracting from (10.1) the equation for the steady-state solution. Notice that $P(z,t) = 0$ for $z \to \pm \infty$, and the same holds for J.

We now consider two successive values of z where $P = 0$ at a certain instant t, and we call these positions z_1 and z_2. Hence

$$P(z_1(t),t) = 0 \quad , \quad P(z_2(t),t) = 0 \quad , \qquad (10.4)$$

and $P \neq 0$ for z-values in between z_1 and z_2. To be specific, we assume that P is positive in the region between z_1 and z_2, and we write

$$P(z,t) > 0 \quad \text{for} \quad z_1(t) < z < z_2(t) \quad . \qquad (10.5)$$

We introduce the quantity

$$G(t) = \int_{z_1(t)}^{z_2(t)} dz' P(z',t) \quad , \qquad (10.6)$$

which is obviously a positive quantity. These definitions are illustrated in figure 4. Consider now the time derivation of G, which obeys the equation

$$\begin{aligned} \frac{d}{dt}G(t) &= P(z_2(t),t) \frac{d}{dt} z_2(t) - - P(z_1(t),t) \frac{d}{dt} z_1(t) \\ &+ w[P(z_2(t),t) - P(z_1(t),t)] \\ &+ D[p(z_2(t),t) + qJ(z_2(t),t) - p(z_1(t),t) - qJ(z_1(t),t)] \quad . \end{aligned} \qquad (10.7)$$

The intensity at any position is determined by the value of M at that position, so that $J(z,t) = 0$ as soon as $P(z,t) = 0$. If we apply (10.4), we notice that the only non-zero terms on the right-hand side of (10.7) are the density deviations p. As will be obvious from an inspection of figure 4, these terms make the time derivative of G negative, and we find

$$\frac{d}{dt}G(t) = D[p(z_2(t),t) - p(z_1(t),t)] < 0 \quad . \tag{10.8}$$

In the same way we may prove that the derivative of the integral of P over a region of negative P-values is positive. We conclude that the quantity

$$H(t) = \int_{-\infty}^{\infty} dz\,|P(z_1 t)| \quad , \tag{10.9}$$

which is a measure of the deviation of M from the steady-state form, can only decrease with time. In this sense, an initial deviation from the steady state can only decrease. This finishes our proof that the solitary-packet solution, discussed in the previous section is stable.

Fig. 4. Sketch illustrating the definition of two successive positions z_1 and z_2, where the integrated density M is equal to the steady-state solution in the moving frame. The solid line indicates M, the steady-state solution is represented by the broken line. The shaded area is the quantity G, defined in (10.6).

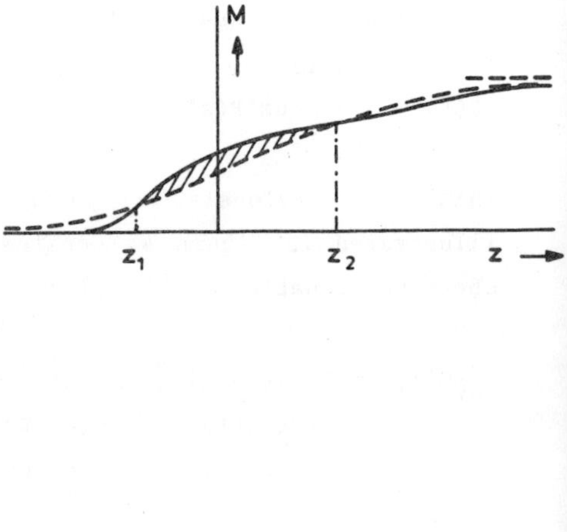

11. Conclusions

We have derived a general equation of motion for the density of atoms in a buffer gas, irradiated by monochromatic velocity-selecting light. This equation is given in (4.10), where the drift velocity $\vec{u}$ is given in (4.14). The absorption equation, that determines the local intensity for a given density distribution, is written in (5.1). When the velocity-changing collisions are described by thermalizing random jumps at a rate ζ_g, and for a long cell with negligible lateral intensity variations, these general results take the special form (7.5) and (7.8). These equations have solitary-packet solutions which are discussed in section 9. We have demonstrated that an arbitrary distribution of atoms with a given integrated density M_∞ always tends to this solitary solution.

References

1. F.Kh. Gel'mukhanov and A.M. Shalagin, Pis'ma Zh. Eksp. Teor. Fiz. 29 (1979) 773 (JETP Lett. 29 (1979) 711).
2. F.Kh. Gel'mukhanov and A.M. Shalagin, Zh. Eksp. Teor. Fiz. 78 (1980) 1674 (Sov. Phys. JETP 51 (1980) 839).
3. A.D. Antsigin, S.N. Atutov, F. Kh. Gelmuk'hanov, A.M. Shalagin and G.G. Telegin, Opt. Commun. 32 (1980) 237.
4. H.G.C. Werij, J.P. Woerdman, J.J.M. Beenakker and I. Kuscer, Phys. Rev. Lett. 52 (1984) 2237.
5. G. Nienhuis, Phys. Rev. A31 (1985) 1636.
6. P.R. Berman, Appl. Phys. 6 (1975) 283.
7. H. Feshbach, Ann. Phys. 5 (1958) 357.
8. R. Zwanzig, J. Chem. Phys. 33 (1960) 1338.
9. U.M. Titulaer, Physica A100 (1980) 234.

GAS KINETICS IN A LASER RADIATION FIELD

F.Kh.Gel'mukhanov and A.M.Shalagin

Institute of Automation and Electrometry, Siberian Branch of the USSR Academy of Sciences, 630090 Novosibirsk, USSR

ABSTRACT

This lecture is devoted to the new physical phenomenon - light-induced drift (LID) and related with LID phenomenons. The discovery of the LID effect causes the qualitative changes in the notions of the nature of light action on a gas state. In the LID phenomenon case the light and the gas exchange only by entropy and in this sense the LID effect can be called entropic. At the initial stage of theoretical investigations it was clear already that class of the physical problems caused by LID wasn't exhausted by LID phenomenon itself. The "nonforce" mechanism of creating of nonequilibrium which has been discovered by us changes qualitatively the character of the all totality of transport phenomena proceeding in radiated gas.

All this allows us to speak about the conceive of the new branch in the physics of interaction of radiation with matter, namely, about the gas kinetics in a laser radiation field.

In 1979 we predicted a new physical phenomenon and called it the phenomenon of light-induced drift (LID) [1,2]. The discovery of this phenomenon caused the qualitative changes in the notions of the nature of a light action on a gas state. It was thought earlier that nonequilibrium conditions (relative to translational degrees of freedom) could be created due to momentum exchange between radiation and matter (direct force action) or dissipation of light energy. In the LID phenomenon case internal flows and density gradients of the gas components are created in the initially equilibrium space-homogeneous gas. It should be noted that for the creating this nonequilibrium situation neither momentum transfer from light to matter nor dissipation of light energy are required. Thus the light and the gas exchange only by entropy and in this sense the light action can be called entropic. It is very important circumstance from the point of view of the academical physics.

Let us remined the physics of the LID phenomenon. We consider the interaction of a plane running monochromatic light wave with an ensemble of absorptive particles in a mixture with a buffer gas. Radiation is absorbed at a transition 0-1 from the ground state 0. The absorptive line is broadened by the Doppler effect. Under these conditions only those particles interact with radiation whose velocity projection on the wave vector $\vec{k}$ is close to the resonant one, i.e. which corresponds to the condition $\vec{k}\vec{v} = \Omega \equiv \omega - \omega_{10}$. Here ω is the radiation frequency, ω_{10} is the frequency of the transition 0-1. If $\Omega \neq 0$ the excited particles appear with nonzero velocity projection on the wave vector, i.e. there is the flow of the excited particles $\vec{j}_1$ which is collinear to $\vec{k}$. In the ground state the opposite flow $\vec{j}_0$ occurs due to the decrease of unexcited particles in the interval of resonant velocities. The radiation initiates the flows $\vec{j}_1$ and $\vec{j}_0$ with opposite directions. Until collisions have manifested themselves the flows $\vec{j}_1$ and $\vec{j}_0$ have equal values, so the gas as a whole is motionless (if the light pressure, which usually can be neglected in the LID theory, is not taken into account). In the presence of a buffer gas each flow $\vec{j}_i$ is impeded and the force density $\vec{F}_i$ is equal

$$\vec{F}_i = - m_a \nu_i \vec{j}_i ,$$

where m_a is the mass of an absorptive particle, ν_i has the meaning of the transport collisional frequency (see below); $i=m,n$. In general the interaction laws of excited and unexcited particles with buffer one are different, so the transport frequences are different too ($\nu_1 \neq \nu_0$). In consequence of this a total nonzero friction force density $\vec{F} = \vec{F}_1 + \vec{F}_0$ arises, which affects the gas and puts it into motion. This is the physical basis of the LID phenomenon.

The drift direction depends on the sign of the difference $\nu_1 - \nu_0$ and changes to the opposite one with the change of the sign of the detuning Ω. For example, if $\nu_1 > \nu_0$ and $\Omega < 0$, the particles drift towards the $\vec{k}$-direction. The momentum conservation law makes the buffer gas to move in the opposite direction

The same effect as the LID is produced qualitatively also by well known light pressure effect, i.e. it produces flows of absorbing particles in the gas and leads to spatial redistribution of the density. We compare now the degrees of manifestation of the LID and of the light pressure effect. Let us consider a gas in a cell with closed ends. In this case $\vec{j} = \vec{j}_1 + \vec{j}_0 = 0$ and therefore $\vec{F} = -(\nu_1 - \nu_0)\,\vec{j}_1$. If $\Omega = k\bar{v}$, we obtain the following rough estimate for the friction force density

$$F \sim m_\alpha (\nu_0 - \nu_1)\, \bar{v}\, \rho_1 ,$$

where $\bar{v}$ is the mean thermal velocity and ρ_1 is the excited particle density. We compare this force with the light pressure force, which density F_L is equal approximatelly to $\hbar k \Gamma_1 \rho_1$. Here Γ_1 is the decay constant of the excited state. For $\Gamma_1 \sim \nu_1$ we obtain

$$\frac{F}{F_L} \sim \frac{m_\alpha \bar{v}}{\hbar k} \left(\frac{\nu_0 - \nu_1}{\nu_1} \right) .$$

The principal factor that determines this relation is $m_\alpha \bar{v}/\hbar k$ - the ratio of the thermal momentum of the particle to the photon momentum. For the optical region of the spectrum and at the room temperature $m_\alpha \bar{v}/\hbar k \sim 10^4$. Thus, the force F causing the LID can exceed by three orders of magnitude the force density of the spontaneous light pressure.

Unlike the light pressure and other effects the LID is not connected with the direct force action exerted by the radiation on the individual particles. In this respect LID occupies a special position. The energy of the directional motion of the particles, produced in the LID case, is drawn from the thermal energy of the gas. This decreases, of course, the entropy of the gas mixture, but this decrease is offset of the entropy of the radiation produced when it is scattered in the gas.

By the present time the existence of the LID phenomenon has been proved in the experiments with Na vapour [3,4,5] , Ne [6] , molecules CH_3F [7-10] and NH_3 [11] . It must be noted specially the work [5]. This work is the first experimental observation of the LID of Na vapour in which the physical adsorption of the Na vapour on cell walls has been eliminated. The problem of interaction between cell walls and Na vapour was solved by coating of cell walls by compressor oil for the high vacuum pumps. The LID of electrons in semiconductors [12,13] and light-induced current in gases [14–17] has been revealed in the experiments [18] and [16], respectively. Recently the effect like LID phenomenon - diffusive suction of particles by light field [19] was observed too [20].

At the initial stage of theoretical investigations it was clear already that class of the physical problems caused by LID wasn't exhausted by LID phenomenon itself. The "nonforce" mechanism of creating of nonequilibrium which has been discovered by us changes qualitatively the character of the all totality of transport phenomena proceeding in radiated gas [21] : besides the particle flows the heat flows appear, the pressure or temperature become anisotropic, the character of diffusion and thermal diffusion changes, the new sound branches appear, the light-induced nonequilibrium of the particle distribution at the magnetic sublevels causes the new mechanisms of gas rotation.

All this allows us to speak about the conceive of the new branch in the physics of interaction of radiation with matter, namely, about the gas kinetics in a laser radiation field.

1. The equations of gas kinetics in a radiation field. Three-liquid hydrodynamics

The state of a radiated gas is defined if the density matrix $\rho_{ij}(\vec{v})$ is known. In a simplest model of two level absorbing particles interacting with the field of a running monochromatic wave $\vec{E}\exp(i\omega t - i\vec{k}\vec{r})$ the next kinetic equations are valid for $\rho_{ij}(\vec{v})$ [22] ($\rho_i(\vec{v}) = \rho_{ii}(\vec{v})$)

$$\left(\frac{\partial}{\partial t} + \vec{v}\nabla + \Gamma_1\right)\rho_1(\vec{v}) = m_\alpha S_1(\vec{v}) - 2\,\mathrm{Re}\left[iG^*\rho_{10}\left(\vec{v} - \frac{\vec{v}_0}{2}\right)\right],$$

$$\left(\frac{\partial}{\partial t} + \vec{v}\nabla\right)\rho_0(\vec{v}) = m_\alpha S_0(\vec{v}) + \hat{\Gamma}_1\rho_1(\vec{v}) + 2\,\mathrm{Re}\left[iG^*\rho_{10}\left(\vec{v} + \frac{\vec{v}_0}{2}\right)\right], \qquad (1.1)$$

$$\left(\frac{\partial}{\partial t} + \vec{v}\nabla + \frac{\Gamma_1}{2} - i(\Omega - \vec{k}\vec{v})\right)\rho_{10}(\vec{v}) = m_\alpha S_{10}(\vec{v}) + iG\left[\rho_0\left(\vec{v} - \frac{\vec{v}_0}{2}\right) - \rho_1\left(\vec{v} + \frac{\vec{v}_0}{2}\right)\right],$$

$$\Omega = \omega - \omega_{10}, \qquad G = Ed_{10}/2\hbar, \qquad \vec{v}_0 = \hbar\vec{k}/m_\alpha .$$

Here d_{10} is the dipole moment matrix element, $\vec{v}_0$ is the recoil velocity, m_α is the mass of absorptive particles, $\hat{\Gamma}_1 \rho_1(\vec{v}) = \Gamma_1 \int d\hat{k}\, \rho_1(\vec{v}+\vec{v}_0)/4\pi$ is the integral operator tacing into account the recoil effect, $S_1(\vec{v})$ and $S_0(\vec{v})$ are the collisional integrals of excited and unexcited particles respectively. The collisional relaxation of the optical coherence between the states is described by the collisional integral $S_{10}(\vec{v})$. Let us make an ordinary assumption that the phase memory in collisions is absent

$$m_\alpha S_{10}(\vec{v}) = -\gamma\, \rho_{10}(\vec{v}), \tag{1.2}$$

where $\Gamma = \gamma + \frac{\Gamma_1}{2}$ is homogeneous half width of the absorption line. The effects of phase memory in the LID problem were taken into account in the works [23-26]. The distribution function of the buffer gas is described by the ordinary Boltzmann equation

$$\left(\frac{\partial}{\partial t} + \vec{v}\nabla\right)\rho_b(\vec{v}) = m_b\, S_b(\vec{v})\,. \tag{1.3}$$

As it is known, even the more simple Boltzmann equation permits approximate solutions only which describe weak nonequilibrium states [27]. We shall obtain the approximate solution of the eqs. (1.1) and (1.3) below using the 13-moment Grad's method [28] well-known in the classical gas dynamics. The following expansion of the distribution function corresponds to the 13-moment method

$$\begin{aligned}
&\rho_\alpha(\vec{v}) = W_\alpha(\vec{c})\Big[\rho_\alpha + \frac{2}{\bar{v}_\alpha^2}\rho_\alpha \vec{w}_\alpha\vec{c} + \frac{2}{\bar{v}_\alpha^4}\sum_{rs}\pi_{\alpha rs}\big(c_r c_s - \tfrac{1}{3}\delta_{rs}c^2\big) + \frac{8}{5\bar{v}_\alpha^6}\vec{h}_\alpha\vec{c}\big(c^2 - \tfrac{5}{2}\bar{v}_\alpha^2\big)\Big],\\
&W_\alpha(\vec{c}) = (\pi\bar{v}_\alpha^2)^{-3/2}\exp(-c^2/\bar{v}_\alpha^2)\,, \qquad \vec{c} = \vec{v} - \vec{u}\,,\\
&\rho_\alpha = m_\alpha n_\alpha = \langle \rho_\alpha(\vec{v})\rangle, \qquad \rho = \sum_\alpha \rho_\alpha\,, \qquad n = \sum_\alpha n_\alpha\,,\\
&\vec{j}_\alpha = \rho_\alpha\vec{u}_\alpha = \langle \vec{v}\,\rho_\alpha(\vec{v})\rangle, \quad \rho\vec{u} = \sum_\alpha \rho_\alpha\vec{u}_\alpha\,, \quad \sum_\alpha \rho_\alpha\vec{w}_\alpha = 0, \quad \vec{w}_\alpha = \vec{u}_\alpha - \vec{u},\\
&p_\alpha = n_\alpha T_\alpha = \tfrac{1}{3}\langle c^2\rho_\alpha(\vec{v})\rangle, \qquad p = nT = \sum_\alpha p_\alpha\,,\\
&\pi_{\alpha rs} = \langle (c_r c_s - \tfrac{1}{3}\delta_{rs}c^2)\,\rho_\alpha(\vec{v})\rangle\,, \qquad \pi_{rs} = \sum_\alpha \pi_{\alpha rs}\,,\\
&\vec{q}_\alpha = \vec{h}_\alpha + \tfrac{5}{2}p_\alpha\vec{w}_\alpha = \tfrac{1}{2}\langle \vec{c}\,c^2\rho_\alpha(v)\rangle\,, \quad \vec{q} = \sum_\alpha \vec{q}_\alpha\,,\\
&\langle \Phi(\vec{v})\rangle = \int d\vec{v}\,\Phi(\vec{v}).
\end{aligned} \tag{1.4}$$

Here $\pi_{\alpha rs}$ and $\vec{q}_\alpha$ are the pressure tensor and the heat flow of the particles of the type $\alpha=0,1,\beta$; $\bar{v}_\alpha=(2T_\alpha/m_\alpha)^{1/2}$ is the average velocity. The heat flow $\vec{h}_\alpha$ differs from $\vec{q}_\alpha$ by the absence of the convective energy transfer $\frac{5}{2}p_\alpha\vec{w}_\alpha$.

The transport equations for the observable values ρ_α , $\vec{j}_\alpha$, $\pi_{\alpha rs}$ and $\vec{h}_\alpha$ are obtained from eqs. (1.1) and (1.2) by multiplying them by the functions

$$\psi_\alpha^{(1)}=1, \qquad \vec{\psi}_\alpha^{(2)}=\vec{c}, \qquad \psi_\alpha^{(3)}=\frac{1}{2}(c^2-\frac{3}{2}\bar{v}_\alpha^2),$$
$$\psi_{\alpha rs}^{(4)}=c_r c_s-\frac{1}{3}\delta_{rs}c^2, \qquad \vec{\psi}_\alpha^{(5)}=\frac{1}{2}\vec{c}(c^2-\frac{5}{2}\bar{v}_\alpha^2)$$

and integrating it over $\vec{v}$. It can be shown the Grad-approximation is well when [29]

$$\frac{\Gamma}{k\bar{v}} \gtrsim 10^{-1}$$

Neglecting by a light pressure ($\vec{v}_0=0$) we obtain the next transport equations for the excited particles ($\frac{d}{dt}=\frac{\partial}{\partial t}+\vec{u}\nabla$) [21,30]

$$\begin{aligned}
&(\Gamma_1+\frac{\partial}{\partial t})\rho_1+\nabla\rho_1\vec{u}_1 = \Pi^{(1)},\\
&(\Gamma_1+\frac{\partial}{\partial t})\rho_1\vec{w}_1+\nabla p_1+\nabla\cdot\pi_1+\rho_1\frac{d\vec{u}}{dt} = \vec{R}_1^{(2)}+\vec{\Pi}^{(2)},\\
&\frac{3}{2}\frac{d}{dt}p_1+\frac{5}{2}p_1\nabla\vec{u}+\nabla\vec{q}_1+(\pi_1\cdot\nabla)\vec{u} = R_1^{(3)}+\Pi^{(3)},\\
&(\Gamma_1+\frac{d}{dt})\pi_{1rs}=R_{1rs}^{(4)}+\Pi_{rs}^{(4)}, \quad (\Gamma_1+\frac{d}{dt})\vec{h}_1=\vec{R}_1^{(5)}+\vec{\Pi}^{(5)}.
\end{aligned} \qquad (1.5)$$

Here

$$R_{\alpha\ldots}^{(i)}=m_\alpha\langle\psi_{\alpha\ldots}^{(i)}\ S_\alpha(\vec{v})\rangle, \qquad \Pi_{\ldots}^{(i)}=\rho_a\langle\psi_{1\ldots}^{(i)}\ 2p(\vec{v})\rangle,$$
$$p(\vec{v})=-\frac{2}{\rho_a}\operatorname{Re} i G^*\rho_{10}(\vec{v}),$$

$\vec{a}\cdot\pi=\sum_{rs}\vec{e}_r a_s\pi_{sr}$, $\vec{e}_r$ is the unit vector along x_r axis, macroscopic variables without α -index relate to the gas mixture as a whole. The absorbing gas as a whole is described by the following equations

$$\begin{aligned}
&\frac{\partial}{\partial t}\rho_a+\nabla\rho_a\vec{u}_a=0, \qquad \frac{d}{dt}\rho_a\vec{w}_a+\nabla p_a+\nabla\cdot\pi_a+\rho_a\frac{d\vec{u}}{dt}=\vec{R}_a^{(2)},\\
&\frac{3}{2}\frac{d}{dt}p_a+\frac{5}{2}p_a\nabla\vec{u}+\nabla\vec{q}_a+(\pi_a\cdot\nabla)\vec{u}=R_a^{(3)},\\
&\frac{d}{dt}\pi_{ars}=R_{ars}^{(4)}, \qquad \frac{d}{dt}\vec{h}_a=\vec{R}_a^{(5)}.
\end{aligned} \qquad (1.6)$$

The evolution of the gas mixture as a whole is described by the transport equations

$$\frac{\partial}{\partial t}\varrho+\nabla\varrho\vec{u}=0,\qquad \varrho\frac{d\vec{u}}{dt}+\nabla p+\nabla\cdot\Pi=0,$$
$$\frac{3}{2}\frac{d}{dt}p+\frac{5}{2}p\nabla\vec{u}+\nabla\vec{q}+(\Pi\cdot\nabla)\vec{u}=0,$$
$$\frac{d}{dt}\Pi_{rs}=R^{(4)}_{rs},\qquad \frac{d}{dt}\vec{h}=\vec{R}^{(5)},\tag{1.7}$$
$$R^{(i)}_{\dots}=R^{(i)}_{\alpha\dots}+R^{(i)}_{\beta\dots},\qquad R^{(i)}_{\alpha\dots}=R^{(i)}_{0\dots}+R^{(i)}_{1\dots}.$$

In the left part of the equations (1.5)-(1.7) the terms unessential for this work are omited. The total equations of gas kinetics in a radiation field are given in the review paper [21]. Linearised moments $R^{(i)}_{\alpha\dots}$ of the Boltzmann collisional integral have the form

$$\vec{R}^{(2)}_{\alpha}=\sum_{\beta}n_\alpha n_\beta\left[\tau^{(1)}_{\alpha\beta}(\vec{u}_\beta-\vec{u}_\alpha)+\tau^{(2)}_{\alpha\beta}\frac{\mu_{\alpha\beta}}{T}\left(\frac{\vec{h}_\beta}{\varrho_\beta}-\frac{\vec{h}_\alpha}{\varrho_\alpha}\right)\right],$$
$$R^{(3)}_{\alpha}=\sum_{\beta}n_\alpha n_\beta\frac{3\tau^{(1)}_{\alpha\beta}}{m_\alpha+m_\beta}(T_\beta-T_\alpha),\tag{1.8}$$
$$R^{(4)}_{\alpha rs}=-\sum_{\beta}\frac{n_\alpha n_\beta}{m_\alpha+m_\beta}\left[\tau^{(3)}_{\alpha\beta}\frac{\Pi_{\alpha rs}}{n_\alpha}+\tau^{(4)}_{\alpha\beta}\frac{\Pi_{\beta rs}}{n_\beta}\right],$$
$$\vec{R}^{(5)}_{\alpha}=\sum_{\beta}\frac{n_\alpha n_\beta}{m_\alpha}\left[-\tau^{(5)}_{\alpha\beta}\frac{\vec{h}_\alpha}{n_\alpha}-\tau^{(6)}_{\alpha\beta}\frac{\vec{h}_\beta}{n_\beta}+\frac{5}{2}\frac{m_\beta T\tau^{(2)}_{\alpha\beta}}{m_\alpha+m_\beta}(\vec{u}_\beta-\vec{u}_\alpha)\right],$$

where $\beta=0,1,B$, the parameters $\tau^{(\ell)}_{\alpha\beta}$ [21,30] are expressed in the terms of Ω -integrals of Chapman-Couling [27],

The expressions (1.8) for $\vec{R}^{(2)}_{\alpha}$ and $\vec{R}^{(5)}_{\alpha}$ describe the collisional interaction of the particle ($n_\alpha\vec{u}_\alpha$) and heat ($\vec{h}_\alpha$) flows. This is the reason of the thermal diffusion and it demonstrates a general phenomenon - the collisional interaction of moments of the distribution functions. However this interaction is weak [27].

The velocity nonequilibrium created by radiation is described by the moments $\Pi^{(i)}_{\dots}$ of the field term $-2\,\mathrm{Re}\,iG^*\varrho_{10}(\vec{v})$. These moments are presented in the work [21]. From the eqs.(1.5) we see directly a nonequilibrium of $\varrho_1(\vec{v})$, which is well known in nonlinear spectroscopy in connection with the Bennet peaks and holes [22]. It is evident that the moments of $\varrho_1(\vec{v})$ are not equal to zero in general case. It means that besides the particle flow $n_1\vec{u}_1$ the heat flow $\vec{h}_1$ of excited particles appear, their temperature changes

($T_1 \neq T$) and their pressure becomes anisotropic ($\pi_{1rs} \neq 0$). As it was shown in the works [1,2] this nonequilibrium of excited particles is transmitted by collisions to a gas as a whole. From the expressions (1.8) for the $R^{(i)}_{a\ldots}$ and $R^{(i)}_{\ldots}$ it follows that this nonequilibrium translation is possible if

$$\sigma_{0\alpha} \neq \sigma_{1\alpha}, \qquad (\alpha = 0, 1, \beta), \tag{1.9}$$

where $\sigma_{i\alpha}$ is the differential cross-section of scattering of particle i on particle α (see table 1).

TABLE 1

Collisional transfer of nonequilibrium on a gas as a whole. T_0 is the temperature of unradiated gas

	excited particles		absorbing particles as a whole	gas mixture as a whole
	$\vec{u}_1$		$\vec{u}_a$	$\vec{u}$
	$\nabla \rho_1 \sim \nabla \mathcal{H}$		$\nabla \rho_a$	$\nabla \rho$
radiation $\longrightarrow$	$T_1 - T$	collisions $\longrightarrow$	$T_a - T$	$T - T_0$
	∇T_1		∇T_a	∇T
	π_1		π_a	π
	$\vec{h}_1$		$\vec{h}_a$	$\vec{h}$

2. LID phenomenon

For simplicity we assume that

$$\frac{n_a}{n_\beta} \ll 1, \qquad \frac{4|G|^2}{\Gamma(\Gamma_1 + \nu_1^{(1)})} \ll 1, \tag{2.1}$$

$$\frac{\Gamma_1}{\nu_1^{(1)}} \ll 1, \qquad \frac{\nu_1^{(i)}}{\nu_1^{(j)}} = \frac{\nu_0^{(i)}}{\nu_0^{(j)}} = \frac{\nu^{(i)}}{\nu^{(j)}}, \tag{2.2}$$

where $\nu_\alpha^{(i)} = \tau_{\alpha\beta}^{(i)} n_\beta / m_a$ is the collisional frequency. The condition of "similarity" (2.2) for the collisional frequencies means that the ratio $\nu_\alpha^{(i)} / \nu_\alpha^{(j)}$ does not depend on the state number α. So the index α will be omitted in this ratio. In the steady-state conditions we obtain from the eqs.(1.5) and (1.6) the next expression for the drift velocity of absorbing gas

$$u_a = \frac{\nu_1 - \nu_0}{\nu_1 \nu_0} \varphi(\Omega) \frac{\bar{v}_a}{n_a \hbar\omega} \frac{dS}{dz} . \tag{2.3}$$

Here z axis is along the wave vector $\vec{k}$, S is the intensity of radiation described by the next equation

$$\frac{dS}{dz} = -\hbar\omega p n_a , \qquad p = \langle p(\vec{v}) \rangle .$$

The so-called φ -function [2,29,31] is expressed in terms of the probability integral $w(z) = \exp(-z^2)[1 + \frac{2i}{\sqrt{\pi}} \int_0^z \exp(t^2)\, dt]$:

$$\varphi(\Omega) = \frac{\langle \vec{k}\vec{v}\, p(\vec{v}) \rangle}{k \bar{v}_a p} \simeq \frac{\mathrm{Re}[z\, w(z)]}{\mathrm{Re}[w(z)]} , \qquad z = \frac{\Omega + i\Gamma}{k\bar{v}_a} . \tag{2.4}$$

In eq.(2.3) the collisional transport frequency ν_α averaged in a proper way was introduced [29,32]

$$\nu_\alpha = \frac{\langle \vec{k}\vec{v}\; S_\alpha(\vec{v}) \rangle}{\vec{k}\vec{j}_\alpha} = \tilde{\nu}_\alpha X(\Omega), \tag{2.5}$$

$$\tilde{\nu}_\alpha = \nu_\alpha^{(1)} [1 - \frac{5}{2} \left(\frac{\mu \nu^{(2)}}{m_a}\right)^2 / \nu^{(1)} \nu^{(5)}] , \qquad \mu = \mu_{ab} ,$$

$$X(\Omega) \simeq \varphi(\Omega) / [\varphi(\Omega) - \frac{\mu}{m_a} \frac{\nu^{(2)}}{\nu^{(5)}} f(\Omega)],$$

$$f(\Omega) = \frac{\mathrm{Re}[z\{z^2 - \frac{3}{2}\} w(z) - i z/\sqrt{\pi}\,]}{\mathrm{Re}\,[w(z)]} .$$

The collisional frequency $\tilde{\nu}_\alpha$ is connected with the ordinary diffusion coefficient $D_\alpha = \bar{v}_\alpha^2 / 2\tilde{\nu}_\alpha$. The Ω -dependence of ν_α is weak in the case $\Gamma/k\bar{v}_a \gtrsim 10^{-1}$ [29].

When the system is optical dense it is possible another regime of LID. It is the so-called optical piston effect [1,4,33] .

The difference of the interaction of excited and unexcited particles with buffer ones changes qualitatively the light pressure effect too, and causes the "negative" light pressure effect [34].

3. Diffusion and thermal diffusion in a radiation field

Let us consider the case $\Omega=0$. If radiation intensity isn't space-homogeneous the radiation produces the nonequilibrium in the space distributions of the excited and unexcited particles. It is evident that collisions transfer this space nonequilibrium to a gas as a whole, if $\sigma_{0b} \neq \sigma_{1b}$. From the eqs.(1.5) and (1.6) the next expression for a density of absorbing particles can be obtained [19]

$$\frac{n_a(r)-n_a(\infty)}{n_a(\infty)} = \frac{\varkappa(r)}{2} \frac{(\nu_1^{(1)}-\nu_0^{(1)})/\nu_1^{(1)}}{1+\varkappa(r)(\nu_1^{(1)}+\nu_0^{(1)})/2\nu_1^{(1)}} . \qquad (3.1)$$

Here

$$\varkappa(r) = \frac{4|G(r)|^2}{\Gamma\Gamma_1}$$

is the saturation parameter, r is the distance between a light beam axis and a point of observation, $n_a(\infty)$ is the density out of a light beam. The expression (3.1) was obtained for the case $\Gamma \gg k\bar{v}$, $\Gamma_1(1+\varkappa) \ll \nu^{(1)}$, $n_a \ll n_b$. When $\nu_1^{(1)} < \nu_0^{(1)}$ the particles are expelled from the light beam $(n_a(r) < n_a(\infty))$ and they are sucked into the light beam if $\nu_1^{(1)} > \nu_0^{(1)}$.

The radiation influences also on the thermal diffusion. This influence is caused by a difference between thermal diffusion coefficients of excited and unexcited particles [21,35] and is connected with the effective averaging (on the states 0 and 1) of the interaction of absorbing and buffer gas.

4. Temperature, pressure anisotropy and heat flows

The temperature change, pressure anisotropy and heat flows appear in the radiated gas due to the velocity selectivity of the interaction between light and particles and due to the difference between scattering cross sections of excited $\sigma_{1\alpha}$ and unexcited $\sigma_{0\alpha}$ particles [21, 30, 36, 37] . The expressions for the temperature, pressure tensor and heat flows of excited particles can be obtained directly from the eqs. (1.5)

$$\frac{T_1-T}{T} = \frac{\Gamma_1}{3\nu^{(1)}}(k\bar{v})^2\frac{(3\Omega^2-\Gamma^2)}{(\Gamma^2+\Omega^2)^2}, \quad \Pi_{1rs}=\delta_{rs}(\delta_{rz}-\frac{1}{3})\frac{\rho_a k^2\bar{v}^4 \varkappa \Gamma_1(3\Omega^2-\Gamma^2)}{2(\Gamma^2+\Omega^2)^2\nu^{(3)}},$$

$$\vec{h}_1 = -\frac{n_1}{n}\lambda\nabla T - \frac{5}{8}T\frac{\varkappa n_a \bar{v}^2\nu^{(2)}}{\nu^{(1)}\nu^{(5)}}\left[\frac{\Gamma_1\Omega\vec{k}}{\Gamma^2+\Omega^2} - \frac{1}{2}\nabla \ln\varkappa\right], \qquad (4.1)$$

$$\varkappa = \frac{4|G|^2\Gamma}{\Gamma_1(\Gamma^2+\Omega^2)} \ll 1, \quad \Gamma_1 \ll \nu^{(1)}, \quad \Gamma \gg k\bar{v}, \quad m_a = m_b = m, \quad \nu_\alpha^{(i)} = \chi_\alpha^{(i)}\frac{n}{m}, \chi_\alpha^{(i)} \simeq \chi_{\alpha\beta}^{(i)}.$$

The collisions transfer this nonequilibrium of excited particles to absorbing gas as a whole if $\sigma_{0\alpha} \neq \sigma_{1\alpha}$

$$T_a - T = (T-T_b)\frac{n_b}{n_a} = \left(\frac{\nu_0^{(1)}-\nu_1^{(1)}}{\nu^{(1)}}\right)\varkappa\frac{n_b}{n}(T_1-T),$$

$$\Pi_{ars} = \Pi_{1rs}\left[\left(\frac{\nu_0-\nu_1}{\nu}\right)n_a + \left(\frac{\nu_0^{(3)}-\nu_1^{(3)}}{\nu^{(3)}}\right)n_b\right]\frac{1}{n}, \qquad (4.2)$$

$$\Pi_{rs} = \Pi_{1rs}\left(\frac{\nu_0-\nu_1}{\nu}\right), \qquad \nu_\alpha = \nu_\alpha^{(3)}+\nu_\alpha^{(4)},$$

$$\vec{h} = -\lambda\nabla T + \frac{5}{8}T\left(\frac{\nu_1-\nu_0}{\nu}\right)\frac{\varkappa n_a\bar{v}^2\nu^{(2)}}{\nu^{(1)}\nu^{(5)}}\left[\frac{\Gamma_1\Omega\vec{k}}{\Gamma^2+\Omega^2} - \frac{1}{2}\nabla\ln\varkappa\right],$$

where $\lambda = 5nT/2m(\nu^{(5)}+\nu^{(6)})$ is a thermal conductivity.

5. Sound branches of radiated gas

There is only one type of longitudinal degenerate sound oscillations in gas in the absence of external fields

$$\omega^2 = g^2c^2, \qquad c^2 = \left(\frac{\partial p}{\partial \rho}\right)_S = \frac{5p}{3\rho},$$

where ω, $\vec{g}$ and c are the frequency, the wave vector and the velocity of a sound wave. The present paragraph is devoted to a theoretical investigation of a spectrum of sound oscillations of a resonantly radiated gas [21,30,38,39]. The investigation which has been made shows splitting of an ordinary branch $\omega^2 = g^2c^2$ and appearance of the new acoustic wave. Let us enumerate them.

1. The effect of light pressure causes transverse sound oscillations which are described by the following dispersion law

$$\omega^2 = \omega_0^2\sin^2\theta, \qquad \omega_0 \sim \Gamma_1\frac{v_0}{c}\frac{\varkappa}{1+\varkappa}. \qquad (5.1)$$

Here Θ is the angle between the wave vectors of sound $\vec{g}$ and light $\vec{k}$.

2. The LID effect causes new sound waves with the following dispersion law

$$\omega = ug \cos\Theta, \tag{5.2}$$

where the velocity $\vec{u}$ is equal approximatelly to that of the light-induced drift. If the masses of the absorptive and buffer particles are equal, the branch (5.2) corresponds to the oscillations of the absorptive gas relative to the buffer one retaining the gas mixture as a whole motionless.

3. As it noted above besides the particle flow radiation initiates the higher velocity moments of a distribution function. The light-induced pressure tensor π_{rs} and heat flow $\vec{h}$ cause sound waves in addition to (5.1) and (5.2). In homogeneous line width limit ($\Gamma \gg k\bar{v}$) the new sound waves for one-component gas are described by the following dispersion law [21,38,40]

$$\omega_{\pm} = \frac{1}{2} g \cos\Theta \left[-s \pm \left(s^2 + 4 s_1^2 \sin^2\Theta\right)^{1/2}\right],$$

$$s = \frac{2}{3c^2}\frac{\partial h}{\partial \rho}, \qquad s_1^2 = -\frac{6\pi}{\rho c^2}\frac{\partial \pi}{\partial \rho}, \qquad \pi \equiv \frac{1}{2}\pi_{zz}. \tag{5.3}$$

If $|s/s_1| \gg 1$ we obtain from eq.(5.3)

$$\omega_+ = -g s \cos\Theta, \qquad \omega_- = g \frac{s_1^2}{s} \sin^2\Theta \cos\Theta. \tag{5.4}$$

Otherwise $|s_1 \sin\Theta / s| \gg 1$

$$\omega_{\pm} = \pm g s_1 \sin\Theta \cos\Theta. \tag{5.5}$$

In these two limiting cases the branches ω_- from (5.4) and $\omega_{\pm}$ (5.5) correspond to transverse oscillations.

6. Macroscopic rotation of gas by light

Radiation causes nonequilibrium in the internal degrees of freedom of absorptive particles. This nonequilibrium is transmited due to collisions to motion of the particles and thus to macroscopic features of gas. We'll consider the effects initiated by the nonequilibrium of the population of magnetic sublevels

[21,41].

1. Magnus phenomenon. Diffusion of oriented particles.

Absorbing circularly polarized radiation the particles receive an angular momentum which is described by the vector of orientation $\vec{\rho}_1$ proportional to the intensity of radiation, i.e. to the saturation parameter $\varkappa$. As the oriented particles arise in the area of localization of radiation they diffuse out of the light beam with the velocity

$$\vec{u} \simeq -D \nabla \ln \varkappa$$

and are affected by the collisions with buffer particles.

In classical hydrodynamics a body moving with the velocity and rotating with the angular velocity $\vec{\omega}$ is affected by the force

$$\delta \vec{F} \backsim [\vec{\omega}\, \vec{u}] .$$

This is the Magnus phenomenon. The microscopic description of the drifting oriented particles [41] reveals an analogy with the motion of the rotating body in gas or liquid. In accordance with this analogy besides an ordinary friction force the absorptive gas affected by the force

$$\delta \vec{F} = \nu' [\vec{u}\, \vec{\rho}_1] = \nu' D [\vec{\rho}_1 \nabla \ln \varkappa] = \nu \rho_1 [\vec{r}\, \vec{\omega}] , \quad \rho_1 = |\vec{\rho}_1| ,$$

which initiates a vortical flow of the absorptive particles

$$\vec{j} \backsim \mathrm{rot}\, \vec{\rho} , \qquad \omega = D \frac{\nu'}{\nu} \frac{\rho_1}{\rho} |\nabla \ln \varkappa| \sim \left(\frac{\ell_0}{a}\right)^2 \frac{\rho_1}{\rho} .$$

The angular momentum associated with this flow is collinear with the wave vector of radiation and changes its direction with the change of the sign of circular polarization.

2. Diffusion of alligned particles. Besides the dipole magnetic moment, i.e. orientation, the radiation initiates hig. her multipolar moments of absorptive partcles. On particular, the quadrupolar magnetic moment , i.e. alignment ρ_2, is initiated. In linearly polarized radiation field the axis of alignment is parallel to the polarization vector $\vec{e}$. The aligned particles

diffusing out of the area of radiation are affected by collisions with buffer particles. The movement of these aligned "ellipsoids" is similar to the movement of a sailing-ship with a keel which is collinear with $\vec{e}$ and a sail which is normal to the wind direction. $\nabla æ$ plays the role of the wind. Thus, there is a nonpotential term in the friction force

$$\delta \vec{F} = -\nu' \varrho_2 \, \vec{e}(\vec{e}\vec{u}) = -\nu' D \varrho_2 \, \vec{e}(\vec{e} \, \nabla \ln æ),$$

which causes the system of whirls in the plane normal to the wave vector of radiation. The lines of these vortical flows are determined by the following equations [21,41]

$$\sin 2\varphi = \Theta(\zeta_0)/\Theta(\zeta), \qquad \zeta = r/a,$$
$$\Theta(\zeta) = \frac{1}{\zeta^2}\left(1 - (1+\zeta^2)\, e^{-\zeta^2}\right),$$

where a is the radius of the light beam, r and φ are the coordinates of the point of observation in the polar system, $a\zeta_0$ is the minimal distance between the beam axis and a flow line. The whirls cause the change in the density of the absorptive particles $\delta\rho(\vec{r})$ both inside and out of the beam

$$\delta \varrho(\vec{r}) = \frac{\nu'}{\nu} \varrho_2 \, \Theta(\zeta) \cos 2\varphi .$$

7. Polarization of a dipolar gas by drift

It is evident that the molecules without a center of symmetry are oriented during their drift with respect to a buffer gas. The physics of this phenomenon is ultimately simple and it can be likened to the orientation of a weathercock by the wind. In our case the polar molecules play a role of this weathercock and the buffer gas is the "wind". The polar molecules have a dipole moment and therefore the constant electric field appears due to their orientation [42].

It is possible an opposite phenomenon that is the drift of radiated polar molecules in a constant electric field [21,43].

The role of radiation is the removal of the gas from the state of a thermodynamics equilibrium.

8. Orientation of stereoisomers by electromagnetic field

In 1978 Baranova and Zel'dovich [44] predicted a propeller effect. Its physics is following. In the field of circular polarization magnetic orientation of particles takes place due to transmision of photon angular momentum to an absorptive particle. As a result the right and the left oriented stereoisomers begin to drift in opposite directions by screwing into a buffer gas like proppelers.

It is possible an opposite phenomenon, that is the magnetic orientation of the right and left steroisomers during their drift. The excited by radiation and unexcited stereoisomers play the role of drifting particles. When a radiational frequency is tuned out of optical resonance ($\Omega = \omega - \omega_{10} \neq 0$), the Doppler effect causes the excitation of the particles with the velocities near $\vec{k}\vec{v} = \Omega$. Thus, the excited and unexcited stereoisomers move in opposite directions when $\Omega \neq 0$. As a result of this motion the latent orientation of the absorptive gas occures due to the orientation of the excited and unexcited particles. A total magnetic moment depending on the sign of the detuning Ω occures with the latent orientation [45]. It is noteworthy the magnetic moment of gas is initiated even by a linearly polarized or nonpolarized radiation [45].

REFERENCES

1 F.Kh.Gel'mukhanov and A.M.Shalagin. JETP Lett.,29,711(1979)

2 F.Kh.Gel'mukhanov and A.M.Shalagin.-Invited Papers 2-nd International Conference on Multiphoton Processes. Budapest, 1980,p.93.

3.V.D.Antsigin, S.N.Atutov, F.Kh.Gel'mukhanov, G.G.Telegin and A.M.Shalagin. JETP Lett.,30,243(1979)

4 H.G.C.Werij, J.P.Woerdman, J.J.M.Beenakker and I.Kuscer. Phys.Rev.Lett.,52,2237(1984)

5 S.N.Atutov.Preprint №288.Institute of Automation and Electrometry, Novosibirsk,1985 (Phys.Lett., in press)

6 S.N.Atutov, P.L.Chapovsky and A.M.Shalagin. Optics Comm., 43,265(1982)

7 P.L.Chapovsky, A.M.Shalagin, V.N.Panfilov and V.P.Strunin. Optics Comm.,40,129(1981)

8 H.Riegler, M.Tacke, H.G.Haefer and E.Skok. Optics Comm., 49,195(1983)

9 V.N.Panfilov, V.P.Strunin and P.L.Chapovsky. JETP,58(3), 510(1983)

10 A.E.Bakarev, A.L.Makas' and P.L.Chapovsky. Preprint №273. Institute of Automation and Electrometry, Novosibirsk,1985 (Sov. J.Quantum Electron., in press)

11 A.K.Folin and P.L.Chapovsky. JETP Lett.,38,№9,549(1983)

12 E.M.Skok and A.M.Shalagin.Pis'ma Zh.Eksp.Teor.Fiz.,32,201(1980)

13 A.M.Dykhne, V.A.Roslyakov and A.N.Starostin.Dokl.AN SSSR, 254,599(1980)

14 F.Kh.Gel'mukhanov and A.M.Shalagin. Sov.J.Quantum Electron., 11,357(1981)

15 A.I.Parkhomenko and V.E.Prokop'ev. Opt.Spectrosc(USSR),53, 597(1982)

16 S.N.Atutov, I.M.Ermolaev and A.M.Shalagin. JETP Lett.,40, №9,1187(1984)

17 I.M.Ermolaev,. Avtometriya,№4,80(1985)

18 A.F.Kravchenko, A.M.Palkina, V.N.Sozinov and O.A.Shegai. Pis'ma Zh.Eksp.Teor.Fiz.,38,328(1983)

19 F.Kh.Gel'mukhanov and A.M.Shalagin.JETP,50,234(1979)

20 S.N.Atutov, S.P.Pod'yachev and A.M.Shalagin. Preprint №228. Institute of Automation and Electrometry.Novosibirsk,1984 (Optics Comm., in press)

21 F.Kh.Gel'mukhanov. Avtometriya,№1,49(1985) (Engl.Transl.: Optoelectronics, instrumentation and data processing)

22 S.G.Raurian,G.I.Smirnov and A.M.Shalagin. Nonlinear Resonances in Atomic and Molecular Spectra. Nauka. Novosibirsk,1979

23 F.Kh.Gel'mukhanov and G.G.Telegin. JETP,53,495(1981)

24 J.Fiutak and S.Zielinska. Optics Comm.,45,392(1983)

25 J.Fiutak, S.Kryszewski and S.Zielinska. Optics Comm.,48, 411(1984)

26 S.Zielinska. J.of Phys.,10B,1333(1985)

27 J.H.Ferziger and H.G.Kaper. Mathematical Theory of Transport Processes in Gases. North-Holland,1972

28 H.Grad. Comm.Pure and Appl.Math.,2,331(1949)

29 F.Kh.Gel'mukhanov, L.V.Il'ichov and A.M.Shalagin. Preprint №286. Novosibirsk,1985 (Physica:A, in press)

30 F.Kh.Gel'mukhanov and L.V.Il'ichov. Khimicheskaya Fizika, 3,№11,1544(1984)

31 V.R.Mironenko and A.M.Shalagin. Izv.AN SSSR, Ser.Fiz.,45, 995(1981)

32 A.M.Shalagin. Synopsis of Thesis prep. by doctor (Institute of Automation and Electrometry, Novosibirsk,1982)

33 G.Nienhuis. Phys.Rev.,A31,1636(1985)

34 F.Kh.Gel'mukhanov. Kvantovaya Elektronika,8,1881(1981)

35 N.V.Karlov. Izv.AN SSSR, Ser.Fiz.,44,2048(1980)

36 F.Kh.Gel'mukhanov, A.I.Parkhomenko,V.E.Prokop'ev and A.M.Shala-

gin. Kvantovaya Elektronika, 7,2246(1980)

37 G.A.Levin and K.G.Folin. Pis'ma Zh.Eksp.Teor.Fiz.,32,160(1980)

38 F.Kh.Gel'mukhanov and L.V.Il'ichov.Phys.Lett.,103A,61(1984); Zh.Eksp.Teor.Fiz.,88,733(1985)

39 F.Kh.Gel'mukhanov. Akusticheskii Zhurnal,31,522(1985)

40 E.E.Zudin and V.N.Sazonov. Zh.Eksp.Teor.Fiz.,88,1600(1985)

41 F.Kh.Gel'mukhanov and L.V.Il'ichov. Zh.Eksp.Teor.Fiz.,88,40(1985); Izv.AN SSSR. MZG, №2,118(1985)

42 F.Kh.Gel'mukhanov and L.V.Il'ichov. Chem.Phys.Lett.,98,349(1983)

43 F.Kh.Gel'mukhanov and G.G.Telegin. Zh.Tekhn.Fiz.,55,1559(1985)

44 N.B.Baranova and B.Ya.Zel'dovich.Chem.Phys.Lett.,57,435(1978)

45 F.Kh.Gel'mukhanov and L.V.Il'ichov. Optics Comm.,53,381(1985)

CROSSED-SECOND-ORDER EFFECTS IN THE ISOTOPE SHIFT OF ATOMIC SPECTRA

A. Steudel

Institut für Atom- und Molekülphysik
Abteilung Spektroskopie
Universität Hannover
Appelstrasse 2, D-3000 Hannover, West Germany

1. Introduction

In recent years the study of the isotope shift (IS) in atomic spectral lines has gained renewed interest and a lot of papers on this field has appeared during the last 10 years. There are two reasons for this development. The first one comes from the nuclear aspect of the IS. The observed shifts consist of two contributions, a mass shift and a field shift, the latter containing the dependence on the nuclear charge distributions. Thus IS can provide information on the structure of the nuclei. For decades this was the main reason for the interest in this field of spectroscopy. Today, special efforts are directed towards the investigation of nuclear properties of long isotopic sequences by optical measurements. Such measurements became possible due to the high sensitivity of optical methods which allows to cope with minute samples of short-lived nuclei and up to now optical IS are the only experimental access used to determine the changes in nuclear charge radii of some highly unstable nuclei (see, e.g., /Ot 81/).

During the last years the emphasis has moved towards the electronic aspects in optical IS and this is the second reason for the renewed interest in this field. The development has been stimulated by the improvement of classical experimental techniques, the availability of new methods like laser spectroscopy, and the advances in theoretical methods.

The increase in experimental accuracy, especially with the laser-atomic-beam and the laser heterodyne techniques has opened a new chapter in optical IS. These shifts can now provide severe tests for the validity of calculation techniques in atomic physics. This paper deals with the electronic aspect of the IS and in particular with the so-called crossed-second-order (CSO) effects. The next chapter reviews first the elementary theory of IS and then the CSO effects. The third chapter describes laser spectroscopic measurements in the arc spectrum of Europium and the fourth chapter gives a survey on the CSO effects in the rare earths region. Finally the last chapter deals with CSO effects in elements of the platinum group.

2. Theory

2.1 Elementary Theory

The IS $\delta\nu^{AA'}$ between two isotopes with mass numbers A and A' observed in a spectral line with frequency ν is given by the difference of the IS $\delta E^{AA'}$ in the two levels belonging to this transition

$$\delta\nu^{AA'} = (\delta E_u^{AA'} - \delta E_l^{AA'})/h.$$

The index u stands for the upper and l for the lower level. In first order perturbation theory the IS $\delta E^{AA'}$ of a level is the sum of three terms, the normal mass shift (NMS), the specific mass shift (SMS), and the field shift (FS) /HS 74, BC 76/

$$\delta E^{AA'} = \delta E_{NMS}^{AA'} + \delta E_{SMS}^{AA'} + \delta E_{FS}^{AA'} . \tag{1}$$

These three IS stem from the operators

$$\hat{N} = \frac{1}{2M} \sum_i \hat{p}_i^2 , \quad \hat{S} = \frac{1}{M} \sum_{i>j} \hat{p}_i \cdot \hat{p}_j , \text{ and } \hat{F} = hC^A \sum_i \delta(\hat{r}_i), \tag{2}$$

respectively, where M is the nuclear mass, $\hat{p}_i$ the momentum of the ith electron, $\delta(\hat{r}_i)$ the Dirac delta function and C^A a constant depending on nuclear properties. The three operators are corrections to the Hamiltonian in which the nucleus is treated as a fixed point mass. The differences of the expectation values of these operators for two isotopes A and A' give the IS $\delta E^{AA'}$ in equation (1).

The mass shift operators $\hat{N}$ and $\hat{S}$ take account of the difference in the nuclear kinetic energy due to the different nuclear masses M of the isotopes. The NMS operator $\hat{N}$ contributes the reduced mass correction to the energy. The SMS operator $\hat{S}$ allows for the influence of correlations in the motion of the electrons on the recoil energy of the nucleus. The non-relativistic effective FS operator $\hat{F}$ takes into account the difference in the finite size and angular distribution of the nuclear charge of the isotopes and thus effects the binding energy of the electrons penetrating the nuleus, that means in the non-relativistic limit the binding energy of the s electrons.

The IS observed in a line is then given by

$$\delta\nu^{AA'} = \frac{m_e}{m} \frac{A'-A}{AA'} (\nu + M_k) + C^{AA'} \Delta D \tag{3}$$

$$A'>A, \quad M_k = k_u - k_l, \quad C^{AA'} = (C^{A'} - C^{A})/4\pi$$

$$D = 4\pi\ |\psi(0)|^2, \quad \Delta D = D_u - D_l,$$

where m_e is the electron mass and m the atomic mass unit. The NMS is proportional to the frequency ν of the line and can easily be calculated exactly. The SMS contains the quantity M_k which is the difference of the Vinti factors k in upper and lower level /Vi 39/. The Vinti factors are linear combinations of products of integrals on the electronic radial wavefunctions and are difficult to calculate. The interest in ab initio calculations of the SMS, that means of the quantity M_k, increased considerably over the last years (see, e.g., /BC 70, Ba 74, MS 82, FS 83, LM 83, FB 83/). But until now the agreement between theory and experiment is varying and only qualitative. This lack of accuracy, however, will not be of importance in this paper, since the IS will be considered in heavy elements only where in suited transitions the SMS is always small and can be estimated with sufficient precision /HS 74/.

The last term in equation (3) is the FS $\delta\nu_{FS}^{AA'}$. D represents the non-relativistic total electron density at the nucleus multiplied by 4π and ΔD is the change of the total electron density at the nucleus in the particular transition. Since, in heavy elements, the FS normally dominates the IS this term has to be considered in more detail. It can be factorized as

$$\delta\nu_{FS}^{AA'} = \frac{a_0^3\, f(Z)}{4Z}\, \Delta D\, \lambda^{AA'}. \tag{4}$$

a_0 is the Bohr radius and f(Z) a known function which increases with Z and which includes relativistic corrections to the electron density and also those due to the finite nuclear charge distribution. $\lambda^{AA'}$ gives the change of the radial nuclear charge parameters

$$\lambda^{AA'} = \delta\langle r^2\rangle^{AA'} + \frac{C_2}{C_1}\,\delta\langle r^4\rangle^{AA'} + \frac{C_3}{C_1}\,\delta\langle r^6\rangle^{AA'} + \ldots\ .$$

The electronic coefficience C_i/C_1 can be taken from /Se 69/. Since, to a good approximation, the electronic wavefunctions can be considered constant over the nuclear volume, the contributions of the higher charge moments are small, often smaller than the errors in the evaluation of λ from optical measurements. With europium, for example, $\lambda^{AA'} = 0.95\ \delta\langle r^2\rangle^{AA'}$ /BBL 69, AKE 85/.

In order to get rid of the frequency dependence of the observed IS it is customary to subtract the NMS from the observed IS. The difference is called the residual IS $\delta\nu_R$

$$\delta\nu_R = \delta\nu_{obs} - \delta\nu_{NMS}\,.$$

In first order perturbation theory the IS is governed by the following rules: (i) The FS is constant over all the levels of a pure configuration. (ii) The SMS is also constant for all levels of a pure configuration provided one is not concerned with a configuration which consists of at least two open sub-shells with angular momenta differing by 1. In this case the mass shift is constant only for all levels of a pure Russell-Saunders (RS) term.

It is, however, very well known, that in many cases these statements are not fulfilled at all. In the various levels of a configuration often very different IS are measured. For a long time it was believed that this is solely caused by the mixing of configurations lying sufficiently close together (close-configuration-mixing). We now know that CSO effects play an important rôle and may cause an IS between the different terms of a configuration and even between the levels of the same term.

2.2 Crossed-Second-Order Effects

The CSO effect of any IS operator $\hat{I}$ and a perturbing Hamiltonian $\hat{O}$ gives an energy contribution to a mono-configurational level x with wavefunction ψ_x which can be written as /BC 76/

$$E_{CSO} = 2 \sum_y \frac{\langle\psi_x|\hat{I}|\psi_y\rangle \langle\psi_y|\hat{O}|\psi_x\rangle}{E_x - E_y} \quad .$$

In this expression E_x and E_y are the zeroth order energies of the configurations C_x and C_y to which the states x and y belong. The sum runs over all states y of all the configurations C_y with large energy separation $E_x - E_y$ to enable a rapid convergence. The CSO effects are therefore sometimes called far-configuration-mixing. Close lying configurations must be allowed for in first order within the intermediate coupling formalism.

The largest CSO effect observed in heavy elements is that between the FS operator and the electrostatic operator $\sum_{i>j} e^2/r_{ij}$. The non-relativistic FS operator is a mono-electronic operator acting solely on s electrons. The only perturbing configuration C_y of interest are therefore those differing from the configuration C_x by an ns-n's excitation. That means a valence or a core ns electron has to be excited to an empty or half-filled n's shell. The CSO contribution of the FS operator and the electrostatic operator can thus lead to different FS in the terms of a pure configuration. There is no J dependence of this effect since the matrix elements of the electrostatic operator do not depend on J.

The general procedure for taking CSO effects into account is to obtain an effective operator. This effective operator then acts inside the configuration of interest, and its expectation values reproduce the CSO contributions. Using the second-quantization angular method Bauche /Ba 69/ has demonstrated that the angular behaviour of the CSO effect of FS and electrostatic operator among the terms of a configuration is that of the Slater integral $G^l(n'l, ns)$ containing at least one s electron. Therefore this CSO effect can be described by one phenomenological parameter called $g^l(n'l, ns)$, whose angular coefficient is the same as that of the Slater integral G^l in the expansion of the energy of the term under consideration.

The CSO effect between the SMS operator $\hat{S}$ and the electrostatic operator may also produce a different IS in different terms of the same configuration like the CSO effect of FS and electrostatic operator. These two CSO effects which result in term dependent IS effects are described in the same way. In heavy elements, however, which in the following will be discussed as examples, the CSO effects of the SMS operator are supposed to be small. The symbol T will be used to characterize these term dependent IS effects.

There is another type of CSO contributions which causes a J dependence of the IS of the levels of a RS term. These are the CSO effects of the FS operator or the SMS operator and a magnetic interaction operator like the spin-orbit interaction. Due to the J-dependence of the matrix elements of the magnetic interaction operators these CSO effects lead to a J dependence in the IS. It has been shown /BC 76/ that all the J dependent effects can be described phenomenologically in the limit of RS coupling by one parameter z_{nl} having the same angular coefficient c_{nl} as the spin-orbit radial integral of the nl electron in the energy expression.

3. Crossed-Second-Order Effects in the Isotope Shift of the Configuration $4f^7(^8S)5d6s$ in Eu I.

3.1 Measurements

There are two basic requirements for investigating CSO effects in the IS of atomic spectra: (i) for the configurations to be studied a correct fine structure analysis and eigenfunctions in intermediate coupling including close-configuration-mixing must be available; (ii) since the J dependence of the IS in the levels of a RS term is usually small, high resolution is needed, that means laser spectroscopy is in most cases the appropriate tool.

As an example we shall now consider term- and J-dependent effects in the IS of the $4f^7(^8S)5d6s$ a ^{10}D and a ^8D terms in the arc spectrum of europium (see Fig.1). It is known from fine structure analysis that, to a good approximation, the levels of these terms represent the pure configuration $4f^7(^8S)5d6s$. Eigenfunctions in intermediate coupling are available /SW 65, GN 71/. All the levels of the a ^{10}D term are metastable. The same is true for the lowest level of a ^8D due to its low J value. All the other levels of a ^8D may decay into levels of the configuration $4f^76s6p$. However, due to the small energy differences the transition probabilities are small, so that these levels are long-lived. Since one wants to get information on the IS between levels of the terms a ^{10}D and a ^8D, the IS in lines has to be studied which connect levels of these terms by a common upper level. Some examples can be seen in Fig. 1.

For the investigation the high resolution laser-atomic-beam technique was applied. In a first step the metastable and the long-lived levels of a ^{10}D and a ^8D have to be populated. This was done by an arc discharge within an atomic beam of Eu. The oven used for this purpose is shown Fig. 2. It is made of stainless steel and is resistance heated by a 0.5 mm tantalum wire which is threaded back and forth through insulator tubes arranged around the crucible. At an oven temperature of 950 K a vapour pressure of about 0.05 torr in the crucible is produced.

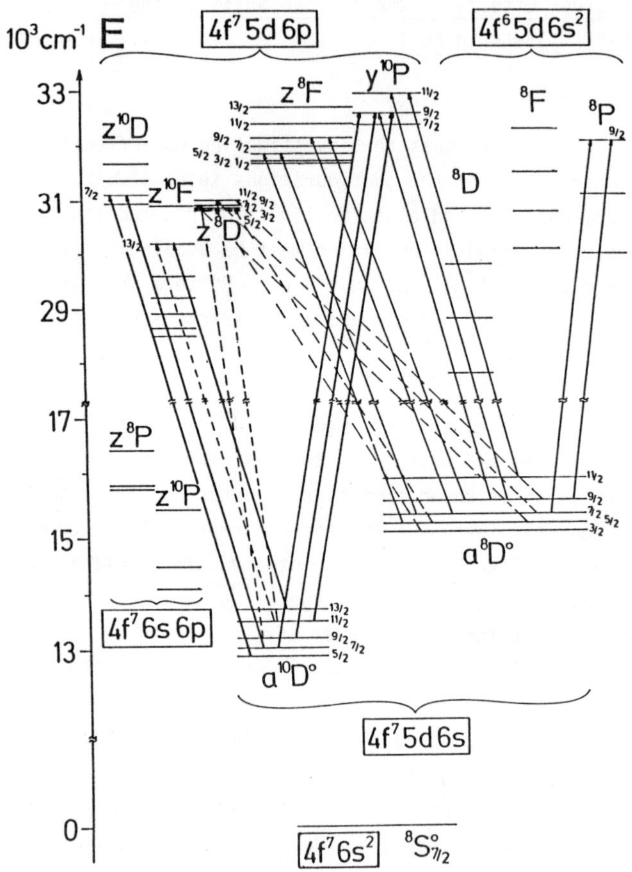

Fig. 1. Part of the energy level diagram of Eu I showing some transitions in which the isotope shift was measured.

The atoms effuse through an orifice of 0.7 mm diameter in the top of the oven and pass a hollow cathode region not shown in the figure. The cathode cylinder was welded directly unto the crucible to achieve automatical heating by conduction. The anode consisted of two parallel tantalum wires at a distance of 7.5 mm from the orifice. The discharge was operated at a current of 300 mA and a cathode-anode voltage of up to 40 V. With this

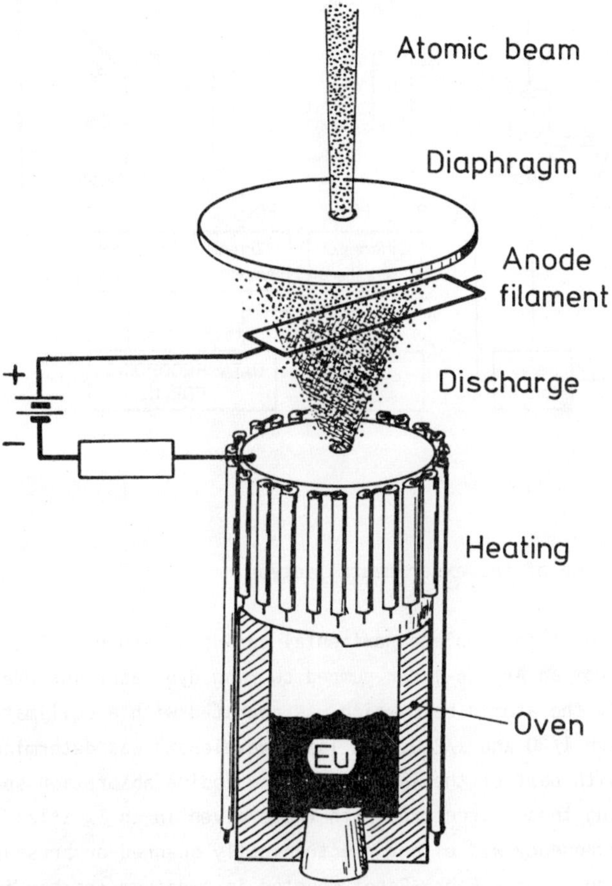

Fig. 2. Scheme of the atomic beam source.

oven arrangement the evaporation rate amounts to less than 0.1 g Eu per hour. This is of importance since europium is rather expensive and the atomic beam has to be operated during many hours. For more details, see /Es 85/.

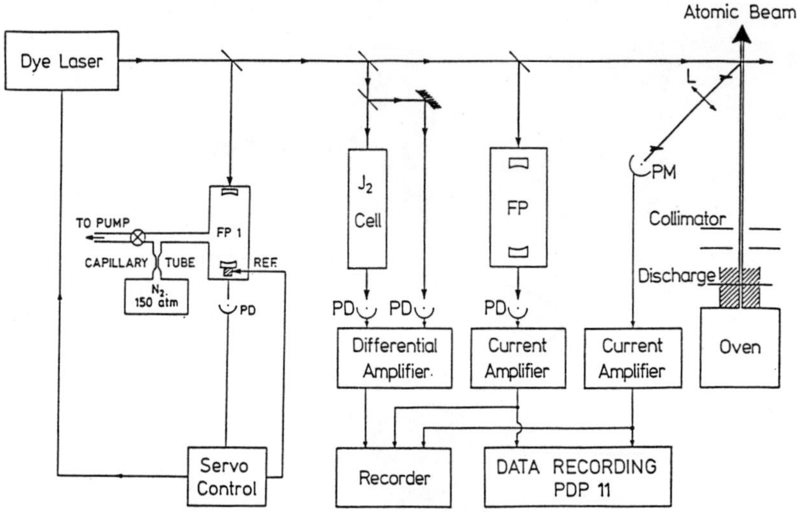

Fig. 3. Scheme of the experimental set-up.

A scheme of the whole experimental set-up is shown in Fig. 3. The light from an Ar-ion-laser pumped cw ring dye laser intersected orthogonally the atomic beam which was operated with a collimation ratio between 1/80 and 1/200. The laser wavelength was determined by measuring with part of the laser light the iodine absorption spectrum and comparing these structures with those given in an I_2 atlas /GL 78/. The laser frequency was either electronically scanned or pressure scanned by means of a Fabry-Perot mounted in a vacuum chamber in which the nitrogen pressure was changed very linearly with time. The scan width amounted to 30 GHz. For calibration of the laser scan part of the laser light was passed through a highly stable Fabry-Perot interferometer having a free spectral range of 150 MHz. The fluorescence light was detected by a photomultiplier perpendicular to both the atomic and the laser beam. To reduce the background from scattered oven light a monochromator, not shown in the figure, was used in front of the photomultiplier and in suited cases the fluorescence was detected on decay channels different from the laser transition. The transmission of the gauge Fabry-Perot and the photomultiplier signal were recorded simultaneously by a strip-chart recorder as well as by a computer.

As an example Fig. 4 shows a recording of the hyperfine structure of the Eu I - line λ 592.65 nm. The hyperfine components belong to the two stable isotopes, ^{151}Eu and ^{153}Eu, which have approximately the same natural abundance. Both isotopes have the nuclear spin 5/2, but the hyperfine splittings are quite different,since the nuclear magnetic dipole moments and the nuclear electric quadrupole moments of the two isotopes differ considerably. The width of the components is due to the natural line width, the remaing doppler width and the width of the laser. The IS is given by the shift of the centers of gravity of the hyperfine splittings of the two isotopes. To get these shifts the experimental recordings were evaluated by means of a least-squares-fitting program which various the hyperfine constants A and B of lower and upper level, the IS and three parameters for the line profile. Thus not only the IS but also the hyperfine constantsare obtained. In this paper, however, only the IS is discussed.

The extensive measurements will not be discussed in detail here. The reader is referred to a forthcoming paper /BEP 86/. As an example Fig. 5 shows some typical results on the residual IS in the levels of a ^{10}D and a ^{8}D derived from the IS measured in the lines (see Fig. 1). In first order perturbation theory one would expect that all the levels of a ^{10}D and a ^{8}D have the same residual IS (zero in Fig. 5). It is quite clear that this is not the case. There is a large term dependence in the IS,since the IS in the a ^{10}D term are about 400 MHz larger than those in the a ^{8}D term. Moreover there is a small but pronounced J dependence of the IS in the levels of each term as can be seen from the differences $\Delta\delta\nu_R$ in Fig. 5. The observed term and J dependencies are caused by CSO effects and will be described in the next section by the phenomenological parameters introduced in section 2.2.

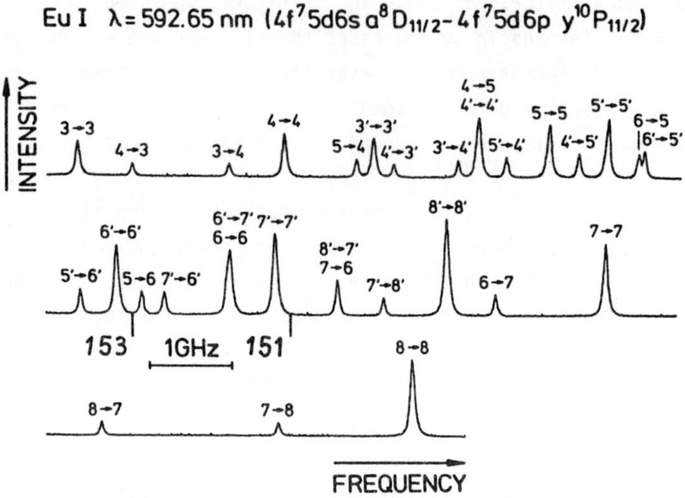

Fig. 4. Fluorescence spectrum of a Eu I - line measured with one laser scan. The hyperfine components are labelled by the F values (lower level - upper level). The F values of ^{153}Eu are primed. The centers of gravity of the hyperfine splittings are indicated.

3.2 Phenomenological Interpretation

For the description of the J-dependent CSO effects in the configuration $4f^7(^8S)5d6s$ a z_{nl} parameter is only needed for the 5d electron, since the 6s electron and $4f^7\ ^8S$ do not give a contribution to the spin-orbit interaction. Hence the CSO effects in the IS of a level of $4f^7(^8S)5d6s$ can be written as (see section 2.2)

$$\delta E_{CSO}^{AA'} = T(^{2S+1}L)^{AA'} + c_{5d}(^{2S+1}L_J)\, z_{5d}^{AA'} .$$

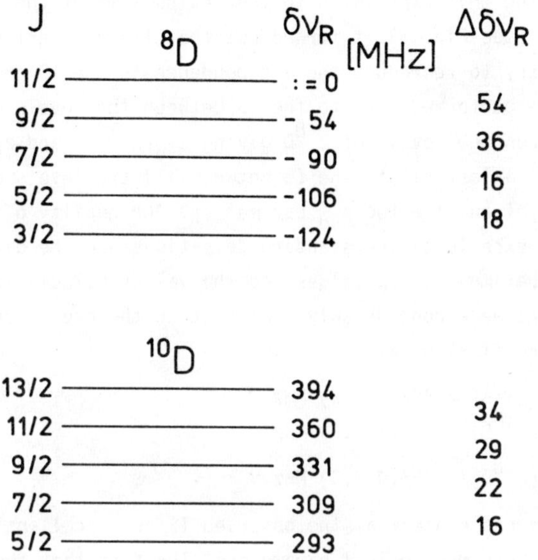

Fig. 5. Some typical results for the residual isotope shift $\delta\nu_R$ in the levels of the Eu I - terms a ^{10}D and a ^{8}D referred to the level a $^{8}D_{11/2}$. $\Delta\delta\nu_R$ residual isotope shift between adjacent levels. Fine structure level scheme not on scale.

The residual IS between two levels of the terms a ^{10}D and a ^{8}D is then given by

$$\delta\nu_R = \Delta T + \Delta c_{5d}\, z_{5d}$$

where $\Delta T = T(a\ ^{10}D) - T(a\ ^{8}D)$
represents the difference of the term dependent CSO effects in a ^{10}D and a ^{8}D and

$$\Delta c_{5d}\, z_{5d} = [c_{5d}(a\ ^{10,8}D_J) - c_{5d}(a\ ^{10,8}D_{J'})]\, z_{5d}$$

the difference of the J dependent CSO effects in the two levels. The coefficients c_{5d} can be calculated, since the wavefunctions are known.

When fitting this expression to the residual IS of the levels of a ^{10}D and a ^{8}D (see Fig. 5) it turned out that two distinct parameters z_{5d} are necessary to reproduce the J dependence in the IS of these terms. Three fits were performed: one to the IS between the levels of a ^{10}D, one to the IS between the levels of a ^{8}D giving z_{5d}(a ^{10}D) and z_{5d}(a ^{8}D), respectively, and a last fit to the IS between all the levels of the two terms yielding ΔT and the two z_{5d} parameters. The quality of all the three fits was excellent; the standard deviations and the differences between the experimental $\delta\nu_R$ values and the values calculated with the three parameters were considerably smaller than the experimental uncertainties. The final parameter values are

$$\Delta T = 408.5(3.2)\ \text{MHz}$$
$$z_{5d}(\text{a }^{10}D) = 44.1(2.6)\ \text{MHz}$$
$$z_{5d}(\text{a }^{8}D) = 55.9(2.3)\ \text{MHz.}$$

As a result one can state that the observed IS are excellently described by these three phenomenological parameters. The fact that two different z_{5d} parameters are necessary to describe the J dependent IS in a ^{10}D and a ^{8}D was quite unexpected. A theoretical interpretation of these findings is still lacking and for further discussion more experimental material on the J dependence of the IS in the various terms of a configuration is needed.

3.3 Ab initio Interpretation

The ab initio interpretation of the parameter ΔT which is spin-independent was performed by the Hartree-Fock method and that of the spin-dependent parameter z_{5d} by the Dirac-Fock method /BEP 86/. The Hartree-Fock method is especially suited for the calculation of FS-CSO effects because it benefits from Brillouin's theorem. Therefore, as concerns these far-configuration-mixing effects, a monoconfigurational Hartree-Fock wavefunction is particularly accurate for the evaluation of total electron densities at the nucleus /BK 72, La 72/.

We discuss first the parameter ΔT, the residual IS between the terms a ^{10}D and a ^{8}D. This IS is in principal due to the CSO effects of the FS and the SMS operator with the electrostatic operator, but the CSO effect of the SMS operator is expected to be small. In order to get the pure FS between the terms a ^{10}D and a ^{8}D the SMS was calculated for these two terms by means of the Hartree-Fock code by Froese Fischer /Fr 69, Fr 78/. Thus $\delta\nu_{SMS}(a\ ^{10}D-a^{8}D)$ = -5 MHz was obtained. It is well-known that SMS evaluated by the Hartree-Fock methods are rather inaccurate, but in the present case the SMS is only a small contribution to the term dependent IS ΔT=408.5 MHz. Therefore it was taken into account to get the term dependent FS

$$\Delta T_{FS}(a\ ^{10}D - a\ ^{8}D) = 414 \text{ MHz}$$

which is due to the CSO effect between FS and electrostatic operator.

Now equation (4) can be used to calculate a theoretical value for ΔT_{FS}. The difference of the total electron densities at the nucleus, ΔD , was evaluated using the Hartree-Fock code by Froese Fischer: $D(a\ ^{10}D)$ = 2234832.8 a.u., $D(a\ ^{8}D)$ = 2234824.1 a.u., ΔD = 8.7 a.u. .

The nuclear parameter $\lambda^{151,153}$ is known from optical, electronic x-ray, and muonic x-ray investigations. The results obtained by these different experimental methods are in excellent agreement. The weighted mean value is for two stable Eu isotopes $\lambda^{151,153} = 0.576(15)\ \text{fm}^2$. The function f(Z) was recently calculated anew /Zi 85, BBP 85/; the value for Eu is f(Z) = 21.67 GHz/fm^2. With these values one obtains the theoretical FS

$$\Delta T_{FS,th}(a\ ^{10}D - a\ ^{8}D) = 430 \text{ MHz}.$$

This is in good agreement with the experimental value of 414 MHz.

According to section 2.2 the parameter ΔT_{FS} can be written as

$$T_{FS} = \alpha g^2(5d,6s) + \beta g^3(4f,6s)$$

$$\Delta T_{FS} = \Delta\alpha\ g^2(5d,6s) + \Delta\beta\ g^3(4f,6s)$$

$$\Delta\alpha = \alpha(a\ ^{10}D) - \alpha(a\ ^{8}D)\ ,\ \Delta\beta = \beta(a\ ^{10}D) - \beta(a\ ^{8}D).$$

α and β are the coefficients of the Slater integral $G^2(5d,6s)$ and $G^3(4f,6s)$, respectively. They can be calculated by means of the known wavefunctions /SW 65, GN 71/ or by the Hartree-Fock method. One obtains $\Delta\alpha = -0.1013$ and $\Delta\beta = -1.1002$.

The phenomenological parameters g^2 and g^3 cannot be determined separately from the experimental data. The Hartree-Fock method, however, enables us to calculate these parameters. For this purpose two fictitious terms T_1 and T_2 have to be introduced, the energies of which are chosen in such a way that the difference $\Delta D(a\ ^{10}D - T_1)$ depends only on $g^2(5d,6s)$ and $\Delta D(a\ ^{10}D - T_2)$ only on $g^3(4f,6s)$. The result of the Hartree-Fock calculation is then

$$\Delta D(a\ ^{10}D - T_1) = 14.5 \text{ a.u.} \sim -0.3\ g^2(5d,6s)$$

$$\Delta D(a\ ^{10}D - T_2) = 1.1 \text{ a.u.} \sim -0.3\ g^3(4f,6s).$$

In order to get the g^1 values the electron densities have to be converted into MHz by means of a calibration factor for which the ratio

$$\frac{\Delta T_{FS}}{\Delta D(a\ ^{10}D - a\ ^{8}D)} = \frac{414 \text{ MHz}}{8.7 \text{ a.u.}} = 47.6\ \frac{\text{MHz}}{\text{a.u.}}$$

can be used. The g^1 values are then

$$g^2(5d,6s) = -2301 \text{ MHz}$$

$$g^3(4f,6s) = -175 \text{ MHz}.$$

With these parameters we get for the FS between the terms $a\ ^{10}D$ and $a\ ^{8}D$ the theoretical value

$$\Delta T_{FS,th}(a\ ^{10}D - a\ ^{8}D) = 426 \text{ MHz}$$

which is in very satisfying agreement with the experimental value of 414 MHz.

We now turn to the interpretation of the z_{5d} parameter. For this purpose it is practical to consider the residual IS between the a $^{10}D_{13/2}$ level and a fictitious level denoted $^{10}D_{av}$ of which the energy is the weighted average energy of the five a $^{10}D_J$ levels. The residual IS between these two levels derived from the measurements is

$$\delta\nu_{FS,exp}(^{10}D_{13/2} - {}^{10}D_{av}) = 44.1(2.6)\ \text{MHz}.$$

This IS is identical with the FS, since there is certainly no SMS between the levels of the same term. The value of 44.1 MHz is proportional to the z_{5d} parameter. In order to reproduce this value Dirac-Fock calculations were made which yield a quantity called c_0 which is proportional to the FS

$$c_o(^{10}D_{13/2}) - c_o(^{10}D_{av}) = 3.1\ \text{a.u.}\ .$$

Again a calibration factor is needed . In this case the ratio of the measured FS between the a $^{10}D_{13/2}$ level and the ground state $4f^7(^8S)6s^2$ and the corresponding difference of the c_0 values was used

$$\frac{\delta\nu_{FS,exp}(^{10}D_{13/2} - {}^8S_{7/2})}{c_o(^{10}D_{13/2}) - c_o(^8S_{7/2})} = \frac{4399}{246.1}\ \frac{\text{MHz}}{\text{a.u.}} = 17.9\ \frac{\text{MHz}}{\text{a.u.}}\ .$$

Multiplying the difference of the c_0 values for the levels a $^{10}D_{13/2}$ and $^{10}D_{av}$ by this calibration factor leads to the theoretical FS

$$\delta\nu_{FS,th}(^{10}D_{13/2} - {}^{10}D_{av}) = 3.1\ \text{a.u.} \cdot 17.9\ \frac{\text{MHz}}{\text{a.u.}} = 55\ \text{MHz}$$

which has to be compared with the experimental value of 44.1 MHz. The deviation of 20% lies within the frame of accuracy of the Dirac-Fock calculations. The result suggests that the J dependence of the IS in the a ^{10}D term can be described as FS-CSO effect. This result is in accordance with the expectation based on the heaviness of the element and experimental results in other rare earths (see next chapter).

4. Comparison of Crossed-Second-Order Effects in the Rare-Earth Elements

4.1 g^1 values

The only point where a comparison between experiment and theory can be made so far is the $g^3(4f,6s)$ parameter in Eu. The ground configuration of Eu II, $4f^7(^8S)6s$, forms the terms ^9S and ^7S. From IS measurements in lines from these ground states to levels of the configuration $4f^76p$ /Gu 68/ the FS between the two terms ^9S and ^7S was derived. This value is connected with g^3 according to the van Vleck - theorem (see, e.g., /CO 80/)

$$\delta\nu_{FS}(^9S_4 - {}^7S_3) = 297(21)\ \mathrm{MHz} = -\frac{8}{7}\ g^3(4f,6s)$$

so that

$$g^3(4f,6s) = -260(18)\ \mathrm{MHz}$$

is obtained for the configuration $4f^7(^8S)6s$ in Eu II. This experimental result is in excellent agreement with the theoretical value of

$$g^3(4f,6s) = -250\ \mathrm{MHz}$$

deduced from Aufmuth's non-relativistic Hartree-Fock calculations /Au 82/. That the $g^3(4f,6s)$ value for $4f^7(^8S)6s$ is larger than the corresponding g^3 value in $4f^7(^8S)5d6s$ in Eu I (-175 MHz) is not surprising, since it is well-known that in the configuration $4f^75d6s$ the 6s electron is screened by the 5d electron so that a smaller electron density at the nucleus results.

Thus one expects that the ratio of the g^3 values in $4f^7 6s$(Eu II) and $4f^7 5d6s$(Eu I)

$$\frac{g^3(\text{Eu II})}{g^3(\text{Eu I})} = 1.49$$

is equal to the ratio of the FS in these configurations referred to the corresponding series limits. (The FS values can be deduced from /BPS 81/)

$$\frac{\delta\nu_{FS}(4f^7 6s - 4f^7)}{\delta\nu_{FS}(4f^7 5d6s - 4f^7 5d)} = \frac{4718 \text{ MHz}}{3138 \text{ MHz}} = 1.50 \ .$$

The two ratios are in excellent agreement. So one can be sure that the different values of the g^3 parameters in Eu I and Eu II are caused by the screening effect of the 5d electron on the s electrons. One can also consider the ratio of the magnetic hyperfine splitting factors of the 6s electron, a_{6s}, in the two configurations /BPS 81/ which describes the screening of only the 6s electron by the 5d electron, since a_{6s} is proportional to the density of the 6s electron at the nucleus.

$$\frac{a_{6s}(4f^7 6s)}{a_{6s}(4f^7 5d6s)} = \frac{12780 \text{ MHz}}{8410 \text{ MHz}} = 1.52 \ .$$

Again the ratio is in good agreement with the ratio of the g^3 values indicating that the 5d electron mainly shields the density of the 6s electron at the nucleus.

$g^3(4f,6s)$ has also been measured for the Eu I - configurations $4f^7 6s6p$, g^3=210(20) MHz, and $4f^7 6s7s$, g^3=230(20) MHz /KKW 85a/.The values demonstrate the weak screening of the 6s electron by 6p and 7s, respectively when comparing with the g^3 value of -260(18)MHz in $4f^7 6s$.

The configuration $4f^7(^8S)6s7s$ forms four terms between which term dependent FS were observed. These FS which are essentially described by $g^3(4f,6s)$ can be well reproduced by Hartree-Fock calculations /Au 85a/.

Table 1. z_{nl} parameters in the lanthanides

Spectrum	Configuration	Isotopes A - A'	nl	z_{nl} (MHz)	$\delta<r^2>$ (fm^2)	$z_{nl}/\delta<r^2>$ (MHz/fm^2)	$z_{4f}/\delta<r^2>\cdot\zeta_{4f}$ ($10^{-6}/fm^2$)	$z_{5d}/\delta<r^2>\cdot\zeta_{5d}$ ($10^{-6}/fm^2$)
Sm I	$4f^6 6s^2$	144-152	4f	50.4(9)[a]	1.243[i]	40.5(7.3)	1.21(18)	
		144-148		19.9(4.2)[b]	0.517[i]	38.5(8.1)		
		148-150		11.7(1.5)[b]	0.303[i]	38.6(5.0)		
		150-152		16.6(2.0)[b]	0.423[i]	39.2(4.7)		
		152-154		8.6(1.3)[b]	0.230[i]	37.4(5.7)		
Eu I	$4f^7 5d6s$	151-153	5d	44.1(2.6)[c]	0.606[c]	72.8(4.7)		7.41(45)
				55.9(2.3)[c]		92.2(4.5)		
Gd II	$4f^7 5d6s$	156-160	5d	42(11)[d]	0.275[k]	152(40)		6.1(1.6)
Dy I	$4f^{10} 6s^2$	160-164	4f	22.8(1.9)[e]	0.267[k]	85.4(8.6)	1.61(16)	
		162-164		11.0(1.4)[e]	0.129[k]	85.3(12.2)		
		161-164		20.5(2.1)[e]	0.233[k]	88.0(9.5)		
		160-163		15.1(2.1)[e]	0.178[k]	84.8(13.2)		
Dy II	$4f^{10} 6s$	160-164	4f	32(12)[f]	0.267[k]	119(47)	2.25(88)	
Er I	$4f^{12} 6s^2$	166-170	4f	40.1(1.3)[g]	0.240[h]	167(16)	2.34(23)	
Er I	$4f^{11} 5d6s^2$	164-170	5d	63.8(6.0)[h]	0.359[h]	178(23)		7.64(85)
		166-170		42.8(4.1)[h]	0.240[h]	178(24)		
		167-170		37.0(3.6)[h]	0.199[h]	186(25)		
		168-170		21.7(1.5)[h]	0.121[l]	179(19)		

a /BCS 77/, b /NGI 81/, c /BEP 86/, d /KKW 85b/, e /PCG 84/, f /Au 78/,
g /PCG 83/, h /Be 85/, i /BSS 80/, k /HS 74/, l /ESV 74/

4.2 z_{nl} Parameters

The experimental material available so far is compiled in Table 1. If the $z_{nl}^{AA'}$ parameters are caused by FS-CSO effects, they should be proportional to $\delta\langle r^2\rangle^{AA'}$. In samarium and dysprosium the z_{4f} parameter and in erbium the z_{5d} parameter was measured for several isotope pairs. Table 1 shows that the ratio $z_{nl}^{AA'}/\delta\langle r^2\rangle^{AA'}$ is constant for the various isotope pairs of an element. This indicates that in the cases studied so far the J dependence can be explained solely by CSO effects of the FS operator.

The parameter z_{nl} is supposed to be also proportional to the spin-orbit radial integral ζ_{nl} so that the ratio $z_{nl}/\delta\langle r^2\rangle\zeta_{nl}$ should be constant for similar configurations. These ratios are given in the last column of Table 1. The configuration $4f^7 5d6s$ was studied in Eu I and Gd II and the configuration $4f^{11}5d6s^2$ in Er I. The values obtained for $z_{5d}/\delta\langle r^2\rangle\zeta_{5d}$ are equal within the limits of error. Concerning the z_{4f} parameter the measurements for configuration of the type $4f^N 6s^2$ in Sm I, Dy I, and Er I and for $4f^{10}6s$ in Dy II have to be considered. In this case the values for the ratio $z_{4f}/\delta\langle r^2\rangle\zeta_{4f}$ do not agree, but seem to indicate a slight increase with the number of 4f electrons. Further accurate measurements in the lanthanides are necessary in order to find out whether this trend is real.

5. Crossed-Second-Order Effects in the 5d-Shell Atoms

In the 5d-shell atoms CSO effects in the IS between levels of terms of the configuration $5d^N 6s$ were studied and in particular the IS between levels belonging to the parent term with the highest multiplicity M and the highest L value. This parent term forms two terms with multiplicity M+1 and M-1, $5d^N(^ML)6s\ ^{M\pm1}L$. The configuration $5d^N 6s$, however, is never pure which is typical for the 5d-shell elements. There is always a more or less strong mixing with the configuration $5d^{N-1}6s^2$. The mixing with $5d^{N+1}$ is weaker and can be neglected in most cases. In order to be able to compare the results obtained in the various 5d-shell atoms one has to evaluate the CSO effects for pure RS terms. Therefore not only the IS of levels which are attributed to the terms $5d^N(^ML)6s\ ^{M\pm1}L$ but also of levels attributed to the configuration $5d^{N-1}6s^2$ has to be studied.

The measurements were made by classical high-resolution spectroscopy. A photoelectric recording Fabry-Perot spectrometer with digital data processing was used and hollow cathodes filled with highly enriched isotopes served as light sources. As an example measurements made in osmium /AW 85/ will be briefly discussed in the next section.

5.1 Crossed-Second-Order Effects in the $5d^7(^4F)6s$ Terms of Os I

The lines in which the IS was investigated are shown in Fig. 6.

Table 2. Residual isotope shifts of low even levels for $^{188-192}$Os referred to the odd level $^5F^o_5$. (1 mK = 10^{-3}cm^{-1})

Levels	$\delta\nu_R$ [mK]
$d^7s\ ^5F_5$	57.3(7)
$d^7s\ ^5F_4$	66.7(6)
$d^7s\ ^5F_3$	63.5(1.0)
$d^7s\ ^3F_4$	98.6(2)
$d^6s^2\ ^3G_5$	210.0(4)
$d^6s^2\ ^3H_6$	215.7(6)

In table 2 are given the experimental residual IS of low even levels referred to the odd level $^5F^o_5$ which is supposed to belong to a mixture of the configurations $5d^76p$ and $5d^66s6p$ /AW 85/. The values are now given in mK, since classical optical spectroscopy was used (1 mK corresponds to 30 MHz). The $\delta\nu_R$ values shown in table 2 are in reasonable agreement with the common views on IS. The levels attributed to the configuration $5d^66s^2$ having a larger s electron density at the nucleus show a much larger IS than the levels attributed to the configuration $5d^76s$. The difference between the various ^5F levels of $5d^76s$ and the ^3F levels can clearly be seen. It is about -35 mK. This, however, is not yet the CSO effect between the pure RS terms ^5F and ^3F we are looking for. The value is strongly influenced by configuration mixing and intermediate coupling.

For further discussion a parametric analysis has to be made in which the residual IS in a level, $\delta\nu_R$, is expressed as a linear combination of phenomenological IS parameters and angular coefficients:

$$\delta\nu_R = \tilde{a}a + \tilde{s}s + \alpha g^2 + \eta h + c_{5d^7}z_{5d^7} + c_{5d^6}z_{5d^6} \cdot \qquad (5)$$

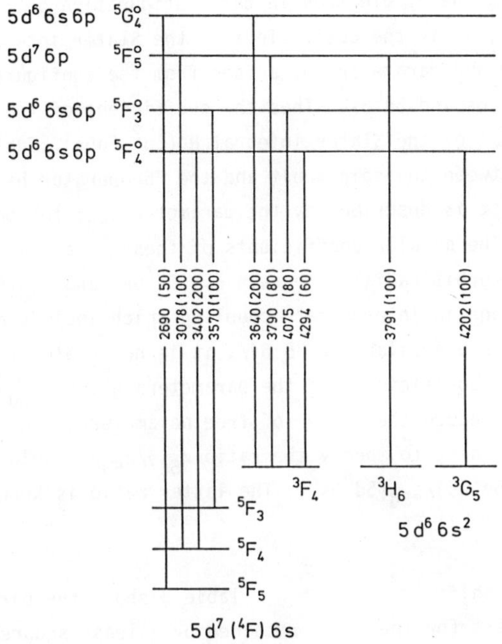

Fig. 6. Part of the energy level scheme of Os I showing the investigated lines, their wavelengths (in nm) and their relative intensities (in parenthesis).

If one takes $\tilde{a} = 1$, the parameter a is the residual IS of the configuration $5d^6 6s^2$ with respect to the reference level. The second parameter s is then equal to the first order IS between the pure configurations $5d^7 6s$ and $5d^6 6s^2$ and the coefficient $\tilde{s}$ gives the percentage of the configuration $5d^7 6s$ in the considered level. Thus the two first terms represent the well-known sharing rule. This rule accounts for all IS contributions which are constant in a configuration. The further terms in (5) represent the influence of second-order effects. Since Os is a heavy element second-order SMS effects should be negligible. Therefore only second-order effects in the configuration $5d^7 6s$ are considered which are caused by the FS operator. For the CSO effect between the electrostatic and the FS operator two parameters, g^2 and h, have to be introduced /Ba 69, BC 76/. The first one refers to the electrostatic exchange integral between the 5d and the 6s electron in the configuration $5d^7 6s$. The coefficient of g^2, α, is the coefficient of the Slater integral $G^2(5d,6s)$. The second parameter, h, arises from the configuration mixing between $5d^7 6s$ and $5d^6 6s^2$. Thus the coefficient of the parameter h, η, is that of the Slater integral $R^2(5d^2,5d6s)$. Further, the CSO effect between the spin-orbit and the FS operator has to be included. This effect is described by the parameter z_{5d7} for $5d^7 6s$ and z_{5d6} for $5d^6 6s^2$. The angular coefficients of these parameters are those of the spin-orbit radial integrals $\zeta_{5d}(5d^7 6s)$ and $\zeta_{5d}(5d^6 6s^2)$. Since eigenfunctions in intermediate coupling which include configuration mixing are available /GBB 64/, it is no problem to calculate the angular coefficients of the parameters g^2, h, z_{5d7}, and z_{5d6}. In order to reduce the number of free parameters, the fact can be used, that according to theory the ratio z_{5d7}/z_{5d6} should be equal to the ratio $\zeta_{5d}(5d^7 6s)/\zeta_{5d}(5d^6 6s^2)$. The latter ratio is known from fine structure analysis.

Table 3. Isotope shift parameters for the isotope pair $^{188-192}$Os. Values in mK.

a	224.1(2.6)
s	-228.2(5.4)
g^2	-67.0(12.0)
h	-2.4(6)
z_{5d7}	2.5(2)
z_{5d6}	2.7(2)

Table 3 shows the parameters obtained by a least-squares fitting procedure. The quoted errors are three times of the standard deviation. The value of the parameter a depends completely on the arbitrarily chosen reference level. The parameter s is the IS between the pure configurations $5d^7 6s$ and $5d^6 6s^2$. This value is of importance for the calculation of the change of the mean-square nuclear charge radius from the optical IS measurements. But this procedure is outside the scope of this paper. The parameter g^2 describes the difference ΔT in the IS between the pure RS terms $5d^7 6s\ ^5F$ and ^3F. This difference is given by the formula

$$\Delta T = T(^5F) - T(^3F) = -\frac{4}{5}\,g^2 .$$

From the g^2 parameter in table 3 the experimental value

$$\Delta T_{exp} = 53.6(9.6)\ \text{mK}$$

is obtained. This value is caused by the CSO effect between the FS and the electrostatic operator. It is in clear contrast to the value of -35 mK mentioned above which one would derive from the measurements at first glance. The z parameters will be discussed in section 5.2 in connection with the results obtained in other 5d-shell atoms.

The experimental value ΔT_{exp} can be compared to a theoretical value obtained by means of formula (4). f(Z) /Zi 85, BBP 85/ and $\delta\langle r^2\rangle$ /AS 85/ are known and ΔD, the difference of the electron density at the nucleus between the pure RS terms $5d^7(^4F)6s\ ^5F$ and ^3F, was calculated by the Hartree-Fock method. But it is well-known that absolute electron densities from non-relativistic Hartree-Fock calculations are often too small. Therefore a scaling factor (experimental FS / theoretical FS) has to be introduced in order to adjust the theoretical FS to the experimental ones (see, e.g., /ACH 78/). For the determination of the scaling factor in Os it is best to use the IS between levels of the low even configurations, since for these levels reliable eigenfunctions are available /GBB 64/. For Os a mean scaling factor of 1.29 is found /Au 85b/.(For Eu the scaling factor is approximately 1.) The theoretical value we obtain for the term dependent IS in Os

$$\Delta T_{th} = 53.9\ \text{mK}$$

is in excellent agreement with the experimental value.

5.2 Comparison of Crossed-Second-Order Effects in 5d-Shell Atoms

Table 4 shows a comparison between experimental values of

$$\Delta T = T[5d^N(^ML)6s\ ^{M+1}L] - T[5d^N(^ML)6s\ ^{M-1}L]$$

and theoretical ones. The error of the experimental values are three times the standard deviation and the errors given with the theoretical values are caused by the uncertainties in the $\delta\langle r^2\rangle$ values used for the calculation. The agreement between experiment and theory is very satisfactory besides for the elements Ir and Pt for which more measurements should be made.

Table 4. Parameters of the term dependent isotope shift, ΔT, and parameters of the J dependent isotope shift, z_{5d}, for 5d-shell atoms. z_{5d} belongs to the terms $5d^N(^ML)6s\ ^{M\pm1}L$. Hartree-Fock values ΔT_{th} without scaling factor.

Isotope pair	Configuration	^{M+1}L	^{M-1}L	$\delta\langle r^2\rangle$ (fm^2)	ΔT_{exp} (mK)	ΔT_{th} (mK)	z_{5d} (mK)	$\frac{z_{5d}7}{\delta\langle r^2\rangle \cdot \zeta_{5d}}$ ($10^{-6}/fm^2$)
$^{178-180}Hf$	$5d^3(^4F)6s$	^5F	^3F	0.078(16)[a]	17.4(2.3)[c]	18.6(3.8)	0.88(2)[c]	10.4(3.0)
$^{184-186}W$	$5d^5(^6S)6s$	^7S	^5S	0.093(8)[b]	34.3(2.8)[c]	37.1(3.0)		
$^{185-187}Re$	$5d^6(^5D)6s$	^6D	^4D	0.109(14)[b]	40.4(4.5)[c]	38.9(5.0)	1.3(9)[c]	5.3(4.2)
$^{188-192}Os$	$5d^7(^4F)6s$	^5F	^3F	0.136(9)[b]	53.6(9.6)[c]	41.5(3.0)	2.6(2)[c]	6.8(9)
$^{191-193}Ir$	$5d^8(^3F)6s$	^4F	^2F	0.043(6)[b]	23(6)[d]	10.6(1.5)	0.2(4)[d]	
$^{194-196}Pt$	$5d^9(^2D)6s$	^3D	^1D	0.045(3)[b]	30(12)[e]	7.7(1.6)	1.9(2.0)[e]	

a /HS 74/, b /AS 85/, c /Wö 85/, d /SW 84/, e /GWB 80/

The ΔT values are on the average about three times larger than the ΔT value found in Eu, whereas the z_{5d} parameters are of the same order of magnitude. The ratio $z_{5d}/\delta\langle r^2\rangle\zeta_{5d}$ is constant within the still rather large limits of error as is expected and moreover it is the same as in the configurations $4f^7 5d6s$ in Eu and Gd and $4f^{11}5d6s^2$ in Er (see table 1).

6. Conclusion

A pronounced term and J dependence is observed in the IS of atomic spectra. These dependencies can be described by phenomenological parameters which represent CSO effects between an IS operator and a perturbing hamiltonian. In heavy elements these CSO effects are mainly caused by the FS operator. This can be proved by ab initio calculations. The term dependence can be explained by the CSO effect between the FS and the electrostatic operator, whereas the J dependence is caused by the CSO effect of the FS operator and a magnetic interaction operator.

The study of CSO effects is on the one hand a useful tool to check the validity of calculation techniques in atomic physics. On the other hand the knowledge of the CSO effects is indispensable for a reliable evaluation of the changes in mean-square nuclear charge radii from the IS observed in atomic spectra.

References

/ACH 78/ Aufmuth, P., Clieves, H.-P., Heilig, K., Steudel, A. Wendlandt, D., Bauche, J.: Z. Phys. A 285, 357 (1978)

/AKE 85/ Ahmad, S.A., Klempt, W. Ekström, C., Neugart, R., Wendt, K.: Z. Phys. A 321, 35 (1985)

/AS 85/ Aufmuth, P., Steudel, A.: unpublished material

/Au 78/ Aufmuth, P.: Z. Phys. A 286, 235 (1978)

/Au 82/ Aufmuth, P.: J. Phys. B 15, 3127 (1982)

/Au 85a/ Aufmuth, P.: Z. Phys. D in press

/Au 85b/ Aufmuth, P.: 2nd ECAMP, Amsterdam 1985

/AW 85/ Aufmuth, P., Wöbker, E.: Z. Phys. A 321, 65 (1985)

/Ba 69/ Bauche, J.: Thêse, Université de Paris, 1969

/Ba 74/ Bauche, J.: J. Phys. (Paris) 35, 19 (1974)

/BBL 69/ Bhattacherjee, S.K., Boehm, F., Lee, P.L.: Phys. Rev. 188, 1919 (1969)

/BBP 85/ Blundell, S.A., Baird, P.E.G., Palmer, C.W.P., Stacey, D.N., Woodgate, G.K., Zimmermann, D.: Z. Phys. A 321, 31 (1985)

/BC 70/ Bauche, J., Crubellier, A.: J. Phys. (Paris) 31, 429 (1970)

/BC 76/ Bauche, J., Champeau, R.-J.: Adv. At. Mol. Phys. 12, 39 (1976)

/BCS 77/ Bauche, J., Champeau, R.-J., Sallot, C.: J. Phys. B 10, 2049 (1977)

/Be 85/ Bernard, A.: Diplomarbeit, Hannover 1985

/BEP 86/ Brüggemeyer, H., Esrom, H., Pfeufer, V., Steudel, A.: Z. Phys. D in press

/BK 72/ Bauche, J., Klapisch, M.: J. Phys. B 5, 29 (1972)

/BPS 81/ Brand, H., Pfeufer, V., Steudel, A.: Z. Phys. A 302, 291 (1981)

/BSS 80/ Brand, H., Seiber, B., Steudel, A.: Z. Phys. A 296, 281 (1980)

/CO 80/ Condon, E.U., Odabasi, H.: Atomic Structure, New York: Cambridge University Press, 1980

/ES 74/ Engfer, R., Schneuwly, H., Vuilleumier, J.L., Walter, H.K., Zehnder, A.: At. Data Nucl. Data Tables 14, 509 (1974)

/Es 85/ Esrom, H.: Thesis, Universität Hannover 1985

/FB 83/ Fonseca, A.L.A., Bauche, J.: Z. Phys. A 314, 275 (1983)

/Fr 69/ Froese Fischer, C.: Comput. Phys. Commun. 1, 151 (1969)

/Fr 78/ Froese Fischer, C.: Comput. Phys. Commun. 14, 145 (1978)

/FS 83/ Froese Fischer, C.: Smentek-Mielczarek, L.: J. Phys. B 16, 3479 (1983)

/GBB 64/ Gluck, G., Bordarier, Y., Bauche, J., Kleef, T.A.M. van: Physica 30, 2068 (1964)

/GL 78/ Gerstenkorn, S., Luc, P.: Atlas du spectre d'absorption de la molecule d'iode: 14800-20000 cm^{-1}, Editions du Centre National de la Recherche Scientifique, Paris 1978

/GN 71/ Goldschmidt, Z.B., Nir, S.: Physica 51, 222 (1971)

/Gu 68/ Guthöhrlein, G.: Z. Phys. 214, 332 (1968)

/GWB 80/ Grethen, H., Winkler, R., Bauche, J.: Physica 98 C, 222 (1980)

/HS 74/ Heilig, K., Steudel, A.: At. Data Nucl. Data Tables 14, 613 (1974)

/KKW 85a/ Kropp, J.-R., Kronfeldt, H.-D., Winkler, R.: Z. Phys. A 321, 365 (1985)

/KKW 85b/ Kropp, J.-R., Kronfeldt, H.-D., Winkler, R.: Z. Phys. A 321, 57 (1985)

/La 72/ Labarthe, J.-J.: J. Phys. B 5, L181 (1972)

/LM 83/ Lindroth, E., Mårtensson-Pendrill, A.-M.: Z. Phys. A 309, 277 (1983)

/MS 82/ Mårtensson, A.-M., Salomonson, S.: J. Phys. B 15, 2115 (1982)

/NGI 81/ New, R., Griffith, J.A.R., Isaak, G.R., Ralls, M.P.: J. Phys. B 14, L135 (1981)

/Ot 81/ Otten, E.W.: Nucl. Phys. A 354, 471 c (1981)

/PCG 83/ Pfeufer, V., Childs, W.J., Goodman, L.S.: J. Phys. B 16, L557 (1983)

/PCG 84/ Pfeufer, V., Childs, W.J., Goodman, L.S.: J. Opt. Soc. Am. B 1, 34 (1984)

/Se 69/ Seltzer, E.C.: Phys. Rev. 188, 1916 (1969)

/SW 65/ Smith, G., Wybourne, B.G.: J. Opt. Soc. Am. 55, 121 (1965)

/SW 84/ Sawatzky, G., Winkler, R.: Verhandl.DPG (VI) 19, 799 (1984)

/Vi 39/ Vinti, J.P.: Phys. Rev. 56, 1120 (1939)

/Wö 85/ Wöbker, E.: Thesis Universität Hannover 1985

/Zi 85/ Zimmermann, D.: Z. Phys. A 321, 23 (1985)

APPLICATIONS OF VUV LASERS IN MOLECULAR PHYSICS

François ROSTAS
Observatoire de Paris, Section de Meudon
Département d'Astrophysique Fondamentale et UA 812 du CNRS
92195 MEUDON Cedex, France

I. Introduction

Coherent, tunable and intense sources of radiation are becoming increasingly available in the vacuum ultra-violet region, thus opening the way to a wealth of new applications in an energy range ($E > 6$ eV) which is of major importance to molecular physics.

The techniques of Laser Spectroscopy which have been developed over the last fifteen years have allowed a very rapid progress of molecular physics. Up to now, however, the wavelength range accessible to tunable lasers ($\lambda > 250$ nm) has severely restricted the choice of molecules which could be studied by these techniques.

During this same period an intense activity has been developing in the far UV which corresponds to a particularly important energy range for molecules :

i) Most of the "simple" molecules, which happen to predominate in natural environments, start absorbing light below 200 nm (1), these are the lighter di- and triatomics (H_2, O_2, N_2, NO, CO, H_2O, CO_2, NO_3, NH ...)

ii) Very elaborate calculations of potential surfaces and collisional dynamics can be performed for these simple molecules, so that experiments can be much more thoroughly interpreted

iii) The highly excited valence states, Rydberg states, ionization and dissociation thresholds of many molecules lie in this energy range. The dynamical processes involving these states are now much better understood especially through Multichannel Quantum Defect Theory (MQDT)(2)(3)

iv) In connection with state to state analysis of reactive collisions, the photodissociation of di- and triatomic molecules in this energy range has been the subject of increased interest (4). The half-collision concept can be used to link these two fields and to understand how high resolution studies of photodissociation, including detailed analysis of the fragments (translational and internal energy, angular distribution, polarization properties) can give access to a detailed understanding of potential surfaces and collision dynamics

v) Photodissociation and beam-beam reactive collision experiments often produce light atomic and molecular fragments which require far UV sources for laser induced fluorescence characterization.

Classical spectroscopy has produced in this spectral region a considerable amount of data which is very useful for preparing and analyzing experiments in molecular dynamics (5). The typical Doppler limited resolution of 0.5 cm^{-1} generally allows rotational analysis but, in most cases, does not provide quantitative predissociation and autoionization profiles.

Due to its adaptability and hitherto unequaled spectral brightness, synchrotron radiation (6) has been the choice source for selective excitation in the far UV in molecular dynamics experiments. However the available brightness is still insufficient to yield a spectral resolution which would allow rotationally selective excitation. Neither does it allow high resolution analysis of the fluorescence emitted by the excited molecules. Orientation of the plane of polarization of the exciting light is also not very practicable.

Laser multiphoton excitation, especially when coupled to ionic detection (MPI) (7) provides an elegant solution to these problems and makes full use of the capabilities of tunable lasers in terms of intensity, high spectral and temporal resolution, polarization and stepwise excitation (REMPI).
A new step forward can be made, now that VUV lasers have become available and allow single photon excitation of the state under study or of the intermediate state in resonant stepwise excitation. Experiments can now be freed from the numerous problems caused by the very high powers needed to obtain the 2 or 3 photon transition which is the first step in the multiphoton excitation experiments.

II. VUV laser sources

The technique most generally used to generate coherent tunable radiation in the VUV region has been the one based on four wave mising (FWM) or Third Harmonic Generation (THG) in rare gases or metal vapours. These methods based on third order non linear effects in gases have been systematically studied and developed in the past decade and have been thoroughly reviewed in several excellent papers (8-11).

Other techniques are available but less frequently used, especially Anti-Stokes Raman Shifting (ASRS) by which fixed frequency or tunable lasers operating in the near UV can be frequency shifted in successive steps by a fixed quantum corresponding to the vibrational spacing of a molecular gas. This method is very much simpler to implement than the four wave mixing technique but does not provide the wide frequency coverage of FWM and is much less efficient below about 150 nm.

The third order mixing technique basically consists in "adding" three photons in the visible or UV range to produce one in the VUV through the third order term in the polarization of the interaction medium produced by the incident electric fields. If a single wavelength is incident on the medium ($\lambda_1 = 2\pi c/\omega_1$) one can obtain an output at $\omega_3 = 3\omega_1$. If two wavelengths are incident, combination such as $\omega_3 = 2\omega_1 \pm \omega_2$ are produced. The power generated at the frequency ω_3 can be expressed as

$$P(\omega_3) \sim N^2 |\chi^{(3)}(\omega_1,\omega_2,\omega_3)|^2 F(L,b,\Delta k) I^2(\omega_1) I(\omega_2)$$

where $\chi^{(3)}$ is the third order susceptibility of the interaction medium, N its number density and $I(\omega)$ is the intensity incident at one of the input wavelengths. F is the phase matching integral, it varies with the length of the interaction region (L), the wavevector mismatch (Δk) between the generated and the incident radiations and with the confocal parameter (b) of the incident beam. The efficiency of the conversion process can be expected to depend critically on the optimization of the parameters which determine F and $\chi^{(3)}$.

The phase matching condition, in focused beam geometry, is easily met for difference frequency mixing (DFM : $\omega_3 = 2\omega_1 - \omega_2$) where F has a broad maximum for $\Delta k = 0$. For sum frequency mixing (SFM : $\omega_3 = 2\omega_1 + \omega_2$) F = 0 if $\Delta k \geqslant 0$ and goes through a maximum for $b\Delta k = -2$. The condition $\Delta k < 0$ implies that the medium must have anomalous dispersion at the frequency ω_3. This happens in limited spectral ranges on the blue side of absorption lines or around autoionizing transitions and makes the choice of the conversion medium depend strongly on the output wavelength which is sought.

The third order susceptibility $\chi^{(3)}$ can be expressed as the sum of terms such as the following :

$$\chi^{(3)}(\omega_1,\omega_2,\omega_3) \sim \sum_{a,b,c} \frac{\langle g|\mu|c\rangle\langle c|\mu|b\rangle\langle b|\mu|a\rangle\langle a|\mu|g\rangle}{(\Omega_{cg} - 2\omega_1 - \omega_2)(\Omega_{bg} - 2\omega_1)(\Omega_{ag} - \omega_1)}$$

where a, b, c stand for excited states of the atom or molecules of the conversion medium, g is the ground state, μ is the dipole transition matrix element, and Ω the complex transition frequency. The term shown exhibits two resonances which play an important part in the four-wave mixing process. Indeed the susceptibility and hence the conversion efficiency is enhanced when either $\omega_3 = 2\omega_1 + \omega_2$ or $2\omega_1$ is in resonance with a real transition of the medium. The two photon resonance is particularly useful since it enhances the conversion process without causing strong absorption.

In order to achieve tunability and benefit from the two-photon resonance enhancement of the conversion efficiency, a typical VUV source will be comprised of two high power pulsed dye lasers, one of which will be set at a two photon resonant frequency ω_1 and the other will be set at a tunable frequency ω_2. The resultant frequency can be tuned in the range $2\omega_1 \pm \omega_{2M}$, excluding however a central gap $2\omega_1 \pm \omega_{2m}$ where ω_{2m} and ω_{2M} are respectively the lowest and highest frequencies on the tuning range accessible to ω_2. The choice of the conversion medium and of the two photon resonance will determine the frequency ranges effectively accessible within the theoretical one defined above and the conversion efficiency.

Greater versatility is obtained at the expense of efficiency if non resonant mixing is produced. Third harmonic generation is also a low efficiency process since it cannot be resonantly enhanced, but it is even simpler to implement since it requires only one input frequency.

A number of conversion media have been studied by now, including metal vapours, rare gases, molecular gases, both in cells and in pulsed jets (9,10). The latter technique looks quite promising since it allows to generate wavelengths below the LiF cutoff (106 nm) without using cumbersome differential pumpings. To take just an example, magnesium vapour which is one of the most popular conversion media can be used to generate VUV in the 140 - 174 nm range by SFM using the $^1D_2 - {}^1S_0$, $\lambda = 430.9$ nm, transition for two photon resonance. Another less efficient range is accessible between 120 and 130 nm. The negative dispersion is provided by very broad autoionizing levels above the ionization limit (162 nm). Typical characteristics of the radiation obtained with dye lasers delivering about 1 MW peak power pulses of visible/UV are summarized in Table 1. For comparison the intensities obtainable from synchrotron sources through a typical 1 m monochromator are also given.

Table 1 - Main characteristics of VUV lasers compared to synchrotron radiation

	Synchrotron radiation		VUV laser	
	without undulator	with undulator	non resonant conversion	resonant conversion
Pulse duration (ns)	1	1	10	10
Repetition rate (Hz)	10^7	10^7	20	20
Line width (nm)	10^{-2}	10^{-3}	10^{-4}	10^{-4}
Numb. of photons per pulse	10^3	10^5	$5\ 10^{10}$	$5\ 10^{12}$
Peak power (W)	$2\ 10^{-5}$	$2\ 10^{-3}$	10	10^3
Spectral brightness (Phot/s/nm)	10^{12}	10^{15}	10^{16}	10^{18}

To summarize, one can say that all wavelengths between 100 and 200 nm are accessible in continuously tunable ranges of 5 to 20 nm. Between 70 and 100 nm most wavelengths are accessible using the rare gases as nonlinear medium but there are gaps and the tuning ranges are shorter. Wavelengths below 70 nm have been produced, generally by using higher order processes (5th, 7th etc...) but at lower efficiency (9) and in short tuning ranges around specific wavelengths.

The spectral width of the radiation produced is generally two or three times that of the dye lasers used. In many applications a resolution of 10^{-4} nm (. 1 cm^{-1}) has been reported (8,12) with possible improvements to 10^{-5} nm.

III. Applications of VUV lasers

A number of applications to atomic and molecular physics have been demonstrated in the laboratories where the conversion processes have been studied.

Table 2 - Main applications of VUV lasers in molecular physics

---

Structure and dynamics of excited states above 6 eV

- High resolution spectroscopy
- Predissociation and autoionization of rotationally resolved states : *cross sections, profiles, lifetimes*
- Collisional complex spectroscopy

Photodissociation of small molecules

- Selective excitation : *rotational resolution, polarization effects*
- Photofragment fluorescence : *spectral resolution, alignment, orientation*
- LIF characterization of fragments : *kinetic energy (Doppler), internal energy, alignment, orientation*

Reactive collisions

- Preparation of reactants by photofragmentation : *state selected molecules, ions or atoms*
- LIF product characterization : *internal states, Doppler profiles*
- Transition state spectroscopy

Photoionization

- Fundamental state (1 photon)
- Excited states
- Product ion state analysis by photoelectron spectroscopy (integrated or threshold)

High sensitivity detection of atoms and molecules

- Laser induced fluorescence
- Resonance enhanced MPI (REMPI)

Molecule/Surface interaction

- Initial state selection
- Product characterization

During the last two or three years the progress of generation techniques has reached a plateau and the VUV laser sources are used more and more in application-oriented experiments. Also a large fraction of the experiments using the MPI technique constitute very good examples of what can be done with VUV lasers. In many cases the interpretation of MPI experiments could indeed be much simplified if the initial multiphoton step was replaced by a single VUV photon excitation. The main fields where VUV lasers have been or would be applied in molecular physics are listed in Table 2 and briefly reviewed in the following sections.

IV. VUV laser spectroscopy

Recently published experiments demonstrate the applicability of the now classical techniques of laser spectroscopy in the VUV region. A forthcoming review paper by C.R. Vidal (12), who has been amongst the pioneers in this field, illustrates this point very convincingly. The molecules which have been studied, the wavelength ranges and the techniques used are summarized below :

H_2	82.8 - 83.3	Absorption	(13)
	110.8 - 111.6	Optical-optical double resonance, ion detection	(14,15)
CO	140 - 160	Rotational level lifetimes	(16,17)
		OODR, selected fluorescence, laser reduced fluorescence (LRF)	(18,19)
NO	179 - 197	Excitation spectra, dispersed and/or selected fluorescence	(20)
		Rotational level lifetimes	(21)
Xe_2	130 - 150	Excitation spectra	(22)
HCN	185 - 192	Excitation spectra, lifetimes	(23)

In the H_2 absorption experiments of Rothschild et al (13) the exceptional spectral resolution (0.005 cm^{-1}) of the frequency locked ArF laser allowed the q and Γ parameters of certain predissociation profiles around 83 nm to be determined.

Vidal and coworkers for CO (18,19), Rottke and Welge for H_2 (14,15) have demonstrated the power of two step excitation techniques (OODR). The "laser reduced fluorescence" signal of the intermediate level (LRF) which is observed when the second excitation step is in resonance is a sensitive probe of predissociations occuring in the final level (19). The combination of two step excitation, photoionization of the final excited state and detection of the ions formed leads to the most extreme sensitivity (14). The optical double resonance technique also has the well known advantages of simplifying molecular spectra by selecting the J values which are excited and allowing sub-Doppler resolution if the laser line width is narrow enough.

Single photon excitation in the VUV, thus opens a vast new field to laser spectroscopy. A comparison of single photon and two or three photon absorption cross sections (7) immediately shows that it is more efficient, starting from a given UV/visible laser power to convert it to a VUV wavelength and do single photon absorption than to use it directly for non resonant multiphoton excitation. Aside from increased efficiency one also benefits by avoiding the numerous secondary effects which are brought about by the high laser powers which have to be used for multiphoton excitation (24). This point will be elaborated further when the state sensitive detection experiments on H_2 will be examined below.

V. Photodissociation

Photodissociation in the VUV is a major factor in the chemistry and energy balance of a number of natural media such as earth and planetary atmospheres, comets, interstellar clouds and circumstellar envelopes. Furthermore spectral analysis of photodissociation processes is a very efficient approach to the study of the potential surfaces and the dynamics of molecular systems. The data obtained from photodissociation experiments is all the more detailed as the state of the initial molecule and that of the fragments produced are better known. The molecules under study should preferably be chosen amongst the di- or triatomic if the experimental results are to be compared to detailed calculations. This choice of small and light molecules implies single photon excitation energies which almost always lie in the VUV.

The most interesting cases are those where the photodissociation spectrum has a rotational structure which implies that the predissociation processes which couple the bound initial state to the continuum are not too fast. The initial angular momentum vector of the molecule can then be selected by means of the wavelength and polarization of the absorbed photon. The state of the fragments is also determined in detail : kinetic, vibrational rotational energy distribution and also their vector properties i.e. angular distribution and alignment. The polarization of the exciting light and of the fragment fluorescence is thus an important tool for probing dynamics of the process.

Until powerful tunable lasers became available for multiphoton excitation and VUV generation, the emphasis in photodissociation studies was mainly on detailed analysis of the fragments. The parent molecule was excited either at fixed frequency by resonance lamps or excimer lasers or at low spectral resolution. The very extensive work already done in this field has been recently reviewed by Leone (25) and Bersohn (26). The new orientations are very clearly indicated in a recent review by J.P. Simons (4) who insists on the importance of combining spectroscopy and dynamics to understand the photodissociation mechanisms and discerns a new era of "high resolution photochemistry" being ushered in. It is perfectly clear that VUV lasers will have a very important part to play in this venture.

The recent work of J.P. Simons and collaborators on H_2O (27), although not done with VUV lasers, illustrates perfectly this new trend. Using a KrF laser tunable for about 1 nm around 248 nm, they have studied the fluorescence of the OH^* $(A^2\Sigma^+)$ fragment produced by two photon rotationally resolved excitation of the $\tilde{C}^1B_1$ state of H_2O. Three processes have been identified in this work :

a) *Direct dissociation through the $\tilde{B}^1A_1$ state.* This process corresponds to the continuum underlying the rotational structure of the $\tilde{C}^1B_1$ state and it produces a strong alignment of the OH^* fragments. However the observed rotational population distribution of OH^* is in partial agreement only with the predictions of Segev and Shapiro (28) which are based on trajectories confined to the $\tilde{B}^1A_1$ state. This work identifies two types of trajectories on the $\tilde{B}^1A_1$ surface : oscillating trajectories on the one hand, which delay the dissociation and destroy alignment and, on the other hand, trajectories leading directly to dissociation which produce alignment and are correlated to higher rotational levels of the OH^* fragment. This correlation between alignment and high J levels is indeed observed experimentally, however the low J levels are underpopulated indicating that the long lived oscillating trajectories favour the passage through a conical intersection from the $\tilde{B}^1A_1$ to the $\tilde{A}^1B_1$ surface which is correlated to OH in the ground state ($X^2\Pi$). This interpretation is in agreement with the observed branching ratio between $OH^*(A^2\Sigma^+)$ and $OH(X^2\Pi)$ which is only 10% in favour of the $OH^*(A)$ output channel.

b) *Heterogeneous predissociation of* $\tilde{C}^1B_1$ by Coriolis coupling with the $\tilde{B}^1A_1$ state confirmed by the absence of fluorescence following excitation of $K_a' = 0$ rotational levels of the $\tilde{C}^1B_1$ state.

c) *Homogeneous predissociation of* $\tilde{C}^1B_1$ leading to OH in the ground state. This process which also affects $K_a' = 0$ levels has been shown to be important in the REMPI spectroscopic work of Ashfold et al (29).

This high resolution photodissociation work sets an example because it is the first one in this excitation energy range where the initial quantum state of the molecule and the state of the product have been completely determined. Comparable work has been done in the visible or near UV on HONO (30) and NO_2 (31) but high resolution photodissociation studies are still quite rare. One can easily predict that further work of this type will increasingly call upon tunable VUV lasers in order to extend the range of accessible excited states.

Single photon direct photodissociation of H_2O has recently been studied by P. Andresen and collaborators (32). Exciting ground state H_2O with a fixed frequency excimer laser (F_2, 157 nm) one can reach a specified point on the potential surface of the $\tilde{A}^1B_1$ state which immediately dissociates into $OH(X^2\Pi) + H$. The distribution of the excess energy between translation, vibration and rotation has been studied by Laser Induced Fluorescence of the OH fragment. Dynamics calculations of Shinke et al (33) on the potential surface of Staemler and Palma (34) reproduce quite satisfactorily the experimental results. Andresen has also shown that this direct photodissociation causes a population inversion in the Λ doublet population of OH in the ground state for high rotational levels. He has been able to show how this inversion was directly connected to the dissociation dynamics and the geometry of the molecular system. This process could be very important in the creation of interstellar OH masers.

In this direct dissociation situation, the initial absorption is structureless and the rotational selection is obtained by cooling the molecules in a supersonic expansion. It is clear, that if the dissociation continuum could be excited at variable wavelength, a larger segment of the upper potential surface could be explored and detailed comparisons with theoretical predictions could be made. One can expect to discover potential surface intersections which would require more elaborate theoretical work.

VI. Reaction product characterization

Visible and UV lasers have been used in many experiments to analyze the internal energy of fragments produced by photodissociation or reactive collisions. The most popular technique is that of Laser Induced Fluorescence (LIF) which is particularly well suited for crossed beam experiments. However, in order to gain access to light fragments such as CO, H_2, H and others, an excitation source in the VUV is needed. Multiphoton excitation is currently used in such cases but recently a few experiments have made use of direct excitation in the VUV.

CO has been detected and rotationally analyzed following the photodissociation of Formaldehyde (35) and Glyoxal (36). The $A^1\Pi \leftarrow X^1\Sigma^+$ transition is excited in the 150 nm region. In the case of Formaldehyde the H_2 fragment has also been analyzed by coherent Anti-Stokes Raman Scattering (CARS)(37).

Atomic Hydrogen produced by the dissociation of HDCO (38) and HI (39) has been detected by fluorescence induced by Lyα photons. The velocity of the H atoms has been determined by Doppler profile analysis.

The Bielefeld group has demonstrated the extreme detection sensitivity which can be obtained through, one photon resonant, two photon ionization (1 VUV + 1 UV). For H, a limit of 10^5 atoms cm^{-3} has been obtained and it can be realistically expected to be reduced to 10 atoms cm^{-3} (40a). For NO a lower detection limit of 10^4 atoms cm^{-3} has been obtained (40b). These detection limits correspond to between 1 and 100 particles in the interaction region.

Br atoms produced by reaction of H or F with HBr or Br_2 have been detected by laser induced fluorescence around 145 nm to determine the reaction cross section and the branching ratio between $Br(^2P_{1/2})$ and $Br(^2P_{3/2})$ (41).

The rotational population analysis of H_2 has been the object of several experiments which allow a comparison of different excitation and detection schemes. Marinero et al (42) and Northrup et al (43) have conducted sensitivity tests of one photon LIF following direct excitation of the $B^1\Sigma_u^+$ state of H_2 between 97 and 107 nm. Rottke and Welge (44) have used a two step excitation scheme with $B^1\Sigma_u^+$ as the intermediate state, followed by ionization through one UV photon. Finally Marinero et al (45) have used multiphoton excitation of the $E^1\Sigma_g^+$ state followed by ionization (REMPI) in the course of an experiment studying the $H + D_2 \rightarrow HD + D$ reaction.

In all these experiments the available laser power was of 1 to 3 MW peak. The VUV conversion had an efficiency of approximately 10^{-6}. In the REMPI experiment the 200 nm photons were produced by Raman Anti-Stokes conversion with an efficiency of about 10^{-3}. In view of the results obtained in these various experiments and assuming realistic improvements for each of them such as VUV conversion efficiency of 10^{-3}, improvement in the ion or photon detection efficiency by factors of 10 or 100 according to the author's indications one can expect the following detection limits :

OODR + ion detection	$2\ 10^2$ mol./cm^3/quant. state		
1 photon excitation + fluorescence	$2\ 10^4$	"	"
2 photon excitation + ion detection	$5\ 10^7$	"	"

For comparison purposes, the CARS technique has a sensitivity of the order of 10^{12} mol./cm^3/quant. state (37) for laser powers similar to the ones used in the above experiments.

It can be seen that changing from fluorescence to ion detection improves the efficiency by two orders of magnitude and that excitation by a single photon improves the sensitivity by about 5 orders of magnitude with respect to two photon excitation. This is quite in accord with the relative cross sections for single and two photon absorption which are in a ratio of about 10^8 at the incident powers used in these experiments ($\sim 10^9$ w/cm^2).

These experiments confirm a previous statement according to which the conversion efficiencies obtainable in VUV generation by third order non linear processes ($10^{-6} < \eta < 10^{-3}$) are such that, for an available laser peak power of about 1 MW, direct one photon excitation in the VUV can be substituted with advantage to two and *a fortiori* three photon excitation.

VII. Photoionization

A strong development of VUV laser applications is to be expected in this field. Indeed high resolution cross sections and profile measurements for single photon ionization from the ground state are of great interest *per se* and in relation with theory. The first measurements of this type have been reported recently on NO (46) with reference to MQDT calculations (2) which take into account the competition between the ionization and dissociation channels.

Numerous studies in the last few years have concerned the photoionization of excited states (47). In these studies Resonant Multiphoton Ionization and Photoelectron Spectroscopy techniques (PES) have been combined to obtain information on the photoionization dynamics of molecules such as H_2 and N_2. Most of these experiments use "one colour" excitation with three photons for resonant excitation of a bound level of the molecule and a fourth photon for achieving ionization. The excitation of the product ion is determined from the excess energy of the released electron (PES). Obviously such experiments would be greatly improved if the bound molecular state was populated by a single VUV photon, or by one VUV and one visible/UV photon in a two step process. The ionization would then be obtained by a variable wavelength photon which would allow the experiment to benefit from the supplementary information about the ionization process provided by Threshold Photoelectron Spectroscopy (TPES).

Similar schemes can be used to produce ions in selected rovibronic states in order to study the state to state dynamics of ion molecule reactions. Such a study has recently been conducted on NH_3^+ ($\tilde{X}$,V = 0-6) at Stanford (48a,b) using 2 photon excitation for the first step. Here again one may expect that single photon VUV excitation will prove advantageous.

Direct VUV ionization of molecules can also be used as a simple and in many cases sufficiently selective means of probing processes occuring in molecular jets (55).

VIII. Other applications

Several experiments use suprathermal Hydrogen atoms in reactions such as $H + H_2$ (45,49,50). These atoms are produced by fixed frequency photodissociation of Hydrides (HI, HBr) using excimer lasers. Replacing these by tunable VUV lasers would allow an increase of the accessible kinetic energy range. The same remark applies to the production of metal atoms from iodides.

Spectroscopy of the transition states of H_3 produced in $H + H_2$ reactions are now possible with VUV lasers and seriously considered (51). H_3 is expected to absorb radiation between the repulsive ground state and an excited state correlated to H (n = 2) + $H_2(^1\Sigma_g^+)$ in the red wing of Lyα above 121 nm.

Laser induced fluorescence and selective photoionization by resonant excitation to a bound intermediate state can be used to characterize traces of atomic and molecular material. The experiments analyzed above concerning the characterization of reaction products demonstrate the high level of sensitivity which can be obtained. This is even truer for atoms which concentrate the absorption oscillator strength on a smaller number of lines. In a feasibility experiment (52), all of the 1000 atoms of a ^{81}Kr sample diluted in Helium have been counted. LIF in the VUV has been proposed for measuring the temperature and density of impurity atoms in Tokamak fusion machines (53).

Molecules desorbed from surfaces can also be studied by VUV Laser Induced Fluorescence. A recent study using the REMPI technique (54) to characterize H_2 molecules having interacted with a copper surface proves the feasibility of such experiments which, as many others, could benefit from the use of VUV lasers.

IX. Conclusion

The present analysis of proven and possible applications of VUV lasers is certainly incomplete, however it shows clearly that a number of laser spectroscopic techniques are now available in the Vacuum Ultraviolet Region thus opening many new possibilities in molecular physics.

Such setups can be put together from commercially available elements at a cost which puts them within reach of many experimental groups. The VUV generation technique has been thoroughly explored by several pioneering laboratories and is now well documented.

The major fields of applications seem to be spectroscopy, high resolution photochemistry and state to state chemistry where the preparation of reactants and characterization of products is of paramount importance. The VUV lasers are thus seen to be an indispensable complement to synchrotron radiation in this particuler range of energies. The same field of applications has already been explored by multiphoton experiments which are simpler to put together. However the advantages of single photon excitation are such that the VUV laser techniques should progressively gain acceptance and replace the MPI methods in many instances.

References

1. Okabe, H., Photochemistry of Small Molecules, Wiley (1978).
2. Giusti-Suzor, A. and Jungen Ch., J. Chem. Phys. 80, 986 (1984).
3. Jungen, Ch., Phys. Rev. Lett. 53, 2394 (1984).
4. Simons, J.P., Journ. Phys. Chem. 88, 1287 (1984).
5 McIllrath, T.J., in Laser Techniques for Extreme Ultraviolet Spectroscopy, T.J. McIllrath and R.R. Freeman Ed., A.I.P. Conference Proceedings No. 90 (1982).
6. Koch, E.E. (Ed.), Handbook on Synchrotron Radiation, North Holland (1983).
7. Johnson, P.M. and Otis, C.E., Ann. Rev. Chem. 32, 139 (1981).
8. Wallace, S.C., Adv. Chem. Phys. 47, 153 (1981).
9. Jamroz, W. and Stoicheff, B.P., in Progress in Optics, E. Wolf (Ed.), North Holland (1983).

10. Hepburn, J.W., Israel J. Chem. 24, 273 (1984).

11. Vidal, C.R., in Tunable Lasers, L.F. Mollenauer and J.C. White (Eds.) Springer (1985).

12. Vidal, C.R., Adv. At. & Mol. Phys. (To be published).

13. Rotschild, M., Egger, H., Hawkins, R.T., Bokor, J. Pummer, H. and Rhodes, C.K., Phys. Rev. A23, 206 (1981).

14. Rottke, H. and Welge, K.H., Chem. Phys. Lett. 99, 456 (1983).

15. Rottke, H. and Welge, K.H., J. de Phys. 46, C1-127 (1985).

16. Provorov, A.C., Stoicheff, B.P. and Wallace, S.C., J. Chem. Phys. 67, 5393 (1977).

17. Maeda, M. and Stoicheff, B.P., Jap. Journ. Appl. Phys. 24, 717 (1985).

18. Klopotek, P. and Vidal, C.R., Can. J. Phys. 62, 1426 (1984).

19. Klopotek, P. and Vidal, C.R., J.O.S.A. B2, 869 (1985).

20. Scheingraber, H. and Vidal, C.R., J.O.S.A. B2, 343 (1985) ; J. Chem. Phys. (to be published).

21. Banic, J.R., Lipson, R.H. and Stoicheff, B.P., Can. J. Phys. 62, 1629 (1984).

22. Lipson, R.H., Laroque, P.E. and Stoicheff, B.P., J. Chem. Phys. 82, 4470 (1985).

23. Hsu, Y.-C., Smith, M.A. and Wallace, S.C., Chem. Phys. Lett. 111, 219 (1984).

24. Girard, B., Billy, N., Vigué, J. and Lehmann, J.C., Chem. Phys. Lett. 102, 168 (1983).

25. Leone, S.R., Adv. Chem. Phys. 50, 255 (1982).

26. Bersohn, R., J. Phys. Chem. 88, 5145 (1984).

27. Hodgson, A., Simons, J.P., Ashfold, M.N.R., Bailey, J.M. and Dixon, R.N., Molec. Phys. 54, 351 (1985).

28. Segev, E. and Shapiro, M., J. Chem. Phys. 77, 5604 (1982).

29. Ashfold, M.N.R., Bailey, J.M. and Dixon, R.N., Chem. Phys. 84, 35 (1984).

30. Vasudev, R., Zare, R.N. and Dixon, R.N., J. Chem. Phys. 80, 4863 (1984).

31. Welge, K.H., in Laser Applications in Chemistry, K.L. Kompa and J. Wanner Eds., NATO A.S.I. ser. B 105, Plenum Press N.Y. p. 123 (1984).

32. Andresen, P., Ondrey, G.S., Titze, B. and Rothe, E.W., J. Chem. Phys. 80, 2548 (1984).

33. Schinke, R., Engel, V. and Staemmler, V., Chem. Phys. Lett. 116, 165 (1985).

34. Staemler, V. and Palma, A., Chem. Phys. 93, 63 (1985).

35. Bamford, D.J., Filseth, S.V., Foltz, M.F., Hepburn, J.W. and Bradley Moore, C., J. Chem. Phys. 82, 3032 (1985).

36. Hepburn, J.W., Sivakumar, N. and Houston, P.L., in "Laser Techniques in the Extreme U.V.", S.E. Harris and T.B. Lucartorto Eds., A.I.P. Conference Proceedings No. 119, p. 126 (1984).

37. Debarre, D., Lefebvre, M., Péalat, M. and Taran, J.P., to be published.

38. Welge, K.H., in Laser Applications in Chemistry, K.L. Kompa and J. Wanner Eds., NATO A.S.I. ser. B 105, Plenum Press N.Y., p. 103 (1984).

39. Schmiedl, R., Dugan, H., Meier, W. and Welge, K.H., Z. Phys. Atoms and Nuclei 304, 137 (1982).

40. a : Zacharias, H., Rottke, H., Danon, J. and Welge, K.H., Opt. Comm. 37, 15 (1981).

b : Zacharias, H., Rottke, H. and Welge, K.H., Appl. Phys. 24, 23 (1981).

41. Hepburn, J.W. et al, J. Chem. Phys. 69, 4311 (1978) ; 74, 6226 (1981) ; 75, 3353 (1981).

42. Marinero, E.E., Rettner, C.T., Zare, R.N. and Kung, A.H., Chem. Phys. Lett. 95, 486 (1983).

43. Northrup, F.J., Polanyi, J.C., Wallace, S.C. and Williamson, J.M., Chem. Phys. Lett. 105, 34 (1984).

44. Rottke, H. and Welge, K.H., Chem. Phys. Lett. 99, 456 (1983).

45. Marinero, E.E., Rettner, C.T. and Zare, R.N., J. Chem. Phys. 80, 4142 (1984).

46. Miller, P.J., Chen, P. and Chupka, W.A., to be published.

47. Pratt, S.T., Dehmer, P.M. and Dehmer, J.L., J. Chem. Phys. 81, 3444 (1984).

48. a : Conaway, W.E., Morrison, R.J.S. and Zare, R.N., Chem. Phys. Lett. 113, 429 (1985).

b : Morrison, R.J.S., Conaway, W.E. and Zare, R.N., Chem. Phys. Lett. 113, 435 (1985).

49. Gerrity, D.P. and Valentini, J.J., J. Chem. Phys. 79, 5202 (1983).

50. Quick Jr., C.R. and Moore, D.S., J. Chem. Phys. 79; 759 (1983).

51. Foth, H.J., Mayne, H.R., Poirier, R.A., Polanyi, J.C. and Telle, H.H., Laser Chem. vol. 2, A.H. Zewail Ed., Harwood, p. 229 (1983).

52. Kramer, S.D., Chen, C.H., Allman, S.L., Hurst, G.S. and Lehmann, B.E., in Laser Techniques in the Extreme U.V., S.E. Harris and T.B. Lucartorto Eds., A.I.P. Conference Proceedings No. 119, p. 246 (1984).

53. Dreyfus, R.W., Bogen, P. and Langer, H., in Laser Techniques for Extreme U.V. Spectroscopy, T.J. McIllrath and R.R. Freeman Eds., A.I.P. Conference Proceedings No. 90, p. 57 (1982).

54. Kubiak, G., Sitz, G.O. and Zare, R.N., J. Chem. Phys. 81, 6397 (1984).

55. Chupka, W.A., private communication.

TRACE ELEMENT DETECTION BY LASER SPECTROSCOPY

Kay Niemax

Institut für Spektrochemie und angewandte Spektroskopie
Bunsen-Kirchhoff-Straße 11, 4600 Dortmund 1, FRG

The principles and the present status of analytical spectroscopy with tunable lasers are presented. The fields of laser-induced fluorescence (LIF) and optogalvanic spectroscopy (OGS) are discussed in detail.
Achievable detection sensitivities of the laser spectroscopy methods are compared with those of approved and current techniques in analytical spectroscopy.
In the second part of the contribution, an outlook is given introducing a new, promising technique in this field which may solve a number of future problems concerning extreme trace element detection in industry, research, medicine and environment.

1. Introduction

It is interesting to note that the introduction of tunable lasers to analytical spectroscopy did not immediately revolutionize the field as it took place e.g. in atomic or molecular spectroscopy. This is mainly due to the necessity of the development of new effective detection schemes which allow to determine trace particles in a host medium in real samples. On the other hand, in the present status, the application of a tunable laser allows the measurement of one element only at a time and is limited to only a few elements because of the limited tunable wavelength ranges.

Laser spectroscopy methods have to compete with powerful and commercially available multielement methods like the Optical Emissison Spectroscopy using e.g. ICP (inductively coupled plasma) or MIP (microwave-induced plasma) sources [1]. In the near future, as long as high-cost laser systems have to be used, the chance to settle laser spectroscopy methods requires high detection sensitivities of the new methods to enter the region of extreme trace detection in the fg (10^{-15}g) or even the ag (10^{-18}g) range. Otherwise the interest to apply those methods will be little, because of the complexity and high costs of the tunable laser systems. As the semiconductor lasers will be improved concerning the power and the extension of the lasing wavelengths towards the visible or even the uv region, these arguments will no longer hold. The author is convinced that in the near future low-cost and simple to handle tunable diode lasers will make laser spectroscopic methods, which are developed now, important and commonly used tools in analytical spectroscopy.

In this contribution, the presented material is restricted to analytical application of lasers in analytical spectroscopy. The active field of laser evaporation of samples [2,3] for analytical purposes will not be discussed here.

The two main spectroscopic techniques using tunable lasers are laser-induced fluorescence (LIF) and laser-enhanced ionisation (LEI) or optogalvanic spectroscopy (OGS). While in LIF, the fluorescence photons of the laser-excited analyte are detected, in LEI (or OGS) enhanced ion rates are measured. Higher excited analyte atoms are more easily ionized by thermal collisions than in low energy levels. A very promising technique is the thermionic diode spectroscopy which basically uses the process of LEI. The main difference compared with the commonly applied LEI techniques is that the additionally, laser-produced ions are detected with a high amplification factor (typically 10^5-10^6). This technique will make extreme trace element detection possible.

2. Selective laser excitation of analyte atoms

The rate equations describing the change of population in two discrete levels 1 and 2 and the ionisation continuum i of analyte atoms (see Fig. 1) in a collisional atmosphere and a radiation field are

$$\dot{N}_1 = -B_{12}\rho_{12}N_1 + B_{21}\rho_{12}N_2 + A_{21}N_2 + C_{21}N_2 - C_{12}N_1 \quad (1)$$

$$\dot{N}_2 = B_{12}\rho_{12}N_1 - B_{21}\rho_{12}N_2 - A_{21}N_2 - C_{21}N_2 + C_{12}N_1 - C_{2i}N_2 \quad (2)$$

$$\dot{N}_i = C_{2i}N_2 \quad (3)$$

where ρ_{12} is the spectral energy density of the radiation field at the transition frequency ν_{12} between level 1 and 2. A_{21}, B_{12} and B_{21} are the familiar Einstein coefficients for spontaneous

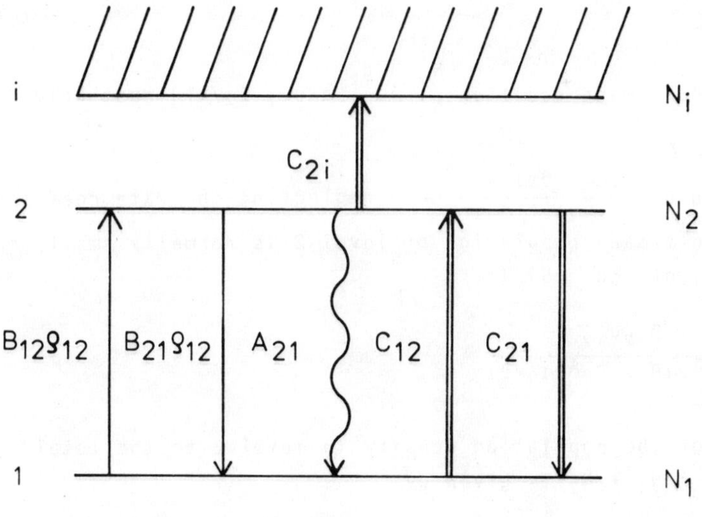

Fig. 1: Energy level scheme of the analyte atom for one-step laser excitation including the optical and collisional rate coefficients.

emission and for induced absorption and emission, respectively. C_{12}, C_{21} and C_{2i} are the collisional rate coefficients for transitions from level 1 → 2, 2 → 1 and from level 2 into the continuum i of the atom. Note that we have neglected the rate coefficients for electronic recombination (i → 2) and (i → 1) and collisional ionization (1 → i) which are very small in practice.

The total number of density of the analyte atom is

$$N_T = N_1 + N_2 + N_i \quad .$$

In a first approach (for LIF), we will neglect the collisional ionization 2 → i ($N_i = 0$). From (1) and (2) we get for the steady-state case ($\dot{N}_1 = \dot{N}_2 = 0$)

$$\frac{N_2}{N_1} = \frac{B_{12}\rho_{12} + C_{12}}{B_{21}\rho_{12} + A_{21} + C_{21}} \tag{5}$$

where the Boltzmann equation $\frac{N_2}{N_1} = \frac{g_2}{g_1} \exp\left[-E_2/kT\right]$ must hold.

Introducing $D_{21} = \frac{A_{21}}{C_{21} + A_{21}}$ and neglecting the rate coefficient C_{12} (the Boltzmann population of level 2 is normally small) the ratio N_2/N_1 of eqn. (5) is

$$\frac{N_2}{N_1} = \frac{B_{12}\rho_{12}}{B_{21}\rho_{12} + A_{21}/D_{21}} \quad . \tag{6}$$

The ratio of the population density in level 2 to the total density $N_t = N_1 + N_2$ is given by

$$\frac{N_2}{N_t} = \frac{N_2}{N_1 + N_2} = \frac{B_{12}\rho_{12}}{(B_{12} + B_{21})\rho_{12} + A_{21}/D_{21}} \quad . \tag{7}$$

With the well known relation between the Einstein coefficients $g_1 B_{12} = g_2 B_{21}$, eqn. (7) changes to

$$\frac{N_2}{N_t} = \frac{B_{12}\rho_{12}}{B_{12}(1 + \frac{g_1}{g_2})\rho_{12} + A_{21}/D_{21}} \quad . \tag{8}$$

In case of optical saturation by the laser field $B_{12}(1 + \frac{g_1}{g_2})\rho_{12} \gg A_{21}$, eqn. (8) simplifies to

$$\frac{N_2}{N_t} \approx \frac{g_2}{g_1 + g_2} \quad . \tag{9}$$

We will assume that the analyte atom after excitation will return to the level 1 in an average lifetime $\tau = 1/A_{21}$. If e.g. $g_1 = g_2$ there is a 50% chance for the atom to be reexcited in case of optical saturation. The number of excitation and fluorescence cycles in the atom can be calculated taking into account the transient time of the moving analyte atoms through the laser beam x/v (x: path length, v: velocity) and the lifetime τ. It is $x/v\tau$. In case of saturation, the number of fluorescence photons emitted by the analyte atom is therefore $x/2v\tau$. If the lifetime is 10 ns (typical for an alkali resonance level) x = 3 mm and v = 2 x 10^4 cm/s, the atom will emit 750 photons which, in principle, should be detectable by a low-noise photomultiplier and an efficient mirror set-up to gather the resonant photons. But background noise (stray light and Rayleigh scattering) limits the detection sensitivity of the LIF technique considerably as discussed below.

If we now take into account collisional ionization from the laser populated level (LEI spectroscopy) and still neglecting electronic recombination of the ions, the following equations hold

$$\dot{N}_i = C_{2i} N_2 \quad \text{and} \tag{3}$$

$$N_T = N_t + N_i \quad . \tag{10}$$

Defining $F = \frac{N_2}{N_t} = \frac{N_2}{N_T - N_i}$ and inserting in eqn. (3) yields

$$\dot{N}_i = C_{2i} F (N_T - N_i) \quad . \tag{11}$$

Note that N_i is only a function of time. Integration of eqn. (10) from t = 0 to t result in

$$N_i(t) = N_T [1 - \exp(-C_{2i} F t)] \tag{12}$$

if the analyte atoms are irradiated by a 'square-formed' laser pulse of Δt_1 length, the ion density is

$$N_i(\Delta t_1) = N_T [1 - \exp(-C_{2i} F \Delta t_1)] \quad . \tag{13}$$

It can immediately be seen that N_i may approach N_T (100% ionization) if

$$\Delta t_1 \gg \frac{1}{C_{2i} F} \quad . \tag{14}$$

In practice one should choose an energy level 2 near to the ionization limit to take advantage of a large collisional ionization rate coefficient C_{2i} . On the other hand, the transition probability for 1 → 2 should be large enough to saturate the transition with the available laser power.

Of course, the atomic level scheme presented in Fig. 1 and discussed above is far from reality and the quantities measured in an experiment are far from the theoretical predictions. Here, one should refer to the more elaborate calculations of Axner et al. [3] who have improved the theory of laser-enhanced ionization by including also intermediate states, well separated from the levels whose polupations are directly influenced by the interaction with the laser radiation field. They found satisfactory agreement between experimental data and values derived by their improved theory.

In experiment, it is advantageous to use a two-step laser excitation scheme (Fig. 2). On the one hand, we can reach highly excited levels near to the ionization limit by two strong dipole transitions and thus, take advantage of an efficient collisional ionization rate; on the other hand, a two step excitation scheme

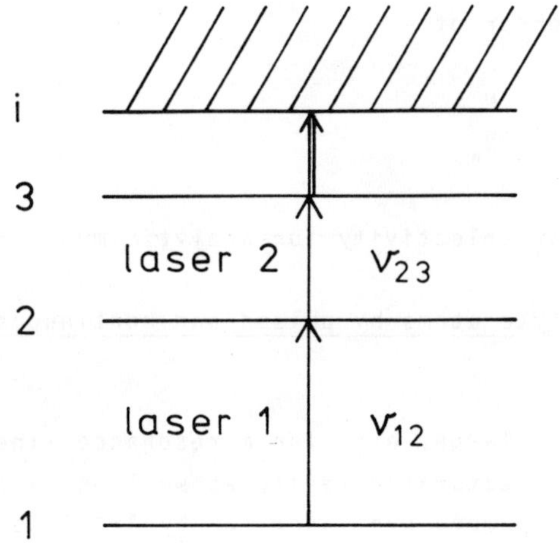

Fig. 2: Energy level scheme of the analyte atom for two-step laser excitation.

improves the selectivity of analyte atoms against wrong atoms of the matrix.

For a one-step excitation scheme, the signal-to-noise ratio (signal of analyte atoms S_a to matrix atoms S_m) is roughly given by $\frac{S_a}{S_m} \sim \left|\frac{\mu_a/\gamma}{\mu_m/\Delta\omega}\right|^2 \frac{N_a}{N_m}$ where μ_a and μ_m are the dipole transition matrix elements of the resonance transition in the analyte and of a nearby resonance in the matrix atom, respectively. γ is the line width of the resonance transition in the analyte, $\Delta\omega$ the frequency detuning of the laser in respect to the resonance transition in the matrix atom. Finally, N_a and N_m are the densities of the analyte and matrix atoms, respectively. Assuming matrix elements μ_a and μ_m of the same order of magnitude ($\mu_a \approx \mu_m$) and a ratio of about 10^4 - 10^5 of $\Delta\omega/\gamma$, we find a signal-to-noise ratio

$$\frac{S_a}{S_m} \sim (10^8 \sim 10^{10}) \frac{N_a}{N_m} \quad .$$

Therefore, the signal-to-noise ratio in a two-step excitation scheme can be of the order of

$$\frac{S_a}{S_m} \sim (10^{16} - 10^{20}) \frac{N_a}{N_m} \quad ,$$

presenting an excellent selectivity for analytic measurements.

3. Excitation of analyte atoms by pulsed and continuous wave lasers

If the dipole moment is large, e.g. for a resonance line, the transition 1 → 2 can be saturated easily as well as by available pulsed, excimer laser, pumped dye lasers as by Ar^+-laser pumped continuous wave dye ring lasers. Taking into account a typical pulse length of about 30 ns and a frequency of 50 Hz, the advantage of a cw system is obvious. Using a pulsed system, the irradiation time of the analyte atoms is the pulse length 3×10^{-8} s only. In case of a cw system, it is the dwell time of the analyte atoms in the laser beam. With the data given above (beam diameter x = 3 mm and velocity v = 2×10^4 cm/s) the dwell time is 1.5×10^{-5} s. Further we will assume that the analyte atoms transverse the laser beam (the laser beams of both systems have the same diameter) only once. To irradiate each transversing atom at least once, the frequency of the pulsed system should be about 67 kHz which is about 1.3×10^3 times higher than the assumed 50 Hz rate. The longer irradiation time as well as the better duty cycle yield a considerable advantage for the cw over the pulsed laser system.

The large difference between a cw and a pulsed system, becomes smaller if the dipole transition $1 \rightarrow 2$ is weak. Here, we can easier saturate with a pulsed laser system because of the higher peak power compared with the cw system. Moreover, if the transition $1 \rightarrow 2$ is situated in the uw spectral range, the light of the high-power pulsed laser systems can be frequency-doubled more efficiently than using a cw system.

If pulsed systems are applied in analytical spectroscopy, the detection of photons in LIF or ions in LEI is gated. This procedure may help to improve the signal-to-noise ratio. In particular in LEI spectroscopy in flames, there is a considerable background due to thermally produced ions (shot noise). A gated detection system reduces the background noise.

4. Analytical laser spectroscopy by laser-induced fluorescence (LIF)

In Chapter 2, a simple estimate of the theoretical detection limit for saturated atoms in the laser beam is given. In an early experiment Fairbank et al. [5] could demonstrate in a cell experiment that nearly 10 Na-atoms /cm^3 are detectable. Later papers by other authors report similar detection limits for pure metal vapours and metal-noble gas mixtures.

The first very important step in analysing the trace elements in a real sample is the evaporation and atomization of the material. Basically, the methods applied in LIF spectroscopy are the same already known in classical analytical spectroscopy. The most common method in analytical atomic spectroscopy are (i) electrothermal atomization in an oven (in most cases made of graphite, like in atomic absorption spectroscopy AAS), (ii) atomization in a flame (see e.g. AAS) and (iii) atomization in a plasma (e.g. the inductively coupled plasma ICP). While the highest evaporation temperatures which can be obtained with graphite ovens and flames are not very different (above 2700 K), the gas temperature in the ICP is much higher (typically about 6000 K).

There is the rough rule that the detection sensitivity of LIF with electrothermal evaporation is superior over flame or ICP techniques. This is mainly due to the large scattering of the exciting laser light by small unburnt particles or molecules in the flame or the ICP. In Table 1 which is based on a compilation by Axner and Magnusson [6] typical current detection limits of the LIF method using flames or ICPs are given. It has to be noted that the value can vary in a wider range depending on the strengths of the laser-induced transitions, the detection line which may differ from the exciting transitions (to reduce scattering) and the power of laser light. The data are comparable to the detection limit which can be reached by the AAS method in flames or the optical emission spectroscopy applying ICPs (also listed in Table 1). On the other hand, the detection limits of LIF are higher than in AAS applying ovens. Only when the LIF method is combined with electrothermal evaporation, significant lower detection limits can be obtained. As examples in Table 1, the detection limits for Co, Fe and Pb publishd by Bolshov et al. [7] are cited. A survey of analytical LIF spectroscopy can be found in two papers by the Ispra-group [8,9].

5. Analytical laser spectroscopy by laser-enhanced ionization SPECTROSCOPY (LEI)

During the last years LEI spectroscopy has beeen developed with some success. Even a commercial LEI instrument is on the market

Table 1: Comparison of detection limits [in ppb] of laser spectroscopy methods (LEI: laser-enhanced ionization, LIF: laser-induced fluorescence) with classical spectroscopy methods (AAS: atomic absorption spectroscopy, ICP: inductively-coupled plasma). *Two-step LEI. (Data are taken from Axner and Magnusson [6] and Bolshov et al. [7] for LIF using electrothermal evaporation out of a graphite cup.

Element	LEI	LIF	LIF (graphite cup)	AAS (oven)	AAS (flame)	ICP
Al	·1	·6		·1	30	2
Bi	·2	3		·4	50	50
Ca	·006	·08		·04	1	·07
Cd	·1*	8		·008	1	1
Co	·06	1000	·002	·2	2	2
Fe	·08	30	·001	1	4	·2
Ga	·03	·9		·1	50	14
In	·001*	·2		·04	30	30
Li	·001	·5		·3	1	
Ni	·02	2		·9	5	2
Pb	·09*	13	2.5×10^{-5}	·2	10	13
Sr	·003	·3		·1	5	·3
Sn	·3*	25		·3	100	25
Tl	·015	4		1	20	4

now [8]. Pioneering work has been done in particular at the NBS in Washington (se e.g. [9, 12] and at Chalmers University in Göteborg (see e.g. [11]). Most of the papers on LEI spectroscopy published so far dealing with studies on flames. Fig. 3 shows schematically a set-up for flame LEI as used by the two groups mentioned. Burners with pneumatic nebulizer which are applied in flame AAS are running with an air-actylene mixture. The introduced droplets of an aerosol are evaporated in the flame (T ≈ 2400 K) and render the analyte atoms which are excited by laser radiation and subsequently ionized by collisions. The ions are

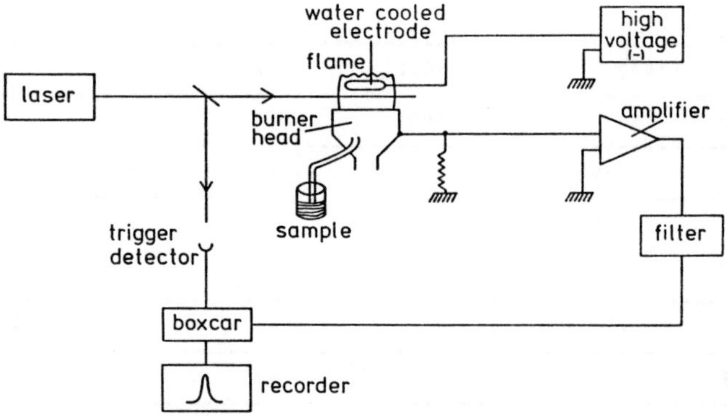

Fig. 3: Typical set-up for flame LEI spectroscopy.

collected by a water-cooled electrode placed in the flame. Electrodes outside the flame may also be used but detailed studies [12] have shown that electrodes inside the flame reduce the dependence on matrix effects.

The detection limits in LEI spectroscopy are also strongly dependent on the oscillation strengths of the laser-induced transition, the laser power, the background signal (thermal ionization and molecular background signal) and the degree of atomization of the analyte element. To improve the detection limit, it is advantageous to use a two-step excitation process (see above), because higher excited atoms are easier collisionally ionized. On the other hand, the selectivity, the discrimination against wrong atoms or molecules, is considerably enhanced (see above).

Detection limits obtained by the NBS and Göteborg group are listed in Table 1 and can be compared with other techniques. Note that the LEI data marked by an asterisk are from two-step excitation experiments. It is likely that the LEI detection limits can still be improved, in particular,if efficient two-step excitation schemes are applied.

It should be noted that, to the author's knowledge, there are two groups who investigated LEI spectroscopy in thermoelectrically heated ovens [13,14]. The results are promising. As already found in the LIF spectroscopy also here in a general the thermoelectrical method seems to be more sensitive than the use of flames.

6. Analytical laser spectroscopy with the thermionic diode

In our laboratories we have started a research project where we will investigate the applicability of thermionic diode detectors for trace element measurement. As the LEI method, thermionic diode spectroscopy (for referencs see e.g. [15]) is based on the detection of collisionally produced ions from laser-excited levels. But in contrast to the LEI method the ions are measured with a large amplification factor (10^5-10^7). This makes the thermionic diode a very sensitive and promising device for analytical spectroscopy.

In Fig. 4, a thermionic diode in its simplest form is shown. There is a cylindrical anode and an axially mounted cathode filament which is heated thermoelectrically. The diode is working in the 'space-charge-limited' mode, the electron cloud in front of the cathode is limiting the diode current. If ions are produced by collisions between laser-excited analyte atoms and other atoms

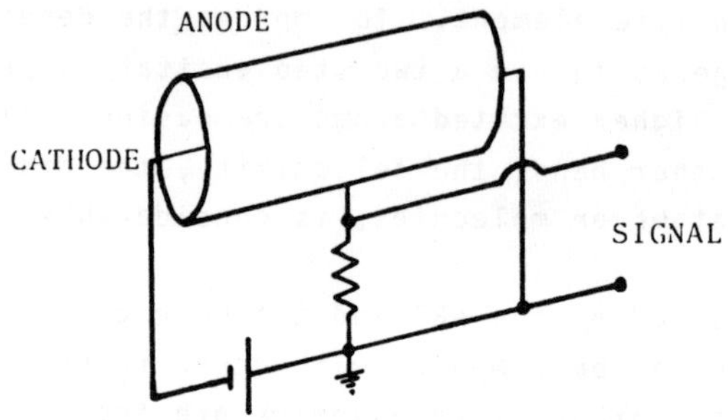

Fig. 4: The thermionic diode

or molecules, the ions are trappped for a long time inside the space charge. The presence of ions N_i reduces the space charge and the diode current j increases by Δj. The ratio $\Delta j/N_i$ represents the amplification of the diode. The detection limit is given by the noise of the system (shot-noise, flicker noise and thermal noise in local resistor and amplifier circuit for signal processing). Having optimum operating conditions, 1 - 10 ions/s may be detected.

If atoms are excited to levels with principal quantum number n > 20, the ionization probability by collision was found to be about unity in a noble gas atmosphere [16]. Theoretically one should be able to detect almost one excited analyte atom per second with the diode; and if the transition can be saturated, nearly all atoms within the laser beam should be detectable (see Chapter 2).

The thermionic diode is known to be an instrument with a large dynamic range. Fig. 5 displays a Doppler-free spectrum of the $4s^2\ ^1S_0$ - $4s10s\ ^1S_0$ 2-photon transition in natural abundant Ca. The dynamic range can be demonstrated plotting the expected ratio of the most abundant ^{40}Ca (96.97%) to the other isotopes ^{M}Ca (e.g. ^{46}Ca: 0.0033%) against the measured ratio (Fig. 6). The data were taken from two different 2-photon lines. As can be seen, the diode is at least linear over more than four orders of magnitude.

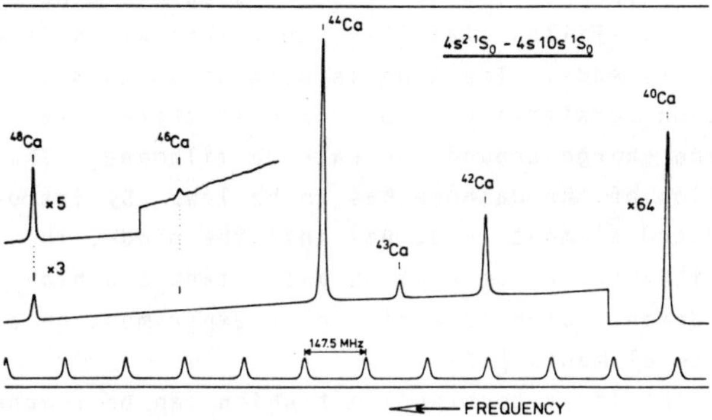

Fig. 5 Off-resonant Doppler-free 2-photon spectrum of the $4s^2$ $^1S_0$ - 4s10s $^1S_0$ transition in Ca and the transmission peaks of a reference cavitiy (from [15]).

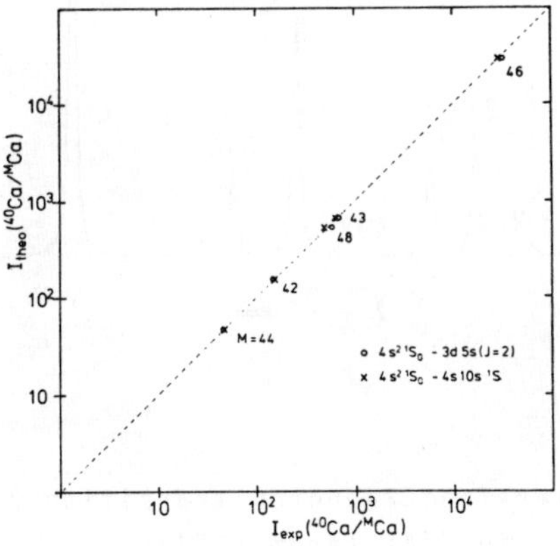

Fig. 6 Plot of the excited against the measured intensity ratios $^{40}Ca/^{M}Ca$ taken from the spectra of two different 2-photon transitions (from [15]).

The spectrum in Fig. 5 has been measured with a thermionic diode of the heat-pipe type [15], where the Ca vapour is generated inside a buffer-gas-filled stainless steel tube which is water-cooled towards its ends. The tube is used as an anode. Optimum detection sensitivity is obtained if there is a pronounced space charge around the cathode filament. Therefore, the work function of the cathode has to be low. By introducing a strongly activated element (e.g. Ba) into the diode, the detection sensitivity can be kept on a constant and high level. This could be demonstrated recently in an experiment on alloying and low-pressure elements [17].
To demonstrate the low detection limit which can be reached with the thermionic diode detector, Fig. 7 shows a directly reproduced

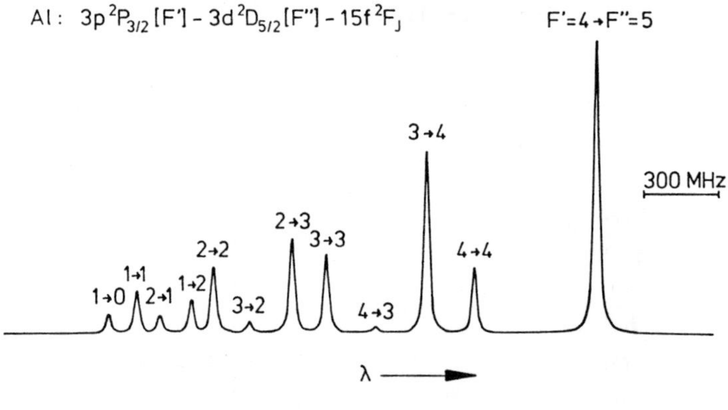

Fig. 7 Resonant Doppler-free 2-photon transition 3p $^2P_{3/2}$ - 3d $^2D_{5/2}$ - 15f ^2F_J in ^{27}Al (from [15]).

Doppler-free resonant 2-photon spectrum of the 3p $^2P_{3/2}$ - 3d $^2D_{5/2}$ - 15f ^2F_J transitions in ^{27}Al. F' and F'' refer to the hyperfine components of the initial (3P) and intermediate state (3D), respectively. The applied laser powers were about 50 nW frequency doubled radiation from a Rh 6 G dye laser (3P-3D) and about 100 mW (3D - 15F). The temperature of the pipe was 1023 K which corresponds to a number density of 6 x 10^8 cm^{-3} of Al $3P_{3/2}$ atoms. Taking into account the experimental signal-to-noise ratio, the detection limit of the strongest transition is about 6 x 10^6 cm^{-3}. But it should be noted that the interaction volume of the laser beams with the vapour was only about .05 cm^3. Therefore the detection limit in the experiment was about 3 x 10^5 atoms. With higher light power for the first step (we were far from saturating the first transition) the detection limit may be improved considerably.

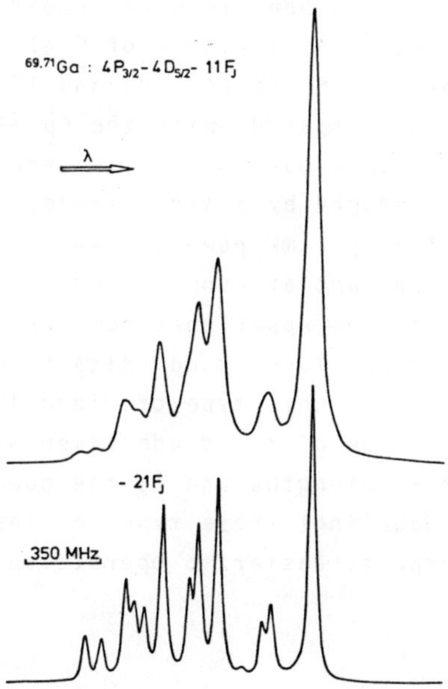

Fig. 8: Resonant Doppler-free 2-photon transitions $4P_{3/2}$ - $4D_{5/2}$ - $21F_J$ and $4P_{3/2}$ - $4D_{5/2}$ - $11F_J$ in ^{69,71}Ga which are taken with two tunable single-mode dye lasers (lower part) and with one dye laser and one diode laser (upper part), respectively.

As can be seen from Fig. 5 or 7, the thermionic diode can be used to measure smallest quantities of isotopic traces, if Doppler-free laser spectroscopy is applied. In case of resonant 2-photon transition (Fig. 7), one takes advantage of the large transition probabilities compared with off-resonant 2-photon transitions (Fig. 5) which, on average, are smaller by a few orders of magnitude. The disadvantage of this method is the high cost of two cw single mode laser systems which have to be used. But is is likely that, in the future, these lasers may be replaced by low-cost solid-state diode lasers. E.g. Fig. 8 shows in the lower part, the spectrum of the 4p $^2P_{3/2}$ - 4d $^2D_{5/2}$ -21f ^2F_J transition in $^{69,71}Ga$ which has been recorded under the same experimental conditions as the Al spectrum of Fig. 7 (the same Ga number density, laser powers: 50 nW [4p-4d] and 100 mW [4D - 21F]. This spectrum can be compared with the 4p $^2P_{3/2}$ - 4d $^2D_{5/2}$ - 11f ^2F_J multiplet upper part of Fig. 8) where the second step 4D - 11F was induced by a single-mode, frequency-stabilized diode laser of only 1 mW power. The light power of the first step was kept constant at about 50 nW. The larger half-widths of the lines in the upper spectrum are due to the larger line width of the diode laser used (Hitachi HL 7801, $\Delta\nu$ = 25 MHz). The current price of this type of diode laser is about $ 50. As the wavelength range of the diode laser will be extended towards shorter wavelengths and as the power will be improved (for frequency doubling) these types of lasers will certainly be parts of compact, easier to operate analytical instruments.

The author gratefully acknowledges financial support by the Deutsche Forschungsgemeinschaft.

References

1. S. Greenfield, H.McD. McGreachin and P.B. Smith, Talanta 22 (1975) 1 and 551 and Talanta 23 (1976) 1.

2. R.H. Scott and A. Strasheim, in Applied Atomic Spectroscopy (Ed. E.L. Grove, Plenum Press, New York 1978), p. 73.

3. K. Laqua, in Analytical Laser Spectroscopy (Ed. N.Omenetto) J. Wiley & Sons, New York 1979 p. 47.

4. O. Axner, T. Berglind, J.L. Heully, I. Lindgren and H. Rubinsztein-Dunlop, J. Appl. Phys. 55 (1984) 3215.

5. W.M. Fairbank, T.W. Hansch and A. Schawlow, J. Opt. Soc. Am. 65 (1975) 199.

6. O. Axner and T. Magnusson, Phys. Scr., in press.

7. M.A. Bolshov, A V. Zygbin and T.I. Smirenkiva Spectrochim. Acta 36B (1981) 1143.

8. Instrument, manufactured by SOPRA (Bais Colombes, France).

9. P.K. Schenck and J.W. Hastie Opt. Engineering 20,(1981) 522.

10. J.C. Travis, G.C. Turk and R.B. Green, Anal. Chem. 54 (1982) 1006A.

11. O. Axner, I. Lindgren, T. Magnusson and H. Rubinsztein-Dunlop, Anal. Chem. 57 (1985) 773.

12. G.C. Turk, Anal. Chem. 53 (1981) 1187.

13. I.V. Bykov, A.B. Skortsov, Yu. g. Tatsii and N.V. Cheklanin, J. Physique 11, (1983) C7-345.

14. O. Axner, I. Lindgren, T. Magnusson and H. Rubinsztein-Dunlop, submitted for publication.

15. K. Niemax, Appl. Phys. B 38 (1985) in press.

16. K. Niemax, Appl. Phys. B 32 (1983) 59.

17. K. Niemax and K.-H. Weber, Appl. Phys. B 36 (1985) 177.

PITCH-DARK ABSORPTION AND WHITE-LIGHT EMISSION IN LASERS

H.H. TELLE
Department of Physics, University College of Swansea
Swansea SA2 8PP, United Kingdom

ABSTRACT

Various aspects of broad-band amplification and coherent transformation of wavelength characteristica in lasers are reviewed in this presentation. Of special interest are broad-band absorption features for the active medium, and multi-color operation of a laser to realize a "white-light" source.

1. INTRODUCTION

In recent months various announcements and advertisements for gas-mixture lasers operating at extended wavelengths could be found in most of the important laser journals. An interesting contribution came from the University of Bochum which reported that the technical development of their three-color helium-selenium laser has been brought to the point of a commercial product /1/. Also in our laboratory there are investigations under way to probe multi-color operation in gas-mixture lasers. This served as a motivation to look into the state-of-the-art and recent developments for broad-band features and multi-color operation of lasers.

By definition the laser is an amplifier for light which in many cases transfers the energy from an input light source at wavelength λ_1 to an output light beam at wavelength λ_2. From the electronic counterpart of amplification theory we know that it is in principle possible to convert and amplify an incoming signal at any frequency to an outgoing signal at probably another frequency. The situation for an idealized light amplifier is shown schematically in Figure 1; a light beam at wavelength λ_1, or containing a sum of wavelengths, enters an active medium which amplifies a certain wavelength component. In an ideal case the optical amplifier is tunable and should also be capable of producing wavelengths differing from those of the incoming light to give either a light beam at a single wavelength λ_2, or again a mixture of wavelengths. One could then speak of a white-light amplifier.

In practise, however, one is faced with a multitude of difficulties, and seldom is it possible to fulfill all the features for such an idealized amplifier. It is well known that dye lasers can be pumped with a light source, e.g., a laser at a shorter wavelength or a broad-band flash lamp, and that their output is tunable over an often wide wavelength range. But not all wavelengths are absorbed sufficiently to result in the necessary population inversion for the active medium, and longer wavelengths than those of the laser output are not absorbed at all. Similar restrictions hold for the output of a laser. Though one often talks of "broad-band" operation of a laser this feature is in general only relative with respect to the mode structure of the laser cavity on one side and the gain profile of the medium on the other side. For example, in dye lasers broad-band operation means at the most a some 10 Å wide output spectrum which is usually narrow in comparison to tha tuning range of the dye. Only for pulsed amplification in picosecond lasers extremely broad output spectra are observed.

Well known but often unwanted is multi-line operation of noble gas ion lasers. For example, Ar^+ lasers operate simultaneously at various wavelengths in the blue-green region of the spectrum, and the introduc-

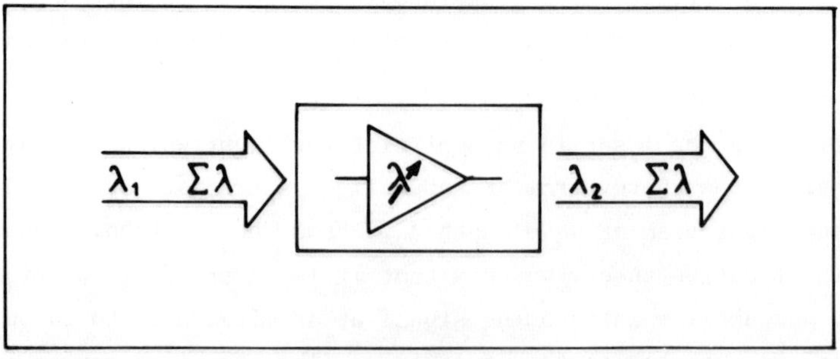

FIGURE 1: Idealized white-light amplifier.

tion of an intra-cavity prism enables the user to select a single-wavelength output. On the other hand, it is very difficult to obtain the additional oscillation of the UV lines of the laser with a single mirror set, and optimised coatings for those wavelengths are introduced which in turn suppress the laser action in the blue-green. "White-light" emission from noble gas ion lasers is however known; the proper mixture of argon and krypton gases gives emission of the Ar^+ lines 488.0 nm and 514.5 nm and the Kr^+ line 647.1 nm /2/. This type of laser is commercially available with high output powers but has the disadvantage that the blue Ar^+ line has a relatively long wavelength; thus, essential parts of the color-triangle (see below) are not covered.

In the following sections we will discuss various aspects of how to realize broad-band absorption in dye lasers, and the concepts of white-light lasers based on gas-mixture lasers (metal vapor and rare gas). Finally, we will mention a few possible applications of light sources described in this presentation.

2. ABSORPTION PROCESSES

The population inversion in an active medium is in many cases produced by the absorption of light. One of the most versatile media is a laser dye which generally has broad absorption bands; in this way optical pumping by various light sources is possible which can be flash lamps or another laser. However, one is faced with certain restrictions. Though the highest output powers for dye lasers are found when they are pumped with high-energy flash lamps the conversion efficiency is in general very low and sometimes the broad spectral distribution of the lamp poses severe problems, i.e., the long-wavelength tail in the spectrum may result in thermal heating of the solvant, whereas a high proportion of UV wavelengths may yield unwanted chemical decomposition of the dye. On the other hand, the wavelengths of pump lasers often do not coincide with the first absorption maximum to the S_1-band though absorption into higher S_n-bands is usually encountered; absorption to these bands is generally much weaker than the fundamental $S_0 \rightarrow S_1$ absorption, and internal conversion processes decrease the efficiency significantly.

2.1. Component mixture in dye lasers

For pulsed aplications the most widely used pump lasers today are the N_2 laser (at 337 nm), the XeCl excimer laser (at 308 nm) and the harmonics of the Nd:YAG laser (at 532 nm and 356 nm). For dyes operating in the red and near IR region of the spectrum the absorption of the UV laser lines is to higher S_n niveaus; as has been pointed out above their absorption cross section is usually rather weak. Of even higher disadvantage is the situation for the second harmonic of the Nd:YAG laser; its wavelength is hardly absorbed at all being right inbetween the first and second absorption band. In such cases an indirect absorption mechanism is chosen which on many occasions is very efficient.

The process in question will be outlined exemplary for a dye laser in the near IR pumped by a N_2 laser. The dye cresyl violet shows laser action around 700 nm; however, its efficiency is rather low when pumped with a N_2 laser and the necessary concentrations of the dye are very high for complete absorption of the 337 nm radiation, and may exceed the

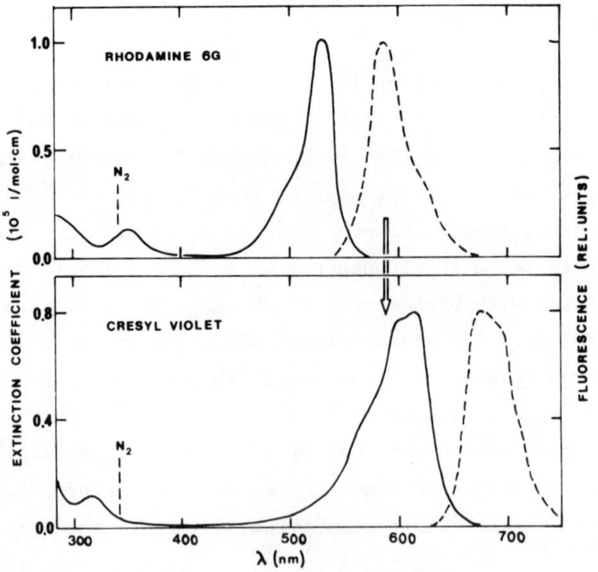

FIGURE 2: Absorption and fluorescence spectra for the laser dyes rhodamine 6G and cresyl violet; the pump wavelength of the N_2 laser is marked, and the fluorescence transfer from one dye to the other is indicated by the downward arrow.

maximum solubility of the dye in its solvant. The efficiency of the dye can be increased dramatically when a second dye is admixed which absorbs the N_2 laser radiation more efficiently and exhibits fluorescence in the same wavelength region as the first absorption band of cresyl violet. Such a dye is the well known rhodamine 6G; its fluorescence around 600 nm is immediately re-absorbed in the first singlet band $S_0 \rightarrow S_1$ of cresyl violet. The situation is shown in Figure 2 where the absorption spectra are drawn in solid lines, and the emission spectra are represented by dashed lines (normalization is with respect to the absorption spectra). In this way the low efficiency of the dye cresyl violet is increased, and the additional benefit is a broader tuning range since the gain profile of the dye is much enhanced.

The method is used for a large variety of dyes in this spectral region, especially when pumped by the 532 nm line of the Nd:YAG laser, and the appropriate mixture of dyes can often be found in the lists of commercially available dyes.

In a similar manner one may improve the efficiency of dyes for laser action in the spectral range just above 500 nm when the pump source is a N_2 laser. Its radiation falls right inbetween the two lowest absorption bands for the according dyes; but the efficiency increases significantly when adding coumarine 1 as a second dye for fluorescence transfer, for example.

It should be pointed out that the procedure of mixing dyes is well established in pulsed applications but that it works usually less satisfactorily in CW operation. This is mainly due to the fact that internal losses via molecular association and interband conversion are of more severe consequence on a longer time scale than is encountered for the short duration of most pulsed dye lasers.

2.2. Synthesis of appropriate dye molecules

The success of the dye mixing procedure has lead to some extensive investigations into the possibility of sythesizing special dye molecules consisting of a chain of two or more components which absorb and fluoresce in various overlapping wavelength bands. Direct energy transfer from one part of the molecule to another with absorption and fluorescence features at longer wavelengths should circumvent the need for emission and subsequent re-absorption of radiation. Such molecules have been produced in past years; however, the great disappointment in all these efforts has been that the molecules readily absorb throughout the spectrum at shorter wavelengths than the anticipated fluorescence, but that no significant fluorescence could be observed. Obviously internal energy transfer quenches the population which is produced in the dye molecule chain.

3. EMISSION PROCESSES

For a variety of experimental applications like for, e.g., color disply and multi-wavelength interference experiments a white light laser with balanced output in the three primary colors would be of great advantage. One approach to realize this concept has been the use of three independent lasers, but a more economic and precise solution should be the use of a single laser resonator. Various different possibilities for white-light operation of a laser have been reported. In general, the proposed and realized solutions base on the principle of gas-mixture lasers.

It has been mentioned above that rare gas ion lasers with proper gas mixtures may produce white-light output but that essential parts in the color-triangle are not covered /2/.

A second solution is a combined $HeCd^+$ and $HeSe^+$ laser tube with a HeNe gas filling. The various Cd^+, Se^+ and Ne emission lines contribute to the mixed-color spectrum /3/.

Another approach uses a hollow cathode discharge tube containing a mixture from cadmium vapor and helium gas, and the Cd^+ lines result in poly-chromatic, nearly white-light emission /4/ even though some difficulties in the experimental realization are still unresolved. A laser design with less operational difficulties than the $HeCd^+$ hollow cathode laser is a $HeSe^+$ laser based on the principle of cataphoresis /5/.

Finally, the successful operation of a white-light dye laser has been reported using a mixture of two laser dyes /6/. In contrast to the previously mentioned lasers which are working continuously this white-light laser with tunable characteristica is a pulsed system.

In the following subsections some of these lasers will be described in more detail.

3.1. White-light emission in hollow-cathode metal-vapor lasers

The hollow-cathode discharge $HeCd^+$ laser operating at transitions of the Cd-ion (in the red at 635.5 nm and 636.0 nm, in the green at 533.7 nm and 537.8 nm and in the blue at 441.6 nm) has been the object of several investigations due to its ability to generate efficient white-light oscillation in a single laser tube. Simultaneous three-color operation in a hollow-cathode $HeCd^+$ laser was reported already more than 10 years ago /7/. Since then further investigations have resulted in new design concepts of the white-light laser and approaches to clarify the excitation mechanisms in such a laser.

The presently known designs may be devided into two basic groups, depending on the position of the anodes and cathodes relative to each other. To the first group belong those tubes in which the configuration of the electrodes establishes an electric field along the axis of the laser tube; this kind of discharge is therefore called a longitudinal hollow-cathode discharge. In the second type the electrode configuration creates an electric field and a discharge current transversly to the laser axis. The different geometrical electrode configurations were investigated thoroughly (see e.g. ref. /8,9/). In the following

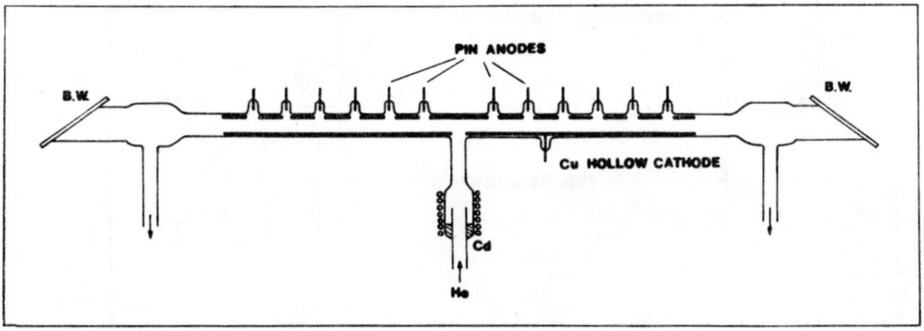

FIGURE 3: Discharge tube for hollow-cathode $HeCd^+$ laser (for details see text).

we like to restrict the discussion to the longitudinal discharge design successfully used in our laboratory /10/.

The laser tube used is shown in Figure 3. It is based on a cylindrical hollow-cathode made from a single copper tube (flute-type discharge tube); the hollow-cathode tube has holes spaced equidistantly above which pin-anodes are placed. The proper dimensions of tube diameter, hole diameter, hole distance and pin distance determine the optimum axial expansion of the negative glow inside the hollow-cathode (for a determination of the dimensions see, e.g., /8/). Cadmium is heated in a reservoir with temperature control and is injected into the discharge region by a flow of helium; a constant flow is maintained by pumping the laser tube at both ends. The hollow-cathode is placed inside a glas tube which is sealed at both ends with brewster-angle windows. They are sufficiently far from the end of the discharge region to prevent their contamination by cadmium deposits; some metal condenses on the cold glas surface at the end of the cathode which is heated by the discharge current. Typical operation conditions for the laser are a total pressure inside the cavity of around 15-30 mTorr with helium flow rates of less than 1 l/min; the reservoir is kept at a temperature to maintain a cadmium pressure of a few mTorr. Total discharge currents can be as high as 1.5 A which are distributed equally via current-limiting resistors to the pin-anodes. The simple cavity consists of a plano and a concave (R=2m) mirror with high-reflectivity dielectric coatings (98%). Under these conditions multi-line output around some 10 mW is obtained.

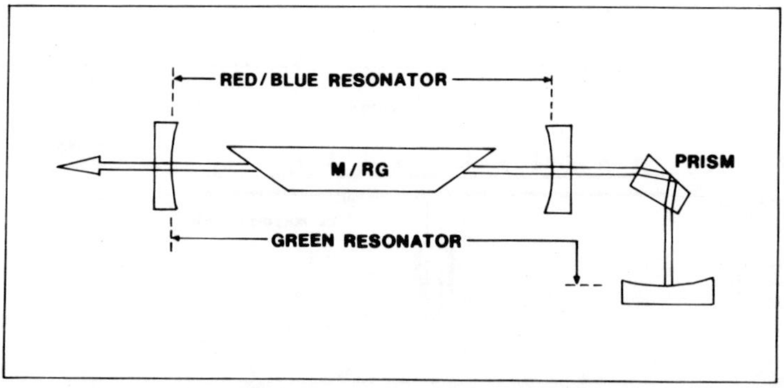

FIGURE 4: Schematics of white-light laser resonator (for details see text).

Such a simple design is by far not the optimum configuration. Different population mechanisms are responsible for the appearence of the red, green and blue lines (see e.g. /11/). This results in different gains for the different transitions, and mirror coatings and operational conditions for the laser have to be chosen carefully to obtain balanced white-light emission. It seems to be of great advantage to use a dispersive resonator in order to enable independent gain variations for the different lines (see ref. /5/ and section 3.2. below). Such a resonator is shown schematically in Figure 4. The laser is formed by two interlocked resonators, one with fixed losses for the red and blue contributions in the emission spectrum, the other contains a low-loss Abbé prism with high dispersion, cut at Brewster angle, with which the loss in the green lines of the laser emission can be adjusted to obtain the wanted white-light emission of the laser. In this way it is possible to cover al spectral shades in the color-triangle formed by the colors of the three laser lines (see the dashed lines in Figure 8a below), and to produce the balanced white-light emission indicated by the cross marked "W" in the same figure.

3.2. White-light emission in cataphoretic metal-vapor lasers

One of the major problems in hollow-cathode discharge lasers has been the homogeneous distribution of the metal vapor and the stable operation of such a laser. For continuously operating metal vapor lasers a homogeneous distribution of metal atoms and ions in the active laser discharge, which is often a positive column plasma in helium buffer gas, is most important. It can be achieved with the technique of cataphoresis /12/.

In this paper we like to describe a cataphoretic $HeSe^+$ laser with a long-life laser tube /13/. The use of selenium in the discharge instead of cadmium has several advantages. In general, much lower heating temperatures are sufficient to produce the equivalent metal vapor density; this in turn makes high-capacity cooling precautions unnecessary. More important seems to be that in the $HeSe^+$ laser a total of more than 46 Se^+ lines between 447 nm and 1.26 um have been found to oscillate both separately and partially simultaneously, depending on the resonator mirrors used. Thus laser lines for white-light operation can be selected from a large ensemble to cover major portions of the color-triangle (see Figure 8b below).

The principles of a cataphoretic laser tube are shown in Figure 5. The active laser discharge capillary is made from quartz (1 mm internal diameter). Losses of helium through its hot walls are avoided by a cooled outer mantle enclosing a large He ballast. Selenium vapor is produced from solid, purified selenium in a side arm; its resistive heating is actively teperature controlled in the range from 240 to 250 °C. The selenium vapor enters the discharge close to the ring anode and is cataphoretically transported down the capillary to the cathode. The latter is placed far back from the exit of the capillary; this reverses the direction of the current, and hence also the cataphoretic flow. This retarding effect on the Se-ions prevents selenium from condensation on the Brewster windows.

The laser is completed by using either a simple resonator with two broad-band dielectric mirrors (reflectivities of 98.5% and 99.8%), or with a dispersive resonator according to Figure 4. With the latter set-up 12 mW white-light output power were obtained from an active length of 700 mm /14/. The mode structure of the laser was found to be TEM_{00} for the red (644 nm and 649 nm) and green (517 nm and 522 nm) primary colors, and TEM_{01} for the blue (460 nm and 464 nm).

With another laboratory set-up of 1000 mm length, 2mm capillary diameter and broad-band dielectric mirrors, e.g., simultaneous oscilla-

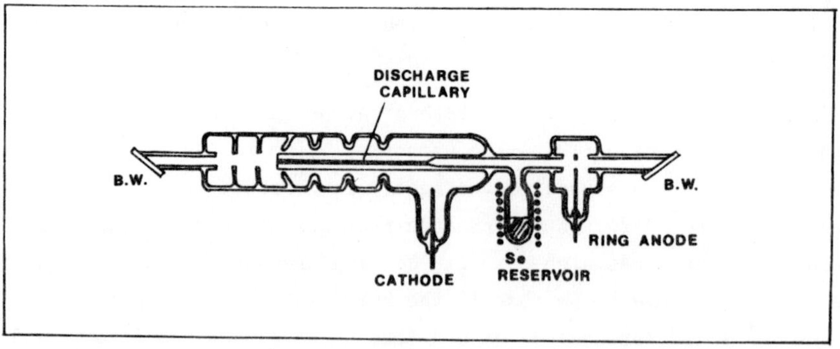

FIGURE 5: Discharge tube for cataphoretic $HeSe^+$ laser (for details see text).

tion of over 20 visible laser lines was observed with a total output power of more than 100 mW /13,14/. Their wavelengths, individual output power and small-signal gains are listed in Table 1.

3.3. White-light emission in dye lasers

While utilization of the $HeSe^+$ laser results already in a more flexible choice of wavelengths for the construction of color-triangles as compared to the $HeCd^+$ laser with only very few laser lines, it still seems desirable to have a more or less free choice of wavelengths. This may be achieved with a multi-wavelength dye laser. Such a laser has been realized recently /6/. The principle of the laser is relatively simple; the multi-color operation of the dye laser could be achieved in the mixture of two or more dyes which all do absorb the pump wavelength but do not have overlapping absorption and emission spectra, and have only low triplet losses. One suitable mixture of dyes is that of coumarine 102 with oxazine 1; their absorption and emission spectra are shown in Figure 6. The radiation of a N_2 laser is readily absorbed in

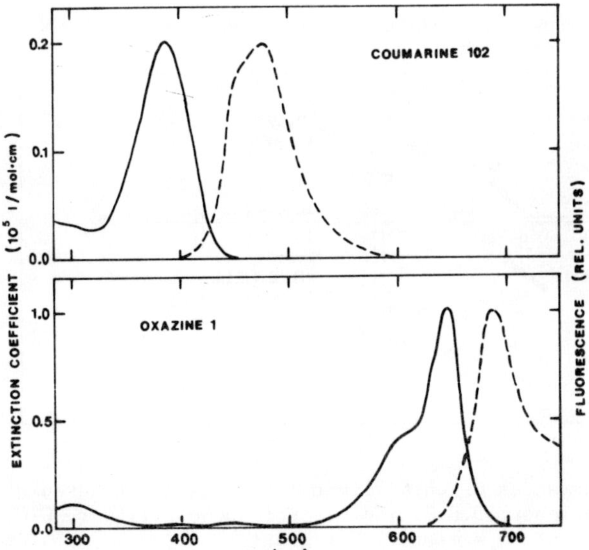

FIGURE 6: Absorption and fluorescence spectra for laser dyes coumarine 102 and oxazine 1.

no.	λ [nm]	P_L [mW]	g_o [% m^{-1}]
1	460.4	8.0	5.1
2	464.8	9.0	
3	484.0	6.0	8.4
4	497.5	8.0	17.8
5	499.3	6.0	18.7
6	506.9	10.0	20.4
7	509.7	1.0	
8	514.2	0.5	
9	517.6	14.0	27.9
10	522.7	15.0	31.5
11	525.3	2.0	
12	527.1	1.0	
13	530.5	6.0	16.6
14	552.2	2.0	3.5
15	556.7	0.5	
16	559.1	1.0	
17	562.3	0.5	
18	574.8	0.5	
19	605.6	3.0	5.4
20	644.4	6.0	4.3
21	649.0	4.0	3.7
22	653.5	0.5	
		total 103.5	

TABLE 1: Wavelengths, individual output poweers and small signal gains of multi-line $HeSe^+$ laser.

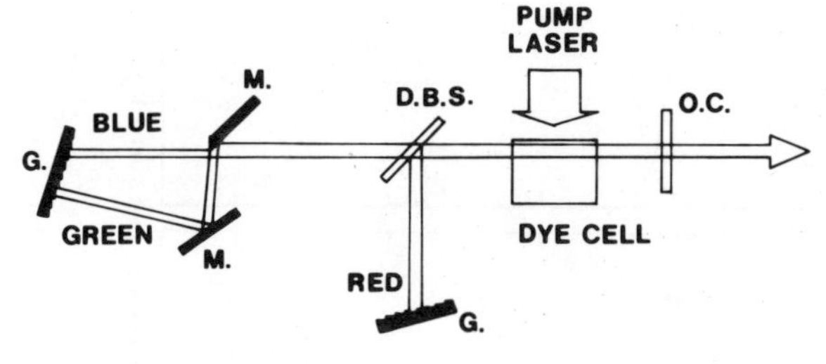

FIGURE 7: Schematics of white-light laser resonator based on a pulsed dye laser; D.B.S. = dielectric beam splitter, O.C. = output coupler, M = mirror, G = grating.

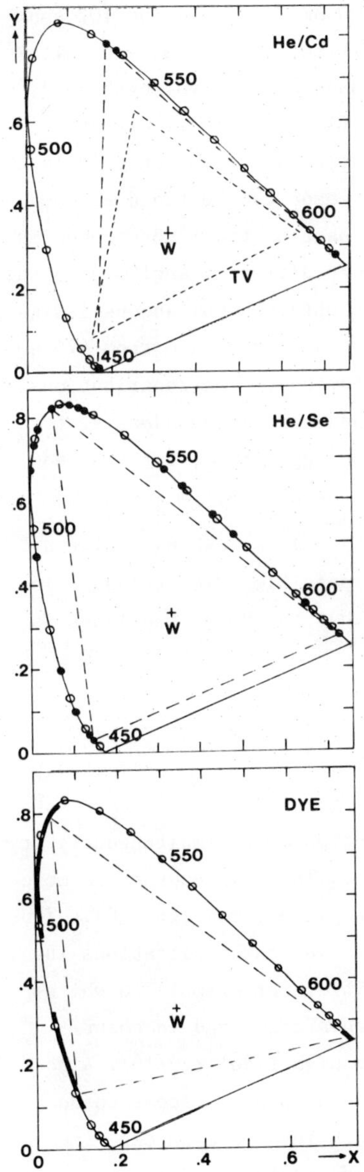

FIGURE 8: Color-triangles for (a) $HeCd^+$ laser, (b) $HeSe^+$ laser and (c) multi-color dye laser; the full circles • indicate output lines of the metal vapor lasers, the fat lines in (c) the tuning ranges of the dye laser; W = balanced white-light point; the dotted line in (a) represents the colors covered by an ordinary color-TV tube.

the S_1 band of coumarine 102, and in a higher S_n band of oxazine 1. On the other hand, nearly no absorption is found for oxatine 1 at wavelengths where coumarine 102 fluoresces. A suitable mixture of the two dyes, both dissolved in methanol, is pumped by a N_2 laser. The two dyes have their optimum of laser action around 490 nm for coumarine 102 and 690 nm for oxazine 1. A special resonator is set up to enable simultaneous operation of the laser at the three fundamental wavelengths red, green and blue (see Figure 7). Part of the fluorescence is directed by a dielectric beam splitter to a first grating which is set to select the red wavelengths; in this branch the dye laser can be tuned between 685 nm and 705 nm in oxazine 1. The second part of the fluorescence is split again, and both beams are directed under different angles onto the same grating. In this way it is possible to obtain simultaneous lasing in coumarine 102 at two wavelengths. The tuning ranges are from 475 nm to 485 nm (blue) and from 495 nm to 520 nm (green). The contributions from the different spectral regions can be adjusted independently, for the blue and green by adjustment of the branching mirror, for the red and blue/green by adjustment of the dye concentrations.

In this way white-light emission of the dye laser can be achieved which can be balanced according to needs; a possible color-triangle is shown in Figure 8c. It should be recalled that the described laser is a pulsed laser.

4. CONCLUSIONS

All lasers described in the previous section emit simultaneously at the three primary colors red, green and blue. The wavelengths are more or less situated in the corners of the color plane (see Figure 8). Thus the advantages for color mixing are obvious. Various applications in the graphics and printing industry are possible, for example, a white-light laser could be used in color-separating systems and in color-negative reproduction from diapositives with high-fidelity /15/. Also an application for color projection and scanning systems seems to be feasible /16/, as well as for multi-wavelength interference experiments and color holography. Another prospective use would be in laser surgery where a changable colored pilot light seems to be of great advantage.

Though some constructural problems remain multi-color metal vapor lasers begin to reach the stage of commercial exploitation, and small-sized lasers with lifetimes of a few thousand hours have already been built for laboratory applications.

REFERENCES

1 "HeSe Laser Finally Looks Like a Winner"; Lasers & Applications (Jan. 1985) p.38

2 E.T. Leonard, M.A. Yaffee, K.W. Billman; Appl. Opt. 9 (1970) 1209

3 V.V. Sabotinov, P.K. Telbizov; Opto-Electron. 6 (1974) 185

4 K.I. Fujji, T. Kakahashi, Y. Asami; IEEE J. Quantum Electron. QE 11 (1975) 111

5 H.P. Popp, E. Schmidt; IEEE J. Quantum Electron. QE 15 (1979) 840

6 Zhang Fu-gen, Yu Chun-lian, Miao Hai-ping, Yang Chen-guang; Acta Phys. Sinica 33 (1984) 1174

7 S.C. Wang, A.E. Siegman; Appl. Phys. 2 (1973) 143

8 K.I. Fujji, S.I. Miyazawa, T. Takahashi, Y. Asami; IEEE J. Quantum Electron. QE 15 (1979) 35

9 J. Mizeraczyk, M. Neiger, J. Steffen; IEEE J. Quantum Electron. QE 20 (1984) 1233

10 K.H. Wong, C. Grey Morgan; J. Phys. D: Appl. Phys. 16 (1983) L1

11 P. Baltayan, J.C. Pebay-Peyroula, N. Sadeghi; J. Phys. B: At. Mol. Phys. 18 (1985) 0000 (in press)

12 J.P. Goldsborough; Appl. Phys. Lett. 15 (1969) 159

13 H.P. Popp, E. Schmidt, M. Neiger, F. Pfeil; Opt. Laser Technol. 13 (1981) 321

14 K.H. Krahn, E. Schmidt; (AEEO Ruhr-Universität Bochum, FRG; private communications)

15 "High-fidelity color negatives produced from slides with 3-beam laser system"; Laser Focus (Sept. 1976) p.16

16 F.T.S. Yu; Optics News 11,5 (1985) 5

SQUEEZED STATES OF LIGHT IN RESONANCE FLUORESCENCE

Ryszard Tanaś

Nonlinear Optics Division, Institute of Physics,
Adam Mickiewicz University, 60-780 Poznań, Poland.

INTRODUCTION

Squeezed states of light have been the subject of intense research in recent years. The literature of the subject have already become quite extensive and I am not going to give a full account of it here. A number of references can be found in the review article by Walls [1]. In this paper I would like to concentrate myself on the problem of production of squeezed states in the process of resonance fluorescence only.

Resonance fluorescence from a coherently driven two--level system is so far the only optical phenomenon in which nonclassical properties of the fluorescent light have been observed experimentally [2,3]. These experiments have given clear evidence of photon antibunching —— another quantum property of light. Photon antibunching is a quantum effect that reveals explicitly particle properties of light. Squeezing on the other hand is a phase sensitive effect and wave properties of light are involved in it. However, the two effects are quantum in nature and cannot be explained within the framework of classical electrodynamics; quantum electrodynamic treatment is needed. The number states /Fock states/

are commonly used to describe theoretically quantum fields. Despite the advantages the namber states present in the theoretical description of light they have one essential disadvantage —— no sources exist that would produce in practice light in such states. Photon antibunching would be most pronounced if the field were in a pure number state. For a nearly monochromatic plane wave of the field the closest quantum counterpart to a classical field is a coherent state. Such a field can be decomposed into two quadrature components with time dependence $\cos\omega t$ and $\sin\omega t$ respectively /the in--phase, and the out-of-phase component/. In a coherent state the fluctuations in the two quadratures are equal and give the minimum value in the Heisenberg uncertainty relation. The quantum fluctuations in a coherent state are equal to the vacuum fluctuations and similarly to a number state are randomly distributed in phase. The vacuum fluctuations that always exist impose the quantum limit to the reduction of noise in a signal. The quantum noise exist even in an ideal laser operating in a pure coherent state. However, it is possible to reduce fluctuations in one quadrature component below the value in a coherent state at the expense of increased fluctuations in the other quadrature component. The states with fluctuations in one quadrature component below the coherent state fluctuations are called squeezed states /see review by Walls [1] and references therein/. The squeezed states may be or may not be minimum uncertainty states. Theoretical predictions show that squeezed states of light can be produced in various nonlinear optical processes [1] /a survey of recent publications and some new results can be found in

[4,5] . One candidate considered as a possible source of light in a squeezed state is a phenomenon of resonance fluorescence from a coherently driven two-level system. I would like to discuss shortly some results concerning the possibility of squeezed states production in the process of resonance fluorescence.

ONE-ATOM RESONANCE FLUORESCENCE

Squeezing in one-atom resonance fluorescence has been predicted by Walls and Zoller [6]. Mandel [7] has made a comparison of photon antibunching and squeezing in resonance fluorescence from a two-level atom showing that detection of squeezed states by phase-sensitive interference with another optical field in a coherent state and measuring the resulting intensity fluctuations, that leads always to sub-Poissonian photon statistics, is at least an order-of-magnitude more difficult than the detection of photon antibunching. Loudon [8] has considered the two-time photon number correlation function resulting in such a homodyne technique. A new homodyne technique that allows elimination of the local oscillator noise has been proposed by Yuen and Chan [9].

Walls and Zoller [6] have shown that in the steady--state resonance fluorescence either quadrature component of the fluorescent field can become squeezed under certain conditions if the Rabi frequency Ω of the exciting field is sufficiently low. When the exciting field is tuned perfectly to the atomic transition the Rabi frequency Ω has to be less than $\sqrt{2}\,\gamma$, with $2\gamma = A$ —— the Einstein coefficient for spontaneous emission. Off resonance excitation further lowers this critical value of Ω. Walls and Zoller have also indicated /without giving the details/ that squeezing can become two times bigger in the transient regime of resonance fluorescence. Arnoldus and Nienhuis [10] have considered the conditions for squeezing in steady-state resonance fluorescence showing that the condition for the presence of squeezed states turns out to be equivalent to the requirement that the intensity of the coherent Rayleigh line contain more than half the total intensity of the fluorescent field. They have also found that squeezed states require small linewidths and either low or moderate Rabi frequencies, or large detunings from resonance. The restriction of Rabi frequencies to low value is lifted in the transient regime of resonance fluorescence as it has been shown by Ficek et al. [11]. These authors have worked out the details of squeezed states generation in the transient regime of resonance fluorescence.

To describe a two-level atom it is customary to use the pseudo-spin operators satisfying the well known commutation relations

$$[S^+(t), S^-(t)] = 2S_3(t)\ ,\ [S_3(t), S^\pm(t)] = \pm S^\pm(t) \quad /1/$$

where $S^+(t)$ $(S^-(t))$ are the raising /lowering/ operators that raise /lower/ the energy of the atom and $S_3(t)$ describes the atomic energy. Defining the Hermitian operators

$$S_1(t) = \frac{1}{2}[S^+(t) + S^-(t)]\ ,\ S_2(t) = -\frac{i}{2}[S^+(t) - S^-(t)] \quad /2/$$

we have

$$[S_1(t), S_2(t)] = i S_3(t), \qquad /3/$$

which implies the Heisenberg uncertainty relation

$$[\langle(\Delta S_1(t))^2\rangle \langle(\Delta S_2(t))^2\rangle]^{\frac{1}{2}} \geq \frac{1}{2} |\langle S_3(t)\rangle| \qquad /4/$$

A squeezed state for the atomic observables $S_1(t)$ or $S_2(t)$ is then defined [6,7] by the condition that either

$$\langle(\Delta S_1(t))^2\rangle < \frac{1}{2} |\langle S_3(t)\rangle|$$

or

$$\langle(\Delta S_2(t))^2\rangle < \frac{1}{2} |\langle S_3(t)\rangle| \qquad /5/$$

If one defines the quantities

$$F_{1,2}(t) = \langle(\Delta S_{1,2}(t))^2\rangle - \frac{1}{2} |\langle S_3(t)\rangle| \qquad /6/$$

the squeezing conditions /5/ mean that either $F_1(t)$ or $F_2(t)$ is negative. In the transient regime all average values of the atomic operators evolve in time and $F_{1,2}(t)$ can take negative or positive values depending on the time that have elapsed since the exciting field was turned on. The formulas for $F_{1,2}(t)$ are quite complicated and will not be given here. The formulas themselves as well as their graphical illustrations can be found in [11]. The main result is that as the laser field increases /Rabi frequency increases/ squeezing in $F_1(t)$ shifts to the region of shorter times, and $F_1(t)$ itself shows an oscillatory behaviour reflecting the Rabi oscillation. The maximum value of squeezing that can be obtained in the transient regime becomes greater than the steady-state maximum and for very strong fields $\Omega = 200\gamma$ can reach its limiting value $F_1(t) = -1/16$ which takes place for very short times.

For steady-state resonance fluorescence, we have

$$F_{1,2}(\infty) = \frac{2\beta^2}{1+\Delta^2+8\beta^2}\left[8\beta^2 \mp (1-\Delta^2)\cos 2\varphi \pm 2\Delta \sin 2\varphi\right] \qquad /7/$$

where $\beta = \Omega/4\gamma$, $\Delta = (\omega_L - \omega_0)/\gamma$ and φ is the initial phase of the exciting field. For $\varphi = 0$ this formula goes over into that derived by Walls and Zoller [6] except for the interchange of the indices 1 and 2 due to different choice of phase. It is evident from /7/ that a change by $\pi/2$ in the initial phase of the exciting field interchanges F_1 and F_2. The maximum of squeezing in steady-state resonance fluorescence turns out to occur for $\Delta = \varphi = 0$ and $\beta^2 = 1/24$, when $F_1 = -1/32$. From /7/ it is clear that F_1 is negative for $8\beta^2 + \Delta^2 < 1$ and F_2 is negative for $\Delta^2 > 8\beta^2 + 1$ which means weak fields.

The above discussion dealt with the problem of squeezing in atomic observables and the relation between this squeezing and the fluorescent field squeezing must be established to answer the question of squeezing in the fluorescent field. In the far-field limit the positive-frequency part of the fluorescent field can be expressed in terms of the atomic operators as follows [12]

$$\vec{E}^{(+)}(\vec{R},t) = \vec{E}_0^{(+)}(\vec{R},t) + \frac{1}{2}\vec{\Psi}(\vec{R})\sum_i S_i^-(t)\, e^{-i\vec{k}_0\cdot\vec{r}_i} \qquad /8/$$

where $\vec{\Psi}^2(\vec{R}) = (6\hbar k_0\gamma/R^2)\sin^2\theta$ with θ being the angle between the observation direction $\vec{R}$ and the atomic transition dipole moment $\vec{\mu}$; $k_0 = \omega_0/c$.

The summation is over all atoms contributing to the fluorescent field. In the case of one atom we have, of course, only one term in the sum. On defining the in-phase component E_1 and the out-of-phase component E_2 of the fluorescent field as

$$E_1 = E^{(+)} + E^{(-)} \quad , \quad E_2 = -i\left(E^{(+)} - E^{(-)}\right) \qquad /9/$$

one can easily calculate the normally ordered variances of the two components using Eq. /8/. It is assumed that the fluorescent field is measured in a direction other than that of the laser beam.

The result is then

$$\langle :(\Delta E_{1,2}(t))^2: \rangle = \vec{\Psi}^2(\vec{R})\left[\langle (\Delta S_{1,2}(t))^2\rangle + \frac{1}{2}\langle S_3(t)\rangle\right] \qquad /10/$$

In the steady state $\langle S_3(\infty)\rangle$ is negative i.e. $\langle S_3(\infty)\rangle = -|\langle S_3(\infty)\rangle|$, and squeezing in the atomic variables given by the condition /6/ implies a negative value of the normally ordered variance of the corresponding component of the fluorescent field. In transient regime, however, $\langle S_3(t)\rangle$ evolves in time and can take positive as well as negative values. It is obvious that intervals of time exist in which the atomic squeezing does not necessarily lead to a nonclassical nature of the emitted field. For more details see [11].

TWO-ATOM RESONANCE FLUORESCENCE

In the case of many identical atoms contributing to the fluorescent field one can introduce the collective atomic operators

$$S^{\pm} = \sum_{i=1}^{N} S_i^{\pm} e^{\pm i \vec{k}_0 \cdot \vec{r}_i} \qquad /11/$$

which inserted into Eqs. /2/ define the collective Dicke's spin variables satisfying the same commutation rules as that for an individual atom. Using expression /8/ for the fluorescent field one can again calculate the normally ordered variances of the two components of the field and the result has the form given by /10/ where the operators should be understood as collective operators. If there are many atoms placed at random positions the sum over the atomic positions in /11/ will be zero and squeezing will not apper. If there are only two atoms placed at definite positions that interact with the resonant laser field and interact with each other via the retarded dipole-dipole interaction the problem requires the solution of a set of fifteen equations describing the evolution of the atomic variables. For the steady state it has been done by Ficek et al [13] who discussed both photon antibunching and squeezing in the two-atom resonance fluorescence. It has been shown that the dipole-dipole interaction between the atoms shifts the maximum of squeezing into the region of finite detuning. The squeezing is most pronounced when the dipole-dipole interaction /in units of frequency/ and the detuning concel out each other. However, the maximum value of squeezing for two interacting atoms is less than that for independent atoms. For very strong dipole-dipole interactions squeezing is washed out completely. More details can be found in [13]. Similar results have been obtained by Richter [14].

CONCLUDING REMARKS

This short review of some results concerning the problem of squeezing in one-atom and two-atom resonance fluorescence is far from being complete. The problem is still under investigation and new results can be expected.
Recently Vogel and Welsch [15] have considered squeezing pattern in resonance fluorescence from a regular system of N atoms. If the atoms are distributed at regular positions the directions exist for which the sum in /11/ does not vanish and squeezing can be observed in such directions. Another possibility is to observe fluorescent light in the direction of the laser beam /forward direction/. In this case squeezing in the many atom resonance fluorescence can be observed, as it has been shown by Heidmann and Reynaud [16]. Although there are many other optical processes that are considered as a possible source of squeezed states the resonance fluorescence is still one of them.

I am grateful to Professor S.Kielich and Dr. Z.Ficek for fruitful discussions.

REFERENCES

1. D.F.Walls, Nature 306, 141 /1983/.
2. H.J.Kimble, M.Dagenais, and L.Mandel, Phys. Rev. Lett. 39, 691 /1977/.
3. G.Leuchs, M.Rateike, and H.Walther, cited by D.F.Walls, Nature 280, 451 /1979/; J.D.Cresser, J.Häger, G.Leuchs, M.Rateike, and H.Walther in Dissipative Systems in Quantum Optics, Vol. 27 of Topics in Current Physics, ed. R.Bonifacio /Springer, Berlin, 1982/.
4. C.M.Caves and B.L.Schumaker, Phys. Rev. A31, 3068 /1985/.
5. C.K.Hong and L.Mandel, Phys. Rev. Lett. 54, 323 /1985/.
6. D.F.Walls and P.Zoller, Phys. Rev. Lett. 47, 709 /1981/.
7. L.Mandel, Phys. Rev. Lett. 49, 136 /1982/.
8. R.Loudon, Opt. Commun., 49, 24 /1984/.
9. H.P.Yuen and V.W.S.Chan, Opt. Lett. 8, 177 /1983/.
10. H.F.Arnoldus and G.Nienhuis, Opt. Acta 30, 1573 /1983/.
11. Z.Ficek, R.Tanaś and S.Kielich, J.Opt.Soc.Am. B1, 882 /1984/; Acta Phys. Polon. A67, 583 /1985/.
12. G.S.Agarwal, Quantum Optics, Vol. 70 of Springer Tracts in Modern Physics, ed.G. Höhler /Springer, Berlin,1974/.
13. Z.Ficek, R.Tanas, and S.Kielich, Opt. Commun. 46, 23 /1983/; Phys. Rev. A29, 2004 /1984/.
14. Th.Richter, Opt. Acta 31, 1045 /1984/.
15. W.Vogel and D.G.Welsch, Phys.Rev.Lett. 54, 1802 /1985/.
16. A.Heidmann and S.Reynaud, J. Phys. /Paris/, in press.

STIMULATED RAMAN SCATTERING IN SILICA OPTICAL FIBERS - light transmission and nonlinear loss

A.Y. Spasov, V.M. Mitev, P.G. Georgiev
Institute of Electronics, Bulgarian Academy of Sciences,
1784 Sofia, Bulgaria

1. Introduction

The use of silica optical fibers for high power transportation in medicine, material processing, scientific research, etc. is limited by the nonlinear losses, due to various nonlinear interactions. The silica optical fibers are an ideal nonlinear media because of the low losses, long length, the possibility to compensate the material dispersion by the waveguide mode dispersion in order to achieve the phase-matching condition. Recent years new coherent light sources in the visible and infrared regions are developed by making use of the Stimulated Raman Scattering (SRS) in optical fibers [1 ÷ 5]. On the other hand, the SRS is considered to be the main source of nonlinear losses in high power transportation [6], since it has low threshold, when working with pump wave having relatively broad spectrum.

In this work an example of nonlinear losses estimation in monomode silica fibers, due to the SRS is presented. The analysis is based on the numerical calculation of the SRS equations. The use of single-mode fibers gives the possibility to concentrate the transmitted power in one mode. Thus, the stability and reproducibility of SRS (or, in general, of any nonlinear interaction) is not complicated by the transverse instability of the pump wave.

II. SRS - basic equations

SRS in a silica fiber leads to arising of a number of Stokes bands, separated by a frequency shift of about 440 cm^{-1}. As these Stokes bands develop successively, the spectral width of each Stokes order becomes broader and, if the pump power is sufficient, the output becomes a continuum.

Here we consider an one-directional propagation of pulses as follows: a pump pulse of wave number k_o and Stokes pulses of wave number k. As we will consider a propagation in not very long distances, the linear losses may be neglected. In case of short pulse propagation, the backward travelling Stokes waves (pulses) may be neglected too. The spectral density of m-th (m=1,2,..) Stokes pulse optical power $p(x, \Theta_o, k, m)$ obey the following equation:

$$\frac{\partial p(x, \Theta_o, k, m)}{\partial x} = \frac{v_k - v_o}{v_k} \frac{\partial p(x, \Theta_o, k, m)}{\partial \Theta_o} +$$

$$+ \int_{k}^{k_{max}} \tilde{g}(k' - k) \left[\tilde{\alpha} + s(x, \Theta_o, k) \right] p(x, \Theta_o, k', m-1)\, dk' -$$

$$- \int_{k_{min}}^{k} \tilde{g}(k - k') \left[\tilde{\alpha} + s(x, \Theta_o, k') \right] p(x, \Theta_o, k, m)\, dk', \qquad (1)$$

$$s(x, \Theta_o, k) = \sum_{m=0}^{\infty} p(x, \Theta_o, k, m). \qquad (2)$$

Here x is a linear coordinate in the direction of fiber length, t is the time, m=0 for the pump power density, v_k - the group velocity of spectral component with wave number k, v_o - the group velocity of pumping wave with number k_o, $\theta_k = v_k t - x$, $\theta_o = v_o t - x$.

Then

$$\left(\frac{\partial p}{\partial x}\right)_{\theta_o = \text{const}} - \frac{v_k - v_o}{v_k}\left(\frac{\partial p}{\partial \theta_o}\right)_{x = \text{const}} = $$

$$= \left(\frac{\partial p}{\partial x}\right)_{\theta_k = \text{const}.}$$

k_{min} and k_{max} are the minimal and maximal value of k; $\tilde{g}(k)$ is the gain coefficient and $\tilde{\alpha}$ is the coefficient of spontaeous Raman scattering - see [2,5]. We consider only down-conversion of the frequency.

The energy conservation law may be written in differential form:

$$\frac{\partial}{\partial x}\left(\int_{k_{min}}^{k_{max}} s\, dk\right) = \frac{\partial}{\partial \theta_o}\left(\int_{k_{min}}^{k_{max}} \frac{v_k - v_o}{v_k} s\, dk\right), \quad (3)$$

or in integral form:

$$\frac{\partial}{\partial x}\int_{-\infty}^{+\infty}\left(\int_{k_{min}}^{k_{max}} s\; dk\right) d\theta_o = 0 \quad (4)$$

The initial conditions for (1) are:

$$p(x,\theta_o,k,m)\Big|_{x=0} = P_o(\theta_o)\,\delta(k - k_o)\,\delta_{m,o}, \quad (5)$$

where $\delta(k)$ is the Dirac δ - function and $\delta_{m,n}$ is the Kronecker deltha.

Dealing with the SRS in case of pulse pumping, one have to take into account the frequency and nonlinear dispersion. The refraction coefficient is equal to:

$$n(\lambda, E) = n_1(\lambda)(1 + n_2 \bar{E}^2), \tag{6}$$

where

$$n_1^2(\lambda) = 1 + \sum_{i=1}^{3} a_i \lambda^2 / (\lambda^2 - b_i) \tag{7}$$

Here λ is the wavelength and the coefficients a_i, b_i are taken from [7]; $n_2 = 3 \times 10^{-22}\ m^2/V^2$ [8].

The field intensity is taken from the formulae:

$$\bar{E}^2 = \int_{k_{min}}^{k_{max}} \frac{s(x,\theta_o,k)}{\varepsilon v_{gr}} dk, \tag{8}$$

where v_{gr} is the group velocity, ε is the dielectric permeability.

III. Numerical calculations

A discretization with respect to k is introduced into the numerical computations. The interval $[k_{min}, k_{max}]$ is divided as follows:

$$k_o = k_{max},\ k_n = k_o - n\Delta k, \quad n = 1,2,\ldots,N,$$

$$k = (k_{max} - k_{min})/N, \text{ i.e. } k_N = k_{min}$$

Let

$$P(x,\theta_o) = \int_{k_i - \frac{k}{2}}^{k_2 + \frac{k}{2}} p(x,\theta_o,k,\emptyset)\, dk, \tag{9}$$

$$S(x,\theta_o,k_i,m) = \int_{k_i - \frac{k}{2}}^{k_i + \frac{k}{2}} s(x,\theta_o,k,m)\, dk, \quad m=1,2,\ldots,M, \tag{10}$$

$$S_{sum}(x,\theta_o,k_i) = \sum_{m=1}^{M} S(x,\theta_o,k_i,m), \tag{11}$$

$$\alpha = \tilde{\alpha}\,\Delta k, \qquad g(k) = \tilde{g}\,(k)/\Delta\, k.$$

Also a discretization with respect to θ_o is introduced:

$$\theta_o = \theta_o^{(1)}, \ldots, \theta_o^{(L)}, \qquad \theta_o^{(l+1)} - \theta_o^{(l)} = \Delta\theta$$

$$\frac{\partial S(\theta_o^{(l)})}{\partial\theta_o} = \frac{S(\theta_o^{(l)} + \Delta\theta) - S(\theta_o^{(l)} - \Delta\theta)}{2\ \Delta\theta} + O(\Delta\theta^2).$$

From the inequality $\left|\dfrac{v_i - v_o}{v_i}\right| \sim |\Delta\theta| \ll 1$,

follows, that this discretization gives an error of order $O(\theta^3)$

Finally:

$$\frac{dP(x,\theta_o^{(l)})}{dx} = -\sum_{i=1}^{N} g(k_i)\,(\alpha + S_{sum}(x,\theta_o^{(l)}, k_i))\, P(x,\theta_o^{(l)}), \tag{12}$$

$$\frac{dS(x,\theta_o^{(l)},k_i,m)}{dx} =$$

$$= \frac{v_i - v_o}{v_i}\ \frac{S(x,\theta_o^{(l)} + \Delta\theta,k_i,m) - S(x,\theta_o^{(l)} - \Delta\theta,k_i,m)}{2\,\Delta\theta} +$$

$$+ \delta_{m,1}\, g(k_i)\,(\alpha + S_{sum}(x,\theta_o^{(l)},k_i))\, P(x,\theta_o^{(l)}) +$$

$$+ (1 - \delta_{m,1}) \sum_{j=1}^{i-1} g(k_i - k_j)\,(\alpha + S_{sum}(x,\theta_o^{(l)}, k_i))$$

$$* \; S(x,\theta_o^{(l)}, k_j,m-1) - \sum_{q=i+1}^{N} g(k_q - k_i)$$

$$* \; (\alpha + S_{sum}(x,\theta_o^{(1)},k_q)) \; S(x,\theta_o^{(1)},k_i,m), \qquad (13)$$

where the initial conditions are:

$$P(\emptyset,\theta_o^{(1)}) = P_o(\theta_o^{(1)}), \quad S(\emptyset,\theta_o^{(1)},k_i,m) = \emptyset. \qquad (14)$$

Here the initial condition $P_o(\theta_o^{(1)})$, $1 = 1,2,\ldots,L$ is expressed by Gaussian curve with fixed energy.

The method of Runge-Kutta may be used for solving of (12) and (13). From physical conditions follows, that

$$P(x,\theta_o^{(1)}) \geqslant \emptyset, \quad S(x,\theta_o^{(1)},k_i,m) \geqslant \emptyset. \qquad (15)$$

When (15) is not satisfied, a numerical instability appears. We impose (15) to avoid this instability, then all negative values will be replaced by zero. The group velocity/in(8)/ v_{gr} is calculated for $k = k_o$ and for small amplitude.

IV. Results and conclusion

The equations (12) and (13) are solved for 50 m monomode fiber with core diameter of 5μm. The step Δx is 5 cm. We consider values of m=∅,1,...,5 , i.e. up to the fifth Stokes. The pump pulse has 10 ps FWHM. The nonlinear loss vs. peak pump power is given in Fig.1 for different pump wavelengths: λ_p = 0.6μm, 1.06μm, and 1.3μm. Since the dispersion for 1.3 μm region is low, the interaction length is longer, and thus, in spite of the lower g, the nonlinear loss is higher. The losses in fibers with different lengths are shown in Fig.2 for λ_p = 1.3 μm. Due to the shorter pulses, the losses calculated in this work are less than the losses, obtained in[5]. The pulse shape of both the pump and Stokes pulses changes substantially during the pulse propagation in the fiber. The energy depletion is not

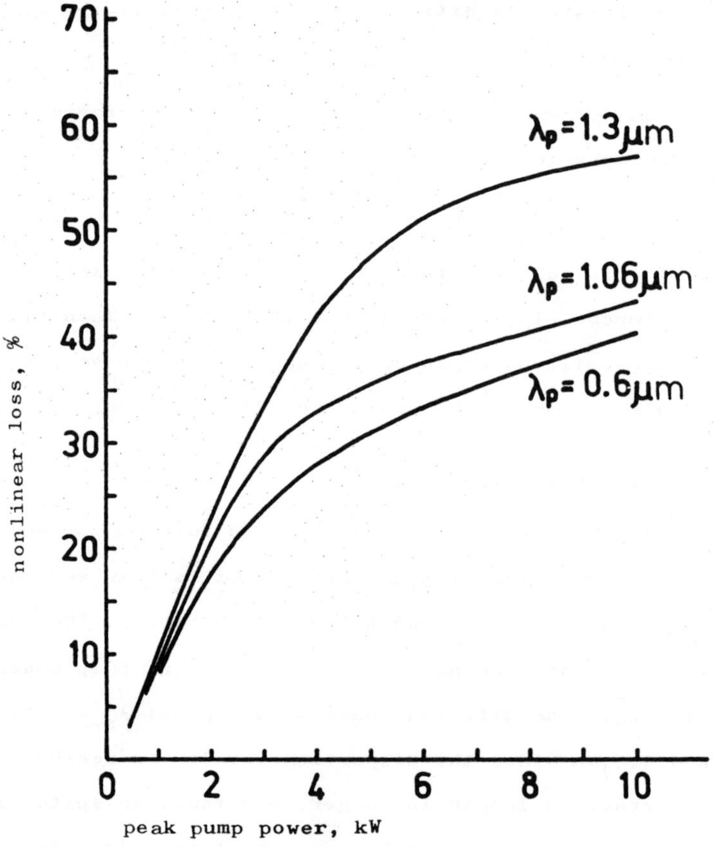

Fig.1

Nonlinear loss dependance on the peak pump power

the same in the peak and off the peak. For some distances the Stokes pulses are substantially shorter.

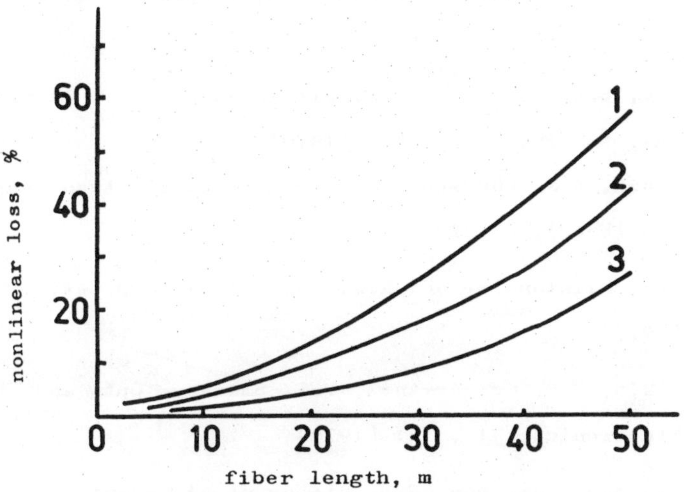

Fig.2

Nonlinear loss dependance on the fiber length

1 - peak pump power 10 kW

2 - peak pump power 4 kW

3 - peak pump power 2 kW

pump wavelength 1.3 m

The procedure developed allows numerical modelling of SRS in optical fibers. In more precise calculations the spectral broadening of the pump pulse, due to the self-phase modulation [9,10] should be taken into account. It will also be interesting to evaluate the nonlinear loss in a soliton pulse propagation in long optical fibers for communication systems [11-13].

References:

1. John An Yeung and Amnon Yariv, J.Opt.Soc.Am., 69, 803 (1979)

2. R.H. Stolen, Clinton Lee and R.K. Jain, J.Opt.Soc.Am.,B,1, 652, (1984).

3. E.M. Dianov, S.K. Isaev, L.S. Kornienko et al, Izv. Akad. Nauk (USSR), ser.Phys., 43, 266 (1979)

4. A.V. Grudinin, A.N. Gurjanov, E.M. Dianov et al, Kvantovaya elektronika (USSR), 8 , 2383 (1981).

5. R.H. Stolen, Clinton Lee and R.K. Jain, Appl.Phys.Lett., 30 , 340 (1977)

6. V.S. Butylkin, V.V. Grigoryants, V.I. Smirnov, Opt. and Quantum Electronics, 11 , 141 (1979)

7. A.V. Belov, A.N. Gurjanov, E.M. Dianov et al, Kvantovaya elektronika (USSR), 5 , 695 (1978)

8. Akira Hasegawa and Frederick Tappert, Appl.Phys.Lett., 23, 142 (1973)

9. W.J. Tomlinson, R.H. Stolen, C.V. Shank, J.Opt.Soc.Am. B, 1, 139 (1984)

10. B. Valk, W. Hodel and H.P. Weber, Opt.Comm., 50, 63 (1984)

11. Akira Hasegawa and Yuji Kodama, Proc.of the IEEE, 69, 1145 (1981)

12. L.F. Mollenauer, R.H. Stolen, M.N. Islam, Opt.Lett.,10, 229, (1985)

13. K.J. Blow and N.J. Doran, Optics Commun., 52, N.5 (1985) 367.

LIST OF PARTICIPANTS

Name	Address
J Czub E Paul-Kwiek S Zielińska W Miklaszewski J Sienkiewicz S Kryszewski M Zukowski	Institute of Theoretical Physics and Astrophysics University of Gdańsk ul. Wita-Stwosza 57 80-952 Gdańsk Poland
Z Konefał J Szczepański R Drozdowski A Sikorska J Heldt	Institute of Experimental Physics University of Gdańsk ul. Wita-Stwosza 57 80-952 Gdańsk Poland
J Musielak T Wujec	Institute of Physics Wyższa Szkoła Pedagogiczna ul. Oleska 48 45-052 Opole Poland
B Pokrzywka	Institute of Physics Jagiellonian University ul. W. Reymonta 4 30-059 Kraków Poland
L Sirko M Łukaszewski Cao Long Van	Institute of Physics Polish Academy of Sciences al. Lotników 32/46 02-668 Warszawa Poland
J Wolnikowski J Szudy P Rudecki B Ziętek P Targowski M Łukaszewicz A Bielski R Bobkowski F Bylicki J Zaremba S Dembiński S Łegowski E Lisicki	Institute of Physics University of N. Copernicus ul. Grudziądzka 5/7 87-100 Toruń Poland

Z Ficek
K Grygiel
W Leoński
M Kazimierczak

Institute of Physics
University of A. Mickiewicz
ul. Grunwaldzka 6
60-780 Poznań
Poland

T Grycuk
M Findeisen

Institute of Experimental Physics
Warsaw University
ul. Hoża 69
00-681 Warszawa
Poland

V Mitev
T Mitsev
V Simeonov
Ch Ghelev

Institute of Electronics
Boulgarian Academy of Sciences
Boul. Lenin 72
1784 Sofia
Bulgaria

J Polecha

Institute of Physics
Czechoslovak Academy of Sciences
Na Slovance 2
180 40 Prague 8
Czechoslovatia